D0712217

SCIENCE AND THE QUIET ART

SCIENCE AND THE QUIET ART

The Role of Medical Research in Health Care

DAVID WEATHERALL

*A Volume of the Commonwealth Fund Book Program
under the editorship of Lewis Thomas, M.D.*

W. W. NORTON & COMPANY / NEW YORK / LONDON

First Edition

The text of this book is composed in Berkeley Book,
with the display set in Onyx and Berkeley Bold Italic.
Composition and manufacturing by
The Maple-Vail Book Manufacturing Group.
Book design by Jam Design.

Library of Congress Cataloging-in-Publication Data

Weatherall, D. J.
Science and the quiet art :
the role of medical research in health care
/ David J. Weatherall.
p.   cm.
"A Commonwealth Fund book."
1. Medicine—Philosophy.   2. Medical sciences.   3. Social
medicine.   I. Title.
R723.W355   1995
610'.1—dc20   94-16483

ISBN 0-393-03744-4

W. W. Norton & Company, Inc., 500 Fifth Avenue, New York, N.Y. 10110
W. W. Norton & Company Ltd., 10 Coptic Street, London WC1A 1PU

1 2 3 4 5 6 7 8 9 0

TO STELLA AND MARK

Contents

Preface

It was his part to learn the powers of medicines and the practice of
healing, and careless of fame, to exercise the quiet art.

Virgil, Aeneid

A few years ago Alick Bearn and Lewis Thomas asked me whether I would
contribute to a series of books being commissioned by the Commonwealth
Fund. The idea was that scientists would write about themselves and their
work in a way that would be comprehensible to the nonspecialist reader.
While flattered, I wondered how a professor of medicine could possibly compete with the other authors in the series, who had devoted their entire lives
to their particular branch of science. Of necessity, clinical professors have
to divide their time between treating patients, teaching, attending mind-
numbing university and hospital meetings, and, at least in my case, trying
with diminishing success to bridge the gap between the clinic and the
increasingly sophisticated world of the medical research laboratory. Though
a life spent moving between the diverse needs of sick people, paranoid scientists, and the bizarre rituals of an ancient university has never been dull, I
doubted whether it is the stuff from which interesting scientific autobiogra-
phies or popular works on science are made.

What finally changed my mind was my deepening concern about the current medical scene. Over recent years there appears to have been an increasing disillusion with modern scientific medicine on the part of the public,
media, politicians, and many doctors themselves. Our apparent failure to
get to the root cause of our major killers, our growing reliance on high-
technology patch-up treatment, the spiraling costs of health care, and concerns about the decline of the pastoral role of doctors in patient care have
combined to make modern medical practice seem both dehumanizing and

ment and guesswork that underlies much of modern high-technology patch-up medicine. After a short survey of what has been achieved in other branches of medical care, we will take a closer look at the origins of our current killers and hazard a guess at the extent to which they result from a clash between the new environment we have created and our genetic makeup. We will then explore some of the exciting new developments in the biomedical sciences and ask when they are likely to make an impact in the clinic and community. And we will also ask how science can be better utilized to improve patient care in the meantime. Finally, we will touch on the lot of patients and some of the choices and ethical issues that will face them and their doctors as medical practice continues its slow evolution from an empiri-cal craft to a scientifically based discipline.

D. J. Weatherall
Oxford
December 1994

PART I

Perceptions of Modern Medicine

O N E

Introduction:
Why Does Modern Medicine
Seem to Have Lost Its Way?

I would suggest that the whole imposing edifice of modern medicine, for all its breath-taking success, is, like the celebrated tower
of Pisa, slightly off balance.

—*Charles, Prince of Wales (1948–)*

In October 1990 the *Economist* carried an article entitled "The Future of Medicine." Under the heading "doctors can damage your wealth" appeared a
cheerless summary of what was to follow: "The costly wonders of modern
medicine have not had much effect on life expectancy or the incidence of
sickness. Should the billions that rich countries spend on high-technology
doctoring be spent on health instead?" To make sure that the reader was left
in no doubt about the slant of its message, the article was also prefaced by a
chilling cartoon depicting a ghoulish-looking doctor, one finger on a button
that was about to plunge a terrified-looking patient into a black tunnel representing a brain scanner. Every part of their anatomy carried price tags: heart
transplant £15,000, brain scanner £500,000, hip replacement £2,200, and
so on; the doctor's rather archaic-looking stethoscope was priced at £10. One
of the patient's legs was encased in a plaster cast across which was scribbled
the message "Get well soon," signed by the minister of health.

This cartoon and article encapsulate a view of medical practice that has
been gathering strength over recent years. The clinician with his finger on
the button that activates the brain scanner, and his undervalued stethoscope
(having just bought a new one, I know for sure that it costs a great deal more
than £10), implies that the art of medicine, in particular the ability of doctors
to care for their patients as individuals, has been lost in a morass of expensive
high-technology investigation and treatment. Even worse, all this ill-directed
activity is doing very little to improve our health. In short, modern scientific
medicine is a failure.

Paradoxically, the origins of the mood of disillusionment with current medical practice can be traced to some of its extraordinary successes earlier this century. In the years after the Second World War, during which the development of new vaccines and the discovery of antibiotics led to the control of many infectious diseases, it appeared that medical science was capable of almost anything. The virtual disappearance overnight of scourges like smallpox, diphtheria, poliomyelitis, and other infectious killers, at least from the more advanced countries, led to the expectation that spectacular progress of this kind would continue. But this did not happen. The diseases that took their place—heart attacks, strokes, cancer, rheumatism, and psychiatric disorders—turned out to be much more intractable. Granted, medical research in the universities and pharmaceutical industry led to some remarkable advances in their symptomatic control: the treatment of heart failure and high blood pressure; powerful analgesic drugs for the management of rheumatic disorders; highly sophisticated diagnostic aids; organ transplantation; life-support systems; the combination of modern surgery and chemotherapy for the cure or palliation of a few forms of cancer; and many other technological feats.

There are now few diseases for which something cannot be done to alleviate symptoms and prolong life. And, by altering our lifestyles, we have made a little progress toward reducing the number of premature deaths from heart disease and cancer. But, overall, it has proved difficult to discover effective forms of prevention or cure. Hence medical science is perceived to have become very effective at expensive, high-technology "patch-up" procedures, and at extending our life spans, but there appears to be little progress toward the definitive control of many of the major killers of Western society. And the spiraling costs of this pattern of medical care are posing serious economic problems, even for the richest of industrialized countries.

The successes of the antibiotic and vaccine era also bred societies that expected much more of their doctors and felt that they had a right to enjoy a state of constant rude health. When this did not happen, there was a move to alternative medicine. If medical science could not cope with chronic backache or lung cancer, why not turn for help to those who claimed that they had the answers? Dietary manipulation, food allergy, herbal remedies, and a variety of other approaches to chronic illness were taken up with enthusiasm. In recent years "health" has become a major industry. Rarely a day passes without the announcement that something that we enjoy eating or drinking, or some other pleasurable activity, is hastening our demise. Bookstalls are packed with information about how to achieve better health. A visitor from another planet, landing at a modern airport and perusing the shelves of its bookshop, might wonder at the sanity of a society that extols the virtues of activities ranging from transcendental meditation to the administration of coffee enemas to sustain a feeling of well-being.

Hopes for a more rational approach to medical care in the future are based on two quite different areas of research and development, which have evolved side by side over recent years. First, there is the growing belief, stemming from studies of the pattern of disease in large populations, a branch of medical science called epidemiology, that many of our common killers can be linked to changes in our environment and lifestyles. And, at the same time, there has occurred a revolution in the biological sciences, particularly in the fields of molecular and cell biology, that may have the potential completely to change the face of health care in the future.

The work of epidemiologists has had a profound effect on the way in which we think about disease. In essence, they are telling us that since many of our problem diseases are the result of the environment and lifestyles that we have created, they should be preventable. This has already led to some serious rethinking about medical education and practice, and to the notion that there should be much more emphasis on learning how to prevent disease than on trying to understand its mechanisms and management. Because prevention is much cheaper than treatment, it is not surprising that this notion is particularly attractive to governments and those who administer our health services. The problem is that we have no idea to what extent many of the current killers of Western society are environmental in origin. There are a few clearcut cases—the relation between lung cancer and cigarette smoking, for example—but many of our intractable diseases seem to reflect complex interactions between our genetic makeup, aging, and the new environment that we have created.

The advent of molecular and cell biology, and the ability to manipulate human genes, has led to a state of tremendous excitement in the medical research world. Rarely a week goes by without a new breakthrough's being splashed over our television screens and newspapers; another human gene has been isolated, and the cause of another disease, which we may not even have heard of, is announced. Remarkable discoveries about the defects in the machinery of our cells are being unearthed, and they seem to be getting at the very heart of the cause of cancer. And information is turning up about aspects of our genetic makeup that make us more or less likely to have heart attacks or to develop diabetes. Whenever these new discoveries are announced, excited scientists or journalists tell us that they will have a major impact on health care "within the near future."

Yet even the atmosphere of expectation generated by these claims has started to sour, for, despite all these promises, as perceived by an impatient, health-conscious society, relatively little of practical value seems to be emerging. Writing in the *Los Angeles Times* recently, U.S. Representative George Brown pleaded, "We've paid for 45 years of discovery: let's start requiring its application to the critical problems in the civilian sector." The notion that much that goes on in science is motivated more by scientists' wish for self-glorification than by any practical end in view is being voiced on both sides of the Atlantic. Furthermore, there is a growing feeling that modern science, whether it involves the manipulation of human genes or inquiries into the origin of the universe, is a debasing activity that is damaging our environment and moving into areas best left alone. In short, science is getting a bad press.

It is not surprising, therefore, that observers of the current medical scene are confused. There is a widespread belief that many of the qualities of good doctoring have been lost in our efforts to understand diseases rather than the problems of sick people. Those who administer our health services wonder where their priorities lie. How can they encourage preventive medicine and try to contain the high-technology practice that is bankrupting their countries at a time when an aging populace is demanding heart transplants, replacement of joints, and open access to intensive care? And those who have the difficult task of evaluating the best way to support medical research are receiving equally conflicting messages. On the one hand, they are being told that modern medical science is contributing very little to the health of the community; on the other, they hear that new developments in the basic sciences have enormous possibilities for our future well-being. Should they be using their limited resources to explore disease in the community and to evaluate the best way to deliver health care, on the basis of current knowledge, or should they take a longer-term view and support those who believe that many diseases will not be controlled until their causes and mechanisms are better understood?

These concerns were highlighted in a recent editorial in the British medical journal *Lancet,* which appeared under the rather unpromising title "Molecular Medicine in the USA: Can Three Paradigms Shift Simultaneously?" As the author, L. S. Rothenberg, explains, the phrase "paradigm shift" is used by historians of science to describe a fundamental change from one set of scientific beliefs to another. How, he asks, will it be possible for the United States to encompass three major and conflicting changes in the pattern of health care in the next century. The molecular sciences, with their reductionist approach to disease, promise to change the whole face of clinical practice and the pharmaceutical industry. But although this may lead to treatments that are much less invasive, brutal, and dehumanizing than current high-technology practice, it could be even more costly. On the other hand, there is also a major shift of emphasis in the United States toward community care and the needs and preferences of patients, with a more holistic approach to patient management. And, perhaps most important, there is a drive to curb the costs of health care, a trend backed strongly by the Clinton administration. Rothenberg wonders whether, in a society in which forty million people have very limited access to medical care, and where some hospitals make it harder to obtain a pair of glasses than an organ transplant, it will be possible to incorporate three such disparate new approaches to medical practice. Although some of his concerns stem from the peculiarities of American health care, I suspect that what he says is relevant to most rich, industrialized countries.

Some of these problems, real though they are, start to look less daunting if they are put into historical perspective and seen against the background of the way in which scientific advances have changed medical practice. In effect, the doctors of today find themselves in much the same position as their predecessors at the beginning of this century. It was already apparent that many of the infectious diseases that were killing their patients could be partly controlled by better housing and hygiene and by other improvements in the environment; a few could even be prevented by vaccination. But it was not clear how far such measures would be successful in controlling these diseases; in the meantime, little could be done for their patients with tuberculosis, meningitis, poliomyelitis, or infections of pregnancy, except to try to control their symptoms and offer comfort. At the same time, however, the doctors were hearing that some exciting developments in the basic biological sciences—in microbiology and immunology, in particular—were just round the corner and might provide the answer to wards full of patients dying of infection, and to the thousands of children permanently disabled by poliomyelitis or tuberculous meningitis. Yet these research fields had been on the move for over half a century and still appeared to be of limited practical value.

The theme that we will explore, therefore, is that many of our current problems are not new; we have lived with uncertainty about the relative roles of environmental improvements and the scientific exploration of disease in determining the state of our health for well over a hundred years. We will see how, as medical care has evolved over the centuries, improvements in living conditions and lifestyles have played an important role. Yet many of our major achievements have also resulted from a genuine understanding of the nature of disease, often stemming from discoveries in many different branches of science, some of which have been made without any particular practical goal in mind. But solutions to medical problems, whatever their source, are rarely complete; because of the complex nature of diseases and those who suffer from them, even seminal advances, such as the control of infectious disease, are often followed by a completely new and unexpected set of questions.

As we look more closely at our current killers, we will see that the main reason for our difficulties, particularly the continued expansion of high-

technology, patch-up practice, is ignorance about their origins and mechanisms, above all because so many of them have multiple causes and are
closely bound up with the mysterious processes of aging. After examining the
relative roles of nature and nurture in their genesis, we will speculate whether
recent developments in the laboratory sciences and epidemiology will solve
our problems, just as advances in our understanding of microorganisms and
the body's reaction to them led to the control of many infectious diseases
earlier this century. It will become apparent that, like their forerunners at the
turn of the century, today's medical sciences may take quite a long time to
affect work in the clinic or the community. Hence, we will also ask how
scientific knowledge that we have already might be applied more effectively
to patient care.

Finally, we will touch briefly on the role of the "old" clinical virtues of
good doctoring. Why do they seem to have been lost in our increasingly
mechanistic world? Might they become completely redundant as we progress
toward a genuine understanding of disease, and as medical practice evolves
from an "art" to a science? As we will see, there is no fundamental contradic-
tion between the scientific approach to the study and management of disease
and the pastoral aspects of medical care. On the contrary, as we develop even
more sophisticated patterns of clinical practice, with all the new ethical issues
that they will pose, the need to treat patients as people, and not diseases, will
be all the greater.

PART II

The Roots
of Medical
Science

T W O

Beginnings

> Physicians are inclined to engage in hasty generalizations. Possessing a natural or acquired distinction, endowed with a quick intelligence, an elegant and facile conversation . . . the more eminent they are . . . the less leisure they have for investigative work. . . . Eager for knowledge . . . they are apt to accept too readily attractive but inadequately proven theories.
>
> —*Louis Pasteur (1822–1895)*

For many of us, an attack of appendicitis is little more than a nuisance that interrupts our lives for a couple of weeks. The diagnosis is based on our symptoms, together with a simple clinical examination: we arrive in the operating room in a pleasantly drowsy state, are put to sleep with an injection, and after a short operation return to the ward and, in a few days, to our home. Yet even this routine event represents the fruit of empirical observations and scientific advances that stretch back for several thousand years and are based on knowledge gained from such disparate fields as anatomy, physics, chemistry, and microbiology. And like that of all modern medical practice, it is an incomplete story. There is still no cast-iron "test" for appendicitis, and the diagnosis relies solely on the surgeon's diagnostic skills. Even the most experienced surgeon will, from time to time, open the abdomen only to find a normal appendix and no explanation for the patient's symptoms.

Largely, I suspect, because of misguided efforts to cram too much knowledge into the minds of our students, the history of medicine is neglected in most medical schools. Recently I introduced Anthony Epstein, the discoverer of the virus that causes infectious mononucleosis (glandular fever) and that bears his name and that of the codiscoverer, to a group of medical students. Afterwards one of them told me that he was surprised to learn that the Epstein of the Epstein-Barr virus had ever existed; I hadn't the heart to ask him whether he thought the virus had received its name by the use of a pin and a telephone directory. The American medical scientist and essayist Lewis Thomas has advanced another reason for the lack of popularity of medical history. It is simply that "it is so unrelievedly deplorable a story." As perceived by Thomas, medicine for century after century consisted of sheer guesswork and the crudest sort of empiricism. Anything that could be dreamed up for the treatment of a disease was tried out. He describes it as the most frivolous and irresponsible kind of human experimentation, based on nothing but trial and error, and usually resulting in precisely that sequence!

An equally irreverent view of the work of doctors of the past animates the recent play *The Madness of George III*, by the English playwright Alan Bennett. Bennett describes the efforts of the leading doctors of the time to control the episodic bouts of bizarre behavior and malaise that afflicted the king throughout much of his reign. Two of them almost kill him in their efforts to control his symptoms by bloodletting, purging, and cupping, an extremely painful procedure in which a hot glass was applied to the skin to encourage blistering and bleeding by suction as it cooled. A third prognosticates on the outcome of his illness by observing, in lingering detail and with evident relish, the shape and quantity of his stools. The only physician who emerges with any credit is a remarkable character called Francis Willis, a leading mad-doctor of the day (psychiatrists had not yet been invented), whose primitive group therapy at least offers some semblance of sense to what otherwise appears to audiences of today as therapeutic madness. At the end of the play a modern, white-coated doctor comes onstage to interpret the king's madness in terms of an inherited biochemical derangement called porphyria. While all this makes the doctors who were practicing at the end of the eighteenth century look silly, it should be pointed out that the retrospective evidence on which the diagnosis of porphyria is based is so tentative that those who have interpreted events in this way may not be much wiser.

Not surprisingly, these views of the past do not please historians, who look on them as no more than misdirected attempts on the part of today's doctors to interpret history in the light of current knowledge, rather than in that available to their predecessors. Since it is unlikely that doctors of the past were less intelligent or compassionate than those of today are, it is essential to try to understand the way in which clinical science and practice have developed in light of what was known at the time. But because medicine is such an all-encompassing activity, reflecting not only sick people and their physicians but also their social, cultural, and religious backgrounds, it is safer to leave the past in the hands of professional historians. However, if we are to appreciate the role of science in current medical practice, it will be worth the risks of a brief journey into this dangerous country to try to understand the thinking of just a few of those whose work led to the birth of modern scientific medicine, sometime around the middle of the nineteenth century.

The bird's-eye view of the development of medical science that follows makes no attempt to compress the extraordinary story of the history of medicine into a few pages. Rather, to help readers better to understand current medical science and practice, and why doctors think and behave as they do, it simply highlights the origins and consequences of a few of the major stepping stones on the way to present-day health care.

EARLY BEGINNINGS

The earliest documentary evidence that has survived from the ancient civilizations of Babylonia, Egypt, China, and India suggests, not surprisingly, that longevity, disease, and death are among humanity's oldest preoccupations. Medical practice, from its beginnings to the present day, has been a mixture of sympathy and kindness allied to well-meaning but often empirical efforts to alter the natural course of events. It has always been an art, practiced against a background of incomplete scientific knowledge about the nature of disease processes. Human beings, like all living things, are immensely complex biological systems. Even today, with all our knowledge of their chemistry and physiology, we have a very limited understanding of the mechanisms that underlie most of the diseases that we encounter in day-to-day practice. Caring for sick people involves making considered judgments on the basis of limited evidence and information. At best, we are slowly reaching the stage at which we are aware of how little we know.

The word "science" stems from the Latin *scientia* (*scire,* "to learn" or "to know"), and hence in its broader sense means learning or knowledge. Currently, it has a more restricted usage, confined to the systematized observation of, and experiment with, natural phenomena. This pithy definition, which appears under "science" in the *Concise Oxford Dictionary,* may not be of much help to readers who wish to understand the meaning of the term "scientific medicine," a problem that, incidentally, seems to worry many historians. As used in this book, the term emphasizes the application of rigorously controlled experiment, measurement, and statistics to try to understand the causes and mechanisms of disease and how it can best be prevented or treated. And it also encompasses careful observational studies at the bedside or in the community, which seek to find recurrent patterns of symptoms and signs that constitute distinct disease entities. The classification of illness in this way—nosology, as it is called—has been an important prerequisite to more sophisticated research in the laboratory or the field.

The first origins of the physical sciences can be traced to observations of the movements of heavenly bodies, and the biological sciences must have started with studies of plants and animals and with some kind of primitive medicine. These early efforts to understand the natural world were sidetracked by animistic interpretations of the control of nature by all-powerful beings, and the intrusion of magic, astrology, and superstition. Attempts to put empirical knowledge into some kind of order by the development of simple arithmetic and the recognition of the periodicity of the seasons were made in ancient Egypt and Babylon, but rigorous submission of knowledge to rational examination—that is, the development of the kind of scientific method used today—probably started with the work of the Greek natural philosophers in Ionia.

From ancient times to the Renaissance knowledge of the living world changed very little, and the distinction between animate and inanimate objects was blurred. Speculations about living things were based, as they still are, of course, on prevailing ideas about the nature of matter. The Babylonians and Egyptians believed that water, air, and earth were the primary constituents of the world; a fourth, fire, was added later. The concept of the all-pervading influence of the four elements was extended to form a theory about how the human body is constituted. In short, it was thought to consist of four humors—blood, phlegm, yellow bile, and black bile. The notion that our patterns of behavior may be modified by different combinations of the humors is reflected in words still used today: "sanguine," "phlegmatic," "choleric," and "melancholic."

The idea that disease results from an imbalance of the humors permeated Greco-Roman medicine. The emergence of Western medical practice as an art, science, and noble profession is usually associated with the name of Hippocrates, a physician who is thought to have lived from about 460 to 361 B.C. Though historians disagree over the relative roles of Hippocrates and his followers as the originators of many of the medical works known as the Hippocratic Collection, and even over whether he ever existed, this remarkable compendium offers a valuable view of medical science and practice of the time. It regarded health as a harmonious balance of the four humors, and disease as the result of an imbalance, or dyscrasia, leading to an abnormal mixture of the humors. Recovery from disease involved the restoration of humoral equilibrium, which proceeded through different stages during which the humors, suitably altered, might be eliminated through secretions or excretions. This concept of pathology, which provided an explanation for both mental and physical illnesses, formed the basis for what, at the time, was a rational approach to treatment by bleeding, by purging, and by modification of diet.

Regardless of their origin, the texts that constitute the Hippocratic Collection became authoritative in Greek, Roman, and medieval times and formed the basis for much of Western medical practice. Apart from the famous Hippocratic oath, which still offers an ethical framework for those few doctors who are familiar with it, they contain detailed descriptions of the natural course of disease, dissertations on diagnosis and prognosis, studies of epi-

FIGURE 1

Galen (129–200). Lithograph by P. R. Vigneron, published 1820–29.

demics, and some very advanced thinking on the management of wounds
and infection. Today's readers who care to browse through the long lists of
aphorisms will find a mixture of acute clinical observations, the validity of
which have stood the test of time, together with many vague impressions,
presumably based on a single patient and akin to some that still permeate
medical practice and teaching.

The doctrine of the four humors continued to be prominent in the work
of Galen of Pergamum (129–200), one of the most remarkable figures in the
history of medical practice and science (figure 1). Galen was part clinician,
part scientist, and part philosopher. His writings were not subjected to serious challenge until the sixteenth and seventeenth centuries. Because the dissection of human bodies was forbidden by the Romans, his anatomy was
based largely, though not entirely, on the study of other species; experience
gained from tending the wounds of gladiators must have provided him with
some grisly insights into living human anatomy. His complex and ingenious
physiology, firmly based on the doctrine of the humors, also incorporated
the Platonic doctrine of the threefold division of the soul. Essentially he
believed that pneuma, or air, undergoes modification in three organs—the
heart, liver, and brain—and is distributed in three groups of vessels—veins,
arteries, and nerves. Pneuma, Galen held, is changed in the liver to produce
the nutritive soul, or natural spirits, which are distributed in the veins. The
heart and arteries are the sites of the vital spirits, which distribute innate heat,
while the brain produces animal spirits, the basis of sensation and movement.
This notion led to the development of a system of anatomy flawed in many
respects; blood from the liver had to move from the right to the left side of
the heart, for example. Although Galen's anatomical descriptions of the heart
and its valves were in many ways correct, his ideas about their function gave
rise to many ambiguities and misinterpretations of his anatomical observations.

But Galen stressed the importance of clinical examination and of the environment in the genesis of illness. He taught the value of bleeding for the
treatment of a wide variety of diseases and studied every conceivable source
of therapeutic agents. The complex drug mixtures that he developed, later
called galenicals, formed the basis for pharmacological treatment for many
centuries.

Galen's teachings dominated medical practice and science throughout late
antiquity and the Middle Ages, and many of his medical and philosophical
writings were translated into Arabic. His work reflects the way in which clinical practice and medical science have evolved over the years. It marks an
attempt to learn more about the workings of the human body through anatomy and physiology, interpreted in light of contemporary ideas about the
nature of living things, and the need to help sick people by means of empiri-

cal treatments that seemed reasonable at the time. Why did his influence last so long? The very extent of his writing, comprising perhaps over four hundred cataloged works, must have played an important part; the communication of ideas, right or wrong, with sufficient vigor and confidence is a valuable prerequisite to the development of a reputation, a problem that still dogs the medical profession today. And there is the matter of authority. As the Johns Hopkins University historian Owsei Temkin has pointed out, medieval men were taught at an early age to accept authority. This was no less true of their attitude toward the prevailing social order, which they believed was divinely established, than of their acceptance of religious belief. Similarly, physicians did not question authority. Truth was believed to have been revealed in the past. Hence Galen's teachings persisted; few authorities challenged them. Faith in authority still pervades the medical profession; in many parts of Europe medical students are discouraged from questioning dogma, and an unhealthy respect for professorial infallibility reigns.

In the period leading up to and including the Middle Ages, medicine in Europe was dominated by the church. Access to works of Hippocrates, Galen, and other ancient writers was limited, and many of their more important works were not translated into Latin until the sixteenth century. This period of European history, during which, except for the expansion of hospitals and the development of simple public health measures to attempt to curb epidemics of infectious disease, there was little progress in the sciences that underpin modern medicine, corresponds roughly to the golden age of Islam. The contribution of Islamic science is, as Ziaudain Sardar notes in a recent review in *Nature,* still a source of controversy. One view is that science began with the Greeks, was preserved by the Arabs, and was finally handed back to its rightful owners at the beginning of the Renaissance. But recent scholarship suggests that Islamic science played a much more important role than that. While there is no doubt that by translating the works of Hippocrates and Galen, it did much to preserve and disseminate the teaching of the Greek medical texts, it also furthered the development of many significant aspects of medical practice. Teaching hospitals were established in the Arab world, and a system of examination and qualification for medical practice was observed. There were improvements in surgical practice, including the cauterization of wounds, and major advances in the use of medicinal herbs and drug therapy. It even appears that the mechanism of the pulmonary circulation—that is, how blood is pumped from the right side of the heart to the lungs and back—was worked out in Egypt as early as the thirteenth century, a fact that was unearthed only as recently as 1924.

The end of the fifteenth century and the sixteenth century saw an awakening of critical thought and challenging of authority in Europe. The work of Copernicus (1473–1543) and, later, Galileo (1564–1642) established the principle of a moving earth and a heliocentric universe, Martin Luther (1483–1546) challenged the domination of the Catholic church, and the medical world started to question some of the teachings of Galen. It was during the Renaissance that Vesalius (1514–1564) and the Padua school firmly established human anatomy as one of the key medical sciences (figure 2).

Vesalius was born in Brussels and became professor of surgery and anatomy in Padua in 1537. He published his great work *De humani corporis fabrica* (On the fabric of the human body) in 1543. This wonderfully illustrated book, seminal in the development of anatomy and physiology, was based on dissections that Vesalius completed before he was twenty-eight years old. His work laid to rest, among other things, at least one of Galen's misconceptions—the idea that blood could travel from the right- to the left-hand side of the heart through the wall, or septum, which separates them. "Not long ago," Vesalius wrote in the second edition of his book, "I would not have dared to diverge a hair's breadth from Galen's opinion. But the septum is as thick, dense and compact as the rest of the heart. I do not see, therefore, how even the smallest particle can be transferred from the right to the left ventricle through it." The apologetic tone of this statement is further testimony to the tenacity of Galen's teaching and the power of authority.

With the exception of some remarkable innovations in surgery, notably in Paris through the work of the surgeon Ambroise Paré (1510–1590), therapeutics advanced very little during the Renaissance. The ideas of the alchemists during later antiquity and the Middle Ages were reflected in the work of a colorful, peripatetic Swiss physician, Theophrastus Bombastus von Hohenheim (1493–1541), who assumed the name Paracelsus. He rejected the notion that health is determined by the balance of the four humors and believed that the body is essentially a chemical system made up of the basic principles of the alchemists. Thus health can be restored by the administration of mineral medicines, not by the organic remedies of Galen. His influence was strong enough to compete with the teachings of Galen through much of the sixteenth and seventeenth centuries. Yet, although he enlivened his lectures on therapeutics in Basel by setting fire to the works of Galen and the influential Islamic physician Avicenna (980–1037) and disregarded much of the medical teaching of the past, he added little of value to therapeutics. However, some of his work, particularly his observations on occupational medicine, the treatment of syphilis, and the relation between goiter and the mineral content of water, was quite advanced for the time. By stressing the importance of chemistry in medicine and by developing a complex but compassionate philosophy of clinical practice, Paracelsus and his many followers undoubtedly opened up an important debate on the nature of disease and therapeutics that was to influence later thinking in scientific medicine.

ANDREAE VESALII .ÆTA. 28

FIGURE 2

Andreas Vesalius (1514–1564), dissecting the forearm of a corpse. Anonymous engraving, ca. 1600.

Considering the slow rate of progress in medicine and biology over the two thousand years from the beginnings of Greek medicine, the rapid pace of developments over the next two hundred years seem all the more remarkable. The advances in science and philosophy in the seventeenth century created an environment that led to equally momentous changes in the medical sciences throughout the seventeenth and eighteenth centuries. Modern physics was founded by Isaac Newton (1642–1727), and the work of Robert Boyle (1627–1691) and Robert Hooke (1635–1703) finally disposed of the basic Aristotelian elements of earth, fire, and water. Hooke, apart from making major contributions to physics, astronomy, and microscopy, showed that animals maintained in a vessel from which air was evacuated died and that those whose chest had been cut open could be kept alive by artificial respiration with bellows placed in the trachea.

Another man whose thinking did much to shape the future of biology and medical practice was the French mathematician, philosopher, and biologist René Descartes (1596–1650). His natural philosophy was very different from anything that had gone before. In particular, he held that all material things, whether animals, plants, or inorganic objects, are ruled by the same mechanical laws. His book *De homine* took a thoroughly mechanistic view of the human body. All living things can be looked on as machines. A sick man is like an ill-made clock; a healthy man, a well-made clock. While this mechanistic approach to human biology and medicine was to be of considerable value and to free much of later thought from the Galenic view of disease, it undoubtedly tended to direct the attention of doctors toward diseases rather than toward the patients who had them. This will be a recurrent theme as we trace the evolution of scientific medicine.

The seventeenth century also brought a major revolution in human physiology, particularly the description of the circulation of the blood by William Harvey (1578–1657) (figure 3). Harvey graduated from Cambridge University and studied in Padua, where he heard about the discovery of valves in veins by Girolamo Fabrici (Hieronymus Fabricius, ca. 1533–1619) and learned about the valves in the heart, which allow the blood to flow in only one direction. After completing his travels, Harvey became a physician to St. Bartholomew's Hospital, London, and to King Charles I, whom he accompanied to the civil war. He mixed a busy practice with the dissection of many different animals. He gradually conceived the idea of the circulation of the blood, which was published in his *Exercitatio anatomica De motu cordis et sanguinis in animalibus* (An anatomical treatise of the motions of the heart and blood in animals) in 1628. The argument of this famous work, which sets out an almost complete description of the circulation, and which applies simple statistical methods to determine the output of blood from the heart, was not accepted by many of Harvey's contemporaries, nor did his research

FIGURE 3

William Harvey (1578–1657). Etching attributed to Richard Gaywood after a hypothetical drawing ascribed to W. Hollar.

have much effect on his own clinical practice; he intended to write another book about its medical applications, but never did so. The only gap in Harvey's description of the circulation was an explanation of how blood could move from the arteries to the veins in the tissues.

Even when interrupted by the remarkable insights of a genius like Harvey, medical research is a continuum, one step forward always depending on what has gone before. The historian Allen Debus has described how the story of Harvey's discovery reflects a succession of pupil-master relationships, starting with Vesalius in Padua. Vesalius's successor Realdo Columbo (ca. 1510–1559) was followed by Gabriello Fallopio (1523–1562), who gave his name to the Fallopian tubes, or oviducts. Fallopio was succeeded by Fabricius, who taught Harvey about valves in veins, a vital factor in the development of his ideas about the circulation. The first European description of the pulmonary circulation appeared in 1553 in a book entitled *Restitution of Christianity*, written by the radical theologian, anatomist, and astronomer Michael Servetus (1511–1553). Columbo of Padua arrived at the same conclusion about the pulmonary circulation, probably independently, in his text on anatomy published in 1559. Interestingly, Vesalius and Servetus both studied under the same teacher in Paris, Johannes Guinter. Thus there is a direct line, starting in Paris, from the great Paduan school of anatomy to William Harvey's discovery of the circulation.

The story of the development of scientific medicine contains many examples of how centers of excellence, like Padua, have spawned long lines of scientists who made major contributions to their fields. The biochemist Hans Krebs has pointed out that modern biochemistry evolved in this way, and more recently several laboratories in the United States and Europe have produced Nobel Prize winners with monotonous regularity. Governments that are suspicious of "centers of excellence," and which would rather spread resources more widely (and thinly), would do well to study the origins of some of the major advances in medical science over the years.

Harvey's work was undoubtedly one of the great landmarks in the development of scientific medicine. In his Harveian Oration, delivered to the Royal College of Physicians of London in 1906, the physician William Osler aptly summed up his contribution as follows: "But at last came the age of the hand—the thinking, devising, planning hand; the hand as an instrument of the mind, now reintroduced into the world in a modest little monograph of seventy-two pages, from which we may date the beginning of experimental medicine."

As we will see as we explore how scientific medicine has evolved, major advances have followed the discovery of new experimental techniques at least as often as the formulation of novel ideas. Harvey's description of the circulation was completed by the researches of an Italian, Marcello Malpighi (1628–

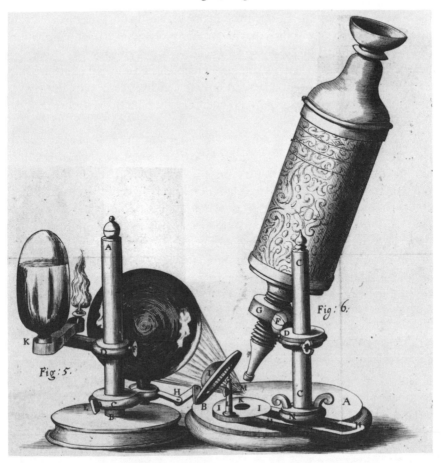

FIGURE 4

Microscope and lighting apparatus. From Robert Hooke, Micrographia *(London: J. Martyn and J. Allestry, 1665).*

1694), who described the capillary circulation for the first time, a discovery made possible by the invention of the microscope. Just as the development of the telescope opened up the universe to Copernicus and Galileo, the microscope had a profound influence on biology and medicine. It is not clear who first invented it; a spectacle maker who worked in Middelburg, Zacharias Jansen (ca. 1588–1631), is a leading candidate. A vivid description of the world as seen through the microscope appeared in the delightful miscellany *Micrographia,* published in 1665 by Robert Hooke (figure 4). But the work that was to give microscopy its central role in the development of biology was carried out by a talented Dutchman, Antoni van Leeuwenhoek (1632–1723), a draper who lived in Delft. His interest in microscopy started with the development of low-power magnifying glasses to inspect the quality of his cloth (figure 5). Over the years he ground more than 550 lenses, of which the best had a magnifying power of over five hundred times. In a series of letters to the Royal Society of London, which elected him a Fellow in 1680, he described a completely new world, which included not only the cells of the blood and spermatozoa but also bacteria, although he did not recognize them as such.

Major advances in many branches of biology in the seventeenth century also stem from the activities of men who dominated the British scientific and medical scene after Harvey. A dozen of them, including four practicing physicians and founder members of the Royal Society, met regularly in Oxford to discuss different aspects of natural philosophy and to describe their experiments. Among them were several physician-scientists who left their mark on medical science. Thomas Willis (1621–1675) carried out a detailed study of the circulation to the brain; his description of the cerebral arterial circle that still bears his name was one of the most important anatomical findings of the period. Richard Lower (1631–1691), Willis's friend and compatriot, studied the pulmonary circulation, deduced how arterial blood arises in the ventricles of the heart, worked on the function of the diaphragm, and, in passing, performed one of the first blood transfusions in man. He was assisted in many of his experiments by the philosopher and physician John Locke (1632–1704), who, though he gave up medical practice, never lost his fascination for physiology.

Lower's first attempt at human blood transfusion mirrors the broad-ranging interests of this group of scholars. Another of them, Christopher Wren (1632–1723), a great architect but also a talented astronomer and anatomist, was fascinated by the problem of cannulating blood vessels and showed that dogs could be given drugs (including alcohol) more effectively intravenously than by mouth. Lower, excited by Wren's work and using his silver cannulas, carried out the first blood transfusions on animals. On November 23, 1667, before a skeptical audience at the Royal Society and unhampered by the attentions of a present-day ethics committee, he injected a small amount of sheep's blood into the veins of a mildly deranged clergyman. Although the rationale for this experiment is not clear, the clergyman recovered sufficiently to give an account of the operation the following week. But, like Harvey's work, this remarkable achievement had to wait for several hundred years, and the discovery of blood groups, before it could make an impact on clinical practice.

It is interesting to compare the sophistication of the experimentation of the Oxford group with their day-to-day clinical practice. John Locke recorded some of Richard Lower's cures in his notebooks, accounts that make extraordinary reading. Lower's prescription for healing wounds precludes the use of

FIGURE 5

Facsimile of the Leeuwenhoek microscope in the University of Utrecht.

lint made from the shirt of a man of strong complexion but advises the use of material made from a woman's smock after she has had a menstrual period. And his prescription for preserving the complexion of fashionable ladies involved cutting up puppies, distilling them over a gentle fire, and applying the resulting fluid morning and evening. He also refused to prescribe Peruvian bark, many years later found to be a source of quinine, for King Charles II's malaria, even though his contemporaries had already demonstrated its value.

When Thomas Willis returned to Oxford after the civil war, in which he had served with the Royalist University Legion, he was accompanied by Thomas Sydenham (1624–1689), another remarkable physician who was also to have a major influence on medical practice, though not in the way his Oxford contemporaries did. Sydenham, who fought as a parliamentary cavalry officer in the war, obtained his medical degree in one year. His view of the role of the universities, and of academe in general, was extremely jaundiced. One of his contemporaries records that he thought that it was better to send a man to Oxford to learn shoemaking than to practice medicine! He disliked the pretentions of doctors who seemed to believe that their scientific research was more important than their day-to-day practice, and he taught that clinical medicine could be learned only at the bedside. His detailed observations of his patients enabled him to describe several diseases for the first time, and his elegant clinical accounts, particularly of gout, from which he suffered for many years, have never been bettered. He helped to define the role and methods of the administration of Peruvian bark for the treatment of fevers. He was equally at home analyzing physical and mental illnesses, believed in the power of the body to cure itself, and was skeptical of some of the standard forms of treatment of the time, especially purging, cupping, and bleeding. And he was a great believer in the value of a healthy life and exercise. One of the many anecdotes about his clinical method tells of his advice to a patient to go on horseback to a wonder-working physician at Inverness, in the north of Scotland. When the patient arrived, he found that this doctor did not exist; Sydenham was convinced that the long horse ride combined with exercise and the anger at finding that his trip was in vain would effect a cure. Accounts of his work suggest that he practiced a very high level of what is now called holistic medicine—that is, he was interested not only in diseases but in every facet of their effects on his patients.

Sydenham has been called the English Hippocrates. He laid the foundation for a view of clinical practice that is still held by many doctors today. It emphasizes the primacy of the doctor-patient relationship, stresses the acquisition of knowledge and skills at the bedside, is suspicious of theory and the value of laboratory research, and, above all, looks on doctoring as a pragmatic art.

There is no doubt, therefore, that the seventeenth century was a remarkable era in the development of science as a whole and of medical science practice in particular. The universal laws of gravity and planetary motion were defined, and electricity, magnetism, chemistry, and many other branches of science were established. With Harvey's work on the circulation of the blood, experimental medicine was put on a firm footing. And, most important, there was a new spirit abroad that questioned the past and that would, in the future, blossom into a more critical and analytical approach to medical practice.

William Harvey's revolutionary work on the circulation made little impact on day-to-day clinical practice. Bloodletting continued with undiminished enthusiasm, resting on what seemed reasonable grounds at the time. Particularly in the case of inflammatory fevers bleeding would have been a logical treatment since the disease process was believed to reflect an excess of hot, feverish blood; patients too weak for bloodletting were treated with milder methods, such as cupping and the application of leeches. A few questioning voices were heard. Jan Baptista van Helmont (1579–1644), for example, suggested that the procedure be subjected to a clinical trial. It was, but not for another two hundred years. An equally long time was to elapse before Harvey's revolutionary ideas, backed by sufficient knowledge of the physiology of the heart and vascular system and the availability of new pharmacological agents, were to form the basis of modern cardiology. Similarly, the remarkable experiments of the Oxford physiologists had little effect on medical care at the time. It will be a recurring theme throughout our story that seminal discoveries in the basic medical sciences may have no practical value for many years but that when major advances in patient care are made, they often rest on fundamental work of this kind.

Another long-standing characteristic of the medical world is the tension that exists between medical scientists and those who confine their activities to the bedside. This division, exemplified by Sydenham's lack of respect for his more experimentally minded colleagues, is still in evidence in our teaching hospitals. Many clinicians regard medical practice as an observational art and believe that the sciences that underpin it are largely irrelevant to their work. They transmit this view to their students, many of whom, because their early training in the basic sciences makes little effort to relate what they are learning to their future role as doctors, are only too receptive to it. In its extreme form this attitude is expressed as a feeling of pride on the part of doctors in not understanding the scientific basis of what they are doing. When the great British physiologist and pharmacologist Sir Henry Dale arrived at St. Bartholomew's Hospital to study medicine in 1900, he was advised by a senior physician, Dr. Samuel Gee, to forget all the physiology he had learned at Cambridge, since medicine was not a science but an empirical art. The dichotomy between medical practitioners and those who try to mix the task of looking after patients with serious medical research can be destructive. At its worst it breeds a state of mind impervious to criticism and retards progress in clinical practice.

The eighteenth century, the Age of the Enlightenment, was, as the historian Thomas Hankins has observed, a period of transition. The scientific revolution continued, and the disciplines that we recognize today gradually began to take shape. There was steady progress in many of the branches of science that later were to have an important role in the development of modern medicine. For the emerging biological sciences it was a time for questioning the divine origin of living things and recognizing that there is order in nature. The spirit of order was epitomized by the work of the Swedish physician and botanist Carl von Linné, or Linnaeus (1707–1778). After publishing his great work on the classification of plants, animals, and minerals, *System of Nature,* in 1735, he went on to write a classification of diseases, which appeared in 1763. Chemistry continued to flourish and began to have an important impact on human physiology. Although thinking was still largely mechanistic, there was increasing insight into the nature of living processes. Joseph Priestley (1733–1804) isolated oxygen in 1777, and the French chemist Antoine Lavoisier (1743–1794) discovered its true nature and function in respiration.

The investigations of Lavoisier, who was to die on the guillotine, led him to believe that "the animal machine is governed by three main regulators: respiration, which consumes oxygen and carbon and provides heating power; perspiration, which increases or decreases according to whether a great deal of heat has been transported or not; and, finally, digestion, which restores to the blood what it loses in breathing and perspiration." Here was the beginning of a genuine understanding of biological functions, because they were becoming accessible to the methods and concepts of physics and chemistry. And it was during the eighteenth century that physiologists began to study the way in which organisms reproduce, a story that we will take up in chapter 8, when we look at the origins of modern genetics. But despite these new insights, knowledge of the basis of life itself remained hazy and developed into little more than a vague form of vitalism.

There was further progress in the sciences that are more directly related to medical practice. Experimental physiologists began to investigate the electrical properties of nerve impulses, and a start was made in working out the processes of digestion. In 1757 the Swiss polymath Albrecht von Haller (1708–1777) published the first volume of his great summary of physiological knowledge of the time, *Elementa physiologiae.* This valuable eight-volume compendium set out what was then known about the function of all the parts of the human body. Giovanni Battista Morgagni (1682–1771), professor of anatomy at Padua, described the results of over seven hundred postmortem examinations in a book published in 1761, *On the Seats and Causes of Disease Investigated by Anatomy*. And later the French surgeon and anatomist Marie François Xavier Bichat (1771–1802) described the microscopic appearances of diseased organs, an important step forward in the development of ideas about the nature of disease. As a result of these seminal studies, pathology and the value of the autopsy were firmly established. And by attempting to correlate the symptoms and signs observed in patients at the bedside with the organ changes found at the postmortem, Bichat and others laid the foundations for a more functional approach to exploring disease mechanisms.

Some genuine therapeutic advances were made in the eighteenth century, notably the discovery by William Withering of Birmingham (1741–1799) that an extract from the foxglove is of value in the treatment of dropsy, or edema, due to heart disease. Edward Jenner (1749–1823), a country practitioner in Gloucestershire, performed the first vaccination, and the London surgeon John Hunter (1728–1793), by initiating the careful pathological study of surgical specimens, started to put surgery on a more rational and scientific footing. There were also the beginnings of a more enlightened attitude toward the mentally sick. Thermometers were invented, although not used in clinical practice, and Benjamin Franklin of Boston (1706–1790), best known for his development of lightning conductors, invented bifocal lenses for treatment of disorders of the eye and, like several of his contemporaries, attempted to treat paralysis and a variety of other diseases with electricity.

The changes in obstetric practice were noteworthy. Before the mid-eighteenth century very few doctors were involved in childbirth; it was almost the exclusive domain of midwives. Very few texts on midwifery existed in the English language, and little was known of the mechanisms of normal labor or of its complications. Yet by the end of the century there were many accomplished accoucheurs, doctors skilled in obstetrics, and in Great Britain almost every surgeon-apothecary, today's general practitioner, had developed skills in the practice of midwifery. The mechanisms of normal labor had been established, and most of its common complications had been described, together with ways of dealing with them. The historian Irvine Loudon has suggested that compared with those of the eighteenth century, the advances in obstetrics in the nineteenth and first third of the twentieth centuries seem almost trivial.

We can also see developments in medical education and clinical practice that were to gain momentum during the nineteenth century and lead to the teaching hospitals of today. One of the most influential physicians of this period, Hermann Boerhaave (1668–1738), set up a hospital designed specifically for teaching. He moved clinical teaching from the lecture theater to the bedside and established Leiden as one of the major centers for the study of medicine. Although, like that of many later teaching hospital clinicians, his approach was rather didactic and uncritical, his emphasis on the importance of clinical method was to be of considerable value in the evolution of medical education. But his views on physiology, and his interpretations of pathology in quasi-chemical terms, reflected the limited understanding of human biology that continued throughout the eighteenth century.

Therapeutics remained largely empirical and, except for observational skills and noninterventionist approaches based on the teachings of Sydenham and for some major improvements in obstetrics, day-to-day practice advanced very little. We noted earlier that George III of England was almost killed by his medical attendants in their efforts to cure his episodic bouts of madness by intensive bleeding, purging, and cupping. In 1799 President George Washington was bled, purged, and blistered until he died, two days after complaining of a sore throat. On the other hand, social conditions improved, and for the first time the birthrate exceeded the death rate in London. Many new hospitals and dispensaries opened, and it was realized that disease could be controlled by better ventilation and sanitation and by better diets. Health became an important political issue, and health education programs were established in several European countries. The scene was set for the great social reforms of the nineteenth century that were to do so much to improve the health of the community.

THE NINETEENTH CENTURY AND THE BEGINNINGS OF MODERN SCIENTIFIC MEDICINE

The nineteenth century was a time of extraordinary movement on most fronts. The volume of scientific knowledge led to the notion of "the scientist" and to specialization; such was the increase of information that the kind of polymath who had driven science in previous centuries could no longer prevail. It was a period of major advances in physics and chemistry, including the development of atomic theory and electromagnetism. Above all, it was the century in which the biological sciences finally achieved a stature equal to that of mathematics and physics. Charles Darwin (1809–1882), with his description of the process of evolution, changed the whole course of biological thinking, and the work of Gregor Mendel (1822–1884) laid the ground for a new science called genetics, which was used later to describe how Darwinian evolution came about. Louis Pasteur (1822–1895) and Robert Koch (1843–1910) founded bacteriology, which, as we will see, was over the next hundred years to lead to the control of many infectious diseases. And Claude Bernard (1813–1878) and his followers in Paris enunciated the seminal principle of the constancy of the internal environment, a notion that profoundly influenced the development of physiology and biochemistry. In the biological sciences the nineteenth century saw the gradual decline of vague theories about "life forces" and a growing belief that living processes can be understood in terms of the laws of chemistry and physics working through complex interactions between the many different types of cells that constitute all living things. This movement was to culminate in the extraordinary achievements in biochemistry and molecular biology in the twentieth century.

It is impossible to overemphasize the importance of the concept of the cell, or cell theory, in the development of the biological and medical sciences during this period. As the French Nobel laureate François Jacob put it in his book *Logic of Life,* "With the cell, biology discovered its atom." The notion of the cell had existed for a long time; the word was used by Robert Hooke in the middle of the seventeenth century to describe the cavities that he observed when examining slices of cork under the microscope. During the eighteenth century observations on cells continued to accumulate, but a general theory about their nature began to crystallize only as microscopes became more efficient in the middle of the nineteenth century. In 1838 the German botanist Matthias Jacob Schleiden (1804–1881) suggested that plants consist of cells. In the following year this notion was extended to animal tissues by another German, the physiologist Theodor Schwann (1810–1882), who wrote that "cells are organisms, and animals as well as plants are aggregates of these organisms, organised in accordance with definite laws." A few years later Rudolf Virchow (1821–1902) pointed out that cells cannot arise spontaneously but are the progeny of other cells. In 1858 he published his most famous work, *Die Cellularpathologie,* which applied cell theory to the study of pathology. All diseases, he held, are diseases of cells. This marked the dawning of modern cellular pathology and of the study of disease at the microscopic and, later, submicroscopic levels.

As we have seen, the work of Morgagni, Bichat, and others during the second half of the eighteenth century set pathology on the road to modern scientific medicine. But in the early nineteenth century pathology was still largely confined to describing the naked-eye appearances of organs at a postmortem, as exemplified by the work of the Viennese Karl von Rokitansky (1804–1878), one of the first physicians to hold a university chair in the subject. The chief problem with this approach was that, when viewed in this way, organs, though clearly the site of disease, often appeared to be normal. To resolve this dilemma Rokitansky formulated a vague and completely erroneous humoral theory that saw these diseases as originating in an imbalance of substances like fibrin and albumin in the blood. But Virchow, by using the microscope, was able to observe pathological changes in apparently normal tissues and organs, and in the process he put the study of pathology on a firm scientific footing. He identified many of the different cells of the blood, described leukemia for the first time, defined the criteria for malignant tumors, and introduced the terms "thrombosis" and "embolism" into medical practice. He was one of the dominant figures in the medical sciences in the second half of the nineteenth century, and his work had a major influence on the way in which they were to develop in Europe and the United States in the future.

Along with these advances in the basic medical sciences came rapid progress in clinical investigation and treatment. René Théophile Hyacinthe Laennec (1781–1826) invented the stethoscope, and an Italian, Scipione Riva-Rocci (1863–1937), designed a device for measuring blood pressure, very similar to those in use today. The popularization of the art of percussion of the chest by Jean Nicolas Corvisart (1755–1821), in which the physician can assess the state of the underlying lung by sensing the character of vibrations generated by gentle taps on the chest wall, greatly facilitated the diagnosis of pneumonia and other respiratory diseases. Remarkably, this simple bedside sign was first described by a Viennese physician, Leopold Auenbrugger (1722–1809), as early as 1761, in a small book, *Inventum novum ex percussione thoracis humani,* which was later translated by Corvisart. The technique is still used today, and if doctors are skilled at interpreting the character of the vibrations, and do not deaden them by heavy-handedness, an effect similar to too much use of the soft pedal on a piano, they can obtain a great deal of information about what is going on in the lungs and in pleural and abdominal cavities.

Up until the nineteenth century a doctor's understanding of disease processes depended almost entirely on his clinical skills at the bedside, backed up by what could be gleaned in the autopsy room. However, the development of methods for the chemical analysis of blood and other body fluids, together with improvements in microscopy, opened up new dimensions for the diagnosis of disease in living patients. These advances culminated in the discovery of x-rays by Wilhelm Röntgen (1845–1923) in the latter half of 1895. By the turn of the century this remarkable ability to visualize what, up to then, had been completely inaccessible parts of the body had started to find application in every branch of medical practice.

During the nineteenth century nitrous oxide and ether anesthetics were developed in the United States, and in Great Britain chloroform anesthesia was used for childbirth. Modern surgery evolved during this time, thanks mainly to the remarkable insights of Joseph Lister (1827–1912), who appreciated the importance of Pasteur's work for the control of wound sepsis. The application of Lister's antiseptic techniques, together with the gradual improvement in anesthesia, led to the flowering of surgery in the latter half of the nineteenth century and early part of the twentieth. And largely through the activities of colonial medical officers the cause of malaria and other parasites was discovered and tropical medicine was born.

It was also during the nineteenth century that psychiatry began to flourish. The importance of mental disorders had, of course, been recognized since the time of Greek medicine, and Willis and Sydenham wrote extensively about melancholia, hysteria, and hypochondria in the seventeenth century. During the eighteenth century hospitals for the mentally ill were established, in Europe and in the United States. But not until the nineteenth century did psychiatry come into its own as an academic discipline, particularly in German-Speaking countries. By careful studies of the natural course of mental illness, the German psychiatrist Emil Kraepelin (1856–1926) was able to develop a comprehensive classification of mental illness, little different from the one used today. The early work of Kraepelin's institute led to the belief that the bulk of psychiatric illness may have a genetic or physical basis. A member of Kraepelin's staff, Alois Alzheimer (1864–1915), described the anatomical changes in the brain in the common form of dementia that still bears his name. But other members later became involved in the Nazi movement and the sterilization of the mentally handicapped, an involvement that brought the notion of the physical basis of mental illness into temporary ill repute.

A very different approach to the study of psychiatric disease was developed in Vienna by Sigmund Freud (1856–1939) and his followers. Their approach started to flourish around the turn of the century. It led to the formation of the psychoanalytical school, which was to gain many adherents both in Europe and in the United States and which until its decline in the middle of the twentieth century, was to have a major influence on psychiatry.

As Jacob Bronowski emphasized in his delightful book *The Common Sense of Science,* scientists of the eighteenth and nineteenth centuries tended to believe that natural phenomena, if taken apart step by step, could be traced to all their causes. But toward the end of the nineteenth century it was becoming clear that chance also plays a key role in natural events. This recognition led to a new approach to the weighing of experimental evidence, one based on statistical methods and destined to have an important influence on biology and medicine in our century.

In the mid-nineteenth century a French clinician, Pierre Charles Alexandre Louis (1787–1872) (figure 6), laid the basis for the statistical analysis of medical practice. One of his earliest successes was to compile sufficient data to prove that bloodletting, which had been practiced for centuries, was not only useless but positively harmful in the management of many diseases. The work of Louis and his followers was to have a profound effect on medical science and clinical practice. Its critical approach to therapy continued to

expand over the next hundred years. Ultimately, it gave rise to modern epidemiology and the use of clinical trials for the assessment of different forms of
treatment. If clinical science, as practiced today, can be said to have started
at any particular time, it was in Paris in the middle of the nineteenth century.
Already earlier, the foundation of the University of Berlin in 1810, had stimulated the expansion of the university system and the establishment of departments for the study of the new sciences that would underpin the growth of
medical knowledge.

Another great contribution of the nineteenth century was the development
of public health and preventive medicine. Despite a steady improvement in
living conditions and control of the environment over the centuries, it was
not until then that public health programs started to evolve. One of the most
colorful episodes in this new awareness involved the British anesthetist and
hygienist John Snow (1813–1858). Snow made careful observations of the
cholera epidemic in London in 1848 and concluded that the infection was
spread by the water supply. His solution to controlling the epidemic was to
remove the handle of the Broad Street pump. Visitors to modern-day London
can visit this scene, now called Broadwick Street, in Soho, and can toast John
Snow in a public house that commemorates his name.

Significant advances in public health occurred on both sides of the Atlantic. In Great Britain several important acts were passed by the government
that brought vast improvements in sanitation and water supplies, and a general cleaning up of the environment. The first, the Public Health Act of 1848,
often known as the Chadwick Act, after Sir Edwin Chadwick (1800–1890),
the civil servant largely responsible for framing its provisions, required the
paving of streets, construction of drains and sewers, collection of refuse, and
procurement of adequate domestic water supplies, together with street cleaning and fire fighting. The 1875 act added provisions regarding the keeping
of animals, public toilets, control of vermin, reporting of disease, registration
and inspection of nursing homes, and maternity and child welfare. Such reg-
ulations were greatly to improve the quality of life and health in Great Britain.
Similar legislation was enacted in the United States at about the same time.

It should be remembered, however, that even in industrialized Western
societies, life was quite different in the middle of the nineteenth century from
what it is now. The first recorded figures for the United States show that the
life expectancy at birth for those who lived in Massachusetts in 1870 was
forty-three years; in 1980 it was seventy-four. The number of deaths in 1870
per 1,000 live births in the same population was 188; in 1980 it was 13.
These figures are a remarkable testimony to advances in both public health
and medical science over the last hundred years.

FIGURE 6

Pierre Charles Alexandre Louis (1787–1872). Lithograph by Maurin. From F. Garrison, History of Medicine *(Philadelphia: W. B. Saunders, 1917).*

LESSONS FROM THE PAST

Mark Twain lamented the improvements in medical practice in the nineteenth and early twentieth centuries because, as he put it, doctors had done so much in former times to keep the population growth in check. Are there any more profound lessons for us from the first few thousand years of man's attempts to get to grips with illness as we consider the role of science in the future of medical practice?

First, there is the matter of authority. The strength of Galen's influence for over a thousand years illustrates how progress may be retarded by the acceptance of everything that has gone before. This unquestioning approach continued to dominate the medical world into the twentieth century and is still entrenched even in some of our great teaching centers. It extends from the traditional belief that there is only "one way" to examine a patient to many forms of investigation, treatment, and prevention that are of unproven value. It is remarkable that, even though over a hundred years have passed since Louis pioneered the statistical approach to medical practice, only in the last few years has some form of structured audit been applied to day-to-day clinical work in Western hospitals.

The medical profession has always been slow to accept the results of work that seems to contradict its long-held prejudices. After Louis had shown that bleeding is an ineffective form of treatment, many doctors continued with this practice. Although it is often assumed that bloodletting had become less fashionable by the end of the nineteenth century, the 1923 edition of William Osler's *Principles and Practice of Medicine,* the standard textbook of medical practice of the time, recommended it for the treatment of heart failure and pneumonia. Indeed, this practice enjoyed a revival in the early part of this century and was used for the management of many disorders. An unwillingness on the part of doctors to change their practice, despite the existence of good evidence that it is not appropriate, remains with us. About twenty years ago it was shown quite unequivocally that the use of drugs to slow the rate of clotting of the blood reduces the number of serious complications after orthopedic surgery, notably the formation of blood clots in veins, which may break off and lodge in the lungs. But there was a long delay before this approach was used widely; even today some patients do not receive anticoagulant drugs before major hip surgery. There are still many other discrepancies between what clinicians perceive as appropriate, on the basis of experience and received wisdom, and evidence founded on solid scientific method.

In some cases the reasons for doctors' not changing their practice have complex roots. The delay in the introduction of chloroform anesthesia in childbirth in the nineteenth century offers an interesting example. In part it may have been based on the biblical teaching "In sorrow thou shalt bring forth children." This view was finally discredited in a remarkable pamphlet written by James Simpson (1811–1870), who introduced ether and chloroform anesthesia to clinical practice in Scotland. He pointed out that the original biblical text was a mistranslation from the Hebrew; the word "sorrow" should have been translated as "labor." But strong opposition also came from London obstetricians who refused to use chloroform because they believed that it caused all sorts of complications, particularly convulsions, which might not appear for a year or more, a belief based on no evidence whatever. As a result many women had to travel from London to Scotland to be delivered under chloroform anesthesia by Simpson. It was not until Queen Victoria asked John Snow to administer chloroform during the birth of Prince Leopold and Princess Beatrice that the London obstetricians finally backed down. They were not pleased, incidentally, when the queen not only wrote a warm letter of congratulation to Simpson but also conferred on him a knighthood. Professional prejudices, suspicion of the new and innovative, and ill-founded dogma have undoubtedly all played a role in delaying progress toward rational and scientifically based medical practice.

But, as with Thomas Sydenham, at least some of the medical profession's apparent disregard for scientific evidence may reflect complex and deeply held convictions about the nature of clinical care. Recently, the medical historian William Bynum has addressed the thorny problem of why the historical partnerships between science and medicine, knowledge and practice, have never been straightforward. He traces a tradition of medical practice, based on admiration for Sydenham's views, that is still a major force in our teaching hospitals of today. It reflects a view of medicine as an individualistic art founded on systematic bedside observations and skills, with its suspicion of theory and laboratory research. But Bynum does not see it as naively antiscientific or dogmatic. In part, it may result from the different attitudes of mind required of the producers of medical knowledge and those who dispense it. The English clinical scientist Thomas Lewis summarized these qualities as follows: "Self confidence is by general consent one of the essentials to the practice of medicine, for it breeds confidence, faith, and hope. Diffidence, by equally general consent, is an essential quality of investigation, for it breeds inquiry. Here then are chief characteristics, each necessary in its own sphere, each unsuited to the other. . . ."

I am sure that Lewis was right; it is extremely difficult for one person to combine the personal qualities required of both a good scientist and an effective clinician. Those who are sick need an air of confidence at the bedside; their recovery may not be hastened by "ifs and buts" and by explanations of the current state of scientific ignorance about their disease. Since clinicians are continually dealing with patients whose diseases do not seem to have a rational explanation based on the science that they learned as students, yet have to do their best to help them, it is not surprising that the ghost of Sydenham still walks the wards of our teaching hospitals.

The best of today's scientifically trained doctors analyze their patients' illnesses and treatment as far as they can with the tools of modern medical science but frequently find themselves in a situation in which knowledge is incomplete and some form of therapy, even if of unproven value, has to be tried. The more medical knowledge increases, the more difficult it is for critical and caring clinicans to dissociate their scientific training from the practical necessity of doing something to relieve suffering, even though they know that they are rarely sure about what they are doing. Medicine has remained an art, but one that has become harder to practice as knowledge of the ignorance that underlies it has increased. Doctors still have to learn to live with uncertainty. For young doctors this can be one of the most difficult and disturbing aspects of their work. At least some of the tension between the science and the practice of medicine, which has undoubtedly impeded medical progress, may reflect a tendency on the part of doctors to rationalize the empiricism of much of what they do.

The history of medical science and practice also emphasizes the importance of pursuing careful scientific experiments for their own sake and the dependence of major advances on a critical mass of knowledge on many different fronts. We have seen how the great seventeenth-century physician-scientists like Harvey, Lower, and Willis carried out elegant experiments and had a sophisticated knowledge of the cardiovascular system, yet were unable to transfer their discoveries from the laboratory to the clinic. As we will see later, it required work from many different disciplines, much of it unrelated to clinical practice, to yield the clinical fruits of Harvey's great concept of the circulation. The development of modern surgery, which led to the routine practice of appendectomy with which we opened this chapter, provides a particularly good example of this pattern of medical progress.

In the days before anesthetics surgery relied on a knowledge of anatomy combined with speed and dexterity on the part of the surgeon. Anecdotes abound, including the story of a surgeon who, while performing a midthigh amputation, removed two fingers from his assistant and both testicles of an observer. But there are better-authenticated cases, such as those of James Syme (1799–1870), who regularly amputated at the hip joint in less than sixty seconds. Surgeons, like present-day athletes and health administrators, became obsessed with setting speed records and, thanks to their remarkable skill, could carry out an immense range of operations. However, such statistics as are available, including a 75 percent mortality rate among Parisian amputees in the 1870s, make it clear that surgery was an unpleasant business (figure 7). The great French surgeon and anatomist Guillaume Dupuytren (1777–1835), who is remembered as the discoverer of the contracture of the

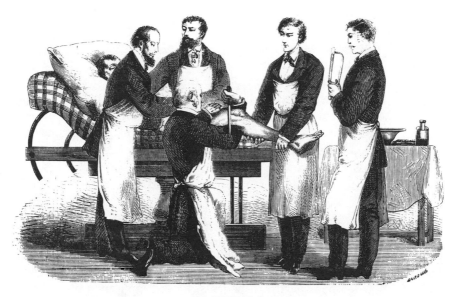

FIGURE 7

Amputation. From Joseph François Malgaigne, Manuel de médecine opératoire, *8th ed. (Paris: G. Baillière, 1874).*

palm that affects some of us as we get older, is quoted as saying that he would "rather die by God's hand than by that of a surgeon."

It was not until the development of anesthesia and Lister's antiseptic approach in the nineteenth century that surgery came into its own. Yet both these advances took quite unexpected and unplanned routes. Humphrey Davy (1778–1829), the British chemist and physicist, described the effects of inhaling nitrous oxide and even observed that it eased the pain of an erupting wisdom tooth, but nobody took up his suggestion that it might be used in surgical operations. Later the inhalation of nitrous oxide and ether formed the basis for a sport called frolics; the prevalence of this pastime in America and astute observations on the effects of these inhalations on pain opened up the field of modern anesthesia. Similarly, as we will see later, Pasteur's development of bacteriology, which led the way to antiseptic surgery, started with observations in the brewing industry.

It appears, therefore, that advances in medical science usually result from the coming together of knowledge at a particular stage of development in a number of fields. Although one area of research may be well developed— Harvey's concept of the circulation, for example—unless work in related fields has led to a critical level of understanding, there will be no practical results. This is the main reason for our current inability to deal with many of the major killers of Western society; we have some knowledge, but in whole areas our level of understanding remains so primitive that we can do no better than offer our patients increasingly sophisticated patch-up procedures to control their diseases.

Until the nineteenth century medical science did not make a great impact on the health of society, and many of the improvements over the centuries resulted from higher standards of living, better hygiene, and other modifications of the environment. The picture has changed dramatically in our own century, although there is still considerable controversy about how much we owe to the impact of scientific medicine and how much to improvements in our environment.

This brief sketch of the development of what we like to call scientific medicine has, I hope, provided some inkling of the enormous scope and breadth of medical research. Good clinical research is often initiated by careful and critical observation at the bedside, which may be the starting point for work in the laboratory encompassing a wide range of disciplines. Or the research may remain in the clinic and ward, and involve careful observations or measurements carried out directly on patients, or the application of statistical methods to assess the value of different types of treatment. On the other hand, many clinical advances have stemmed from curiosity-driven science in completely different fields, whose findings have often required the insight of clinicians able to see their medical application. Such is the rich and variegated background for our picture of modern clinical research and practice.

T H R E E

Early Successes Breed
New Problems

So it is something of a homiletical commonplace to say that the
outcome of any serious research can only be to make two questions
grow where only one question grew before.
— *Thorstein Veblen (1857–1929)*

One of the main problems for anybody who tries to understand the potentials
and limitations of medical research and practice is to appreciate the sheer
complexity of disease processes. Paradoxically, the more medical science has
told us about some of our intractable diseases, the more uncertain clinical
practice has become. Improvements in patient care do not, as the media
would have us believe, stem from a series of "breakthroughs." Rather, they
reflect a steady accumulation of knowledge; a few diseases are prevented,
cured, or more effectively controlled along the way, but each step forward
uncovers another layer of ignorance. This pattern of medical progress, and
the way in which medical research works, is well exemplified by some of its
major achievements during the first half of the twentieth century.

THE DEVELOPMENT OF THE MEDICAL SCIENCES
DURING THE TWENTIETH CENTURY

The latter half of the nineteenth century saw increasing fragmentation and
specialization in medical care and research. The sciences that underpin
human biology and medicine broke away from the mainstream of day-to-day
practice to form disciplines of their own. The oldest, anatomy, was already
firmly established. Human physiology expanded and, with the recognition of
the electrical activity of the heart and ways to monitor it through the development of the electrocardiograph, formed the basis for our present-day investi-

gation and management of heart disease. Equally rapid advances were made in understanding the functions of the lungs and digestive tract and in describing the composition and functions of the blood. Neurophysiology, which explores how the nervous system works, developed on several fronts—the electrical properties of nerve conduction, reflex activity of the spinal cord, and the way in which different parts of the brain are involved in particular functions, for example. And around the turn of the century there was remarkable progress in biological chemistry, including the characterization of proteins, enzymes, and, later, nucleic acids, and in understanding the body's handling of oxygen and the mechanisms of energy production and nutrition. This led to the emergence of another new discipline, biochemistry, an amalgamation of physiology and chemistry.

During the first half of this century advances in biochemistry and physiology gave rise to the notion that all living organisms are made up of lifeless molecules assembled to form cells and organs. Biochemistry described the complex pathways by which cells obtain and utilize energy and how they are driven and controlled by hundreds of different chemical catalysts, or enzymes. Physiology, on the other hand, concerned itself with how the functions of groups of cells and organs are controlled and coordinated, by hormones or reflex pathways in the nervous system, for example. It gradually became clear that all living processes can be studied, and to a large degree interpreted, by the methods and laws of chemistry and physics. That resulted in the emergence of a new science, molecular biology.

At the same time more clinically directed research was on the move. In the preceding chapter we saw how a critical, analytical approach to clinical practice was developed by Louis in Paris in the midnineteenth century. The Paris Hospital school was the major center for the development of clinical science in the first half of the nineteenth century and became the mecca for physicians in Europe and the United States. In addition to excellence in clinical method, those who worked with Louis learned, for the first time, how to apply the statistical method to clinical investigation. In the later nineteenth century the main site of medical science moved from France to Germany, where laboratories were developed in which men and women could devote their time to research in the blossoming basic sciences—anatomy, physiology, and, later, biochemistry. In this atmosphere matured a new generation of clinical scientists, who became interested in physiological medicine, that is, in understanding the fundamental mechanisms of disease.

The influence of these developments in Germany led to the establishment of several university medical schools in the United States. In 1910 the American educator Abraham Flexner (1866–1959) wrote a withering critique of medical education and science in North America, on the basis of his visits to German universities. This attack stimulated the development of specialist clinical departments in many American medical schools and the creation of full-time posts for academic clinicians. Flexner's revolutionary report recommended that medical education begin with a strong foundation in the basic sciences, followed by the study of clinical medicine in an atmosphere of critical thinking and with adequate time and facilities for research. His plan was widely accepted not only in North America but also in many British medical schools.

The appearance of university clinical academic departments in the period between the two world wars and, particularly, after the Second World War led to the emergence of "clinical science," experimentation on patients or laboratory animals on problems that stemmed directly from observations made at the bedside. This eventually brought a remarkable improvement in our understanding of disease mechanisms. In effect, it set the scene for the appearance of modern high-technology medical practice. Not surprisingly, it also had a profound effect on medical education and on the way doctors think about disease. Indeed, those who criticize modern methods of teaching doctors—in particular, its Cartesian approach to the study of human biology and disease—believe that the organization of clinical academic departments along Flexner's lines may have done much to concentrate their minds on diseases rather than on those who suffer from them. More on this later.

The new breed of clinical scientists were quite different from the polymaths of previous centuries. Their clinical work became so specialized that they spent their time dealing with one system—the heart or the lungs, for example. They studied their patients' illnesses in great depth by means of new tools of clinical measurement, and the boundaries between patient care and research became blurred. Today clinical research workers can detect minute bioelectric potentials in nerve, heart muscle, or other tissues, can follow chemical reactions or other functions with the use of radioactive isotopes, have a precise understanding of how drugs work, can study the minutiae of anatomy with the electron microscope or with sophisticated scanning techniques, and can dissect the body's responses to infection by separating the various parts of the immune system. And in the last few years it has been possible for them to start to study disease processes at the level of individual molecules.

Although the development of scientific medicine has been supported by advances in the basic sciences and by the establishment of university departments in the major clinical disciplines, we should not forget the other, and arguably more important, influences on the progress of twentieth-century medicine. First, there was the rapid expansion of the pharmaceutical industry. This industry has a long and turbulent history. In the United Kingdom the first body set up for manufacturing drugs was the Worshipful Society of Apothecaries of London. Established in 1671, it still stands today, although not for drug manufacturing. Throughout the eighteenth century this laboratory had the monopoly for the supply of medicines to the military services, and it also supplied the East India Company. During the nineteenth century many individual drug manufacturers set up businesses, and by the beginning of the twentieth several major drug houses had been founded. More recently the industry has spawned many large international companies with enormous research programs, and their products have had an impact on every branch of clinical practice. In 1990 it was estimated that, between them, the major international drug companies spent in excess of twenty billion dollars on research and development.

But as we saw in the preceding chapter, the main factor behind the better standards of health in industrialized societies in this century is the improvement in social conditions, particularly housing, nutrition, hygiene, and education. Long before the advent of antibiotics and vaccination, these changes, together with the implementation of more effective public health measures, did much to control epidemics and to contain outbreaks of infectious disease. Opinions vary about the importance of research in the universities and in industry, as compared with social change, in improving our health. We will return to this important theme, which dominates much current thinking about the future of medical practice, in later chapters. Here we will take a closer look at how medical research works and at the roots and consequences of three of its most notable achievements in the first half of the twentieth century.

SOME EARLY ACHIEVEMENTS OF
MEDICAL RESEARCH IN THE TWENTIETH CENTURY

The first half of this century brought steady progress toward an understanding of the mechanisms of many common diseases and genuine improvements in both clinical diagnosis and management. Among the major success stories are the isolation of insulin for the treatment of diabetes, the discovery of a cure for pernicious anemia, and the prevention and treatment of infectious disease. The tale of how these advances came about has been told many times. However, to appreciate our current state of knowledge of disease and its management, and the reasons for our increasingly high-technology medical practice, we must trace the aftermaths of seminal developments of this kind As we will see, even these three great discoveries were only stepping stones in the acquisition of knowledge—in effect, incomplete stories that led to new sets of problems. For this reason they are a particularly good place to start in our attempt to describe the mixture of progress and ignorance that underlies the present medical scene.

The Control of Diabetes Mellitus

Sir Peter Medawar (1915–1987), Nobel laureate, immunologist, and philosopher of science, has suggested that the discovery and isolation of insulin for the control of diabetes marked the first major success for scientific medicine. But it is still only a partial success story and a good example of our ability to control though not eradicate a disease, a theme that will often recur as we examine the development of modern medicine.

Diabetes mellitus, diabetes for short, is a serious disease that affects about 4 percent of us in Western societies at some time during our lives. The basic problem is the body's inability to burn glucose to produce energy. Glucose accumulates in the blood and spills into the urine, and, in the body's effort to gain energy from other sources, tissue is broken down, which leads to a loss of weight. The abnormal products of this disordered body chemistry accumulate and may ultimately lead to coma and death. Even if partially controlled, the disease may cause complications, which, together with an increased susceptibility to infection, combine to produce a crippling illness common enough to pose a major public health problem.

Descriptions of what was almost certainly diabetes first appeared over three thousand years ago in Egypt. Physicians in India around 400 B.C. noted not only the sweetness of the urine but also the disease's tendency to run in families. Diabetes was recognized again early in the Christian era, and the Romans gave it the name *diabetes mellitus;* the Greek *diabetes* means "syphon" and *mellitus* "honey" or "sweet." The sweetness of the urine was also rediscovered in the seventeenth century by Thomas Willis of Oxford. About a hundred years later it was found that it is due to sugar, and in the mid-1880s Claude Bernard showed that the sugar that appears in the urine of diabetic patients is not produced in the kidney, the role of which is to remove excess sugar that accumulates in the body.

At the end of the nineteenth century the Austrians Joseph von Mering (1849–1908) and Oskar Minkowski (1858–1931) discovered that dogs whose pancreas was removed became diabetic and soon died. A few years later it was determined that in diabetics there are changes in a particular type of cell in the pancreas that had been described in 1869 by the German physician and pathologist Paul Langerhans (1847–1888), when he was working in Rudolf Virchow's laboratory. These cells, subsequently called the islets of Langerhans, were identified as being different from the other cells of the pancreas, which produce a mixture of powerful digestive juices that are secreted into the gastrointestinal tract. Several attempts were made to isolate the product of the islet cells that might be missing in diabetics, but there was no success until the remarkable researches of Banting and Best in Canada in the 1920s.

FIGURE 8

Frederick Grant Banting (1891–1941).

Frederick Banting (1891–1941) (figure 8) studied medicine in Toronto and, after serving in the First World War in the Canadian Army Medical Corps, returned to Toronto and was appointed a surgeon to the Hospital for Sick Children. Clinical practice was not very lucrative for a young doctor, and so he obtained a part-time job as a demonstrator in surgery and anatomy in London, Ontario. While preparing his student lectures on sugar metabolism, Banting read a paper describing the obstruction of the duct of the pancreas by a stone, an event that had led to the destruction of the cells that secrete digestive juices but that left the islets of Langerhans intact. Hitherto the main problem in extracting the putative antidiabetic principle from the pancreas had been how to protect it from attack by digestive enzymes secreted by the other cells. It struck Banting that this chance clinical observation might provide the answer. On October 31, 1920, during a sleepless night, he wrote down his idea.

Banting asked the professor of physiology in Toronto, John Macleod (1876–1935), whether he could have the facilities to try out an experiment to reproduce this effect. With the help of a medical student, Charles Best (1899–1978) (figure 9), he removed the pancreas of dogs in two stages. First he tied the pancreatic ducts; a few weeks later he removed the pancreas and attempted to isolate the antidiabetic factor, a task he thought would be easier now that the cells that produced digestive juices had withered. After a series of dismal failures things started to improve, and the extract reduced the blood sugar of dogs that had been rendered diabetic by the removal of their pancreas.

As Peter Medawar suggested, what followed would have made a good libretto for a Puccini opera. Banting and Best worked night and day, cooking their meals over a Bunsen burner. With the help of a visiting biochemist, J. B. Collip (1892–1965), they made more pancreatic extracts, and their diabetic dogs started to survive. By now Macleod, who had initially been skeptical about these activities, was beginning to appreciate the importance of their work and set the whole of his laboratory onto the problem and elicited the help of an American pharmaceutical company to develop the extraction program on a large scale. Soon Banting and Best felt ready to test their product, which they called insulin, on a diabetic boy who was a patient at the Toronto General Hospital. The results were extremely gratifying, though the first, crude preparations of insulin caused fevers and severe local inflammation at the site of injection.

Over the following years the pharmaceutical industry isolated insulin from the pancreas of different animals for therapeutic use and modified it so that it could be given in rapid- or slow-acting forms, or in mixtures. The chemical structure of insulin was finally worked out in the 1950s by the English protein chemist Frederick Sanger.

FIGURE 9

Charles Herbert Best (1899–1978).

The isolation of insulin for the control of diabetes by Banting, Best, Collip, and Macleod, and its commercial development by the staff of the pharmaceutical company, Eli Lilly, undoubtedly ranks as one of the major achievements of scientific medicine. But progress since their discovery has been much less spectacular. It soon became clear that there is more than one type of diabetes. The two main forms are now called insulin-dependent and non-insulin-dependent. While there is good evidence that the insulin-dependent variety is due to the self-destruction of the pancreatic islet cells, it is still not clear why this happens. Nothing is known about the cause of insulin-resistant diabetes, although the likelihood of developing it is strongly influenced by genetic factors.

At present the omens for the control of diabetes in the future are not good. The frequency of the insulin-dependent form has increased in recent years and has doubled in young children under the age of five years. This is not an artifact attributable to more refined methods of diagnosis but seems to mark a genuine increase in the prevalence of the disease, presumably due to the action of environmental factors. And even more worrisome is the fact that although it can be controlled with insulin, serious complications still occur. Diabetic eye disease is the major cause of blindness in the United Kingdom for people between the ages of twenty and sixty-five. Diabetes also leads to damage to blood vessels. This increases the likelihood of coronary artery disease or narrowing of the vessels to the limbs, with destruction of the tissues, particularly in the feet. Infected and ulcerated toes cause considerable pain and distress, and cost to the health services. Kidney disease is common; kidney failure due to diabetes is one of the most frequent reasons for patients' having to be maintained on kidney machines or receiving transplants of new kidneys. It is still not clear whether accurate control of the sugar level in the blood in diabetic patients will reduce the prevalence of these complications, though the results of recent large-scale trials suggest that this may be possible.

When we come to non-insulin-dependent diabetes, the position is even less clear. This extremely common disorder occurs in later life and seems to make us more prone to disease of our blood vessels. Although it is associated with obesity, we have no idea why, as has become apparent in recent years, it is particularly common in certain races. It is reaching epidemic proportions in Indian populations, especially those who have emigrated to the West. It is equally common in the Amerindian populations of the American continent, in inhabitants of parts of Melanesia and Polynesia, and in some peoples of African origin. With increased knowledge, it has become apparent that there are probably several other types of diabetes, some of which are also confined to certain racial groups. The more we have learned about this disease, the more complex it has turned out to be.

The story of diabetes is therefore incomplete. As we will see, this is true of many chronic diseases; it is our partial success in the management of conditions like diabetes that has spawned much of the high-technology medical practice found in our hospitals today. By controlling diabetes and allowing diabetic patients to survive indefinitely, but by not learning how to prevent the disease or to control its complications, we have produced a heavy load for our health services. This includes the establishment of clinics and community services for the regular monitoring of the control of the blood sugar of diabetics, the development of sophisticated techniques for dealing with the complications that affect the eye, the utilization of a large part of our facilities for maintaining patients with diseased kidneys on dialysis programs, kidney transplantation, and—the most recent addition to our armamentarium—the transplantation of pancreatic islet cells. Clearly, major advances in scientific medicine can lead to completely unforeseen problems for the provision of medical care in the future. One of our major goals must be to discover the underlying cause of the different forms of diabetes and how they can be prevented.

The "Cure" of Pernicious Anemia

The stories of the discovery of treatments for pernicious anemia and diabetes are linked by a quirk of timing and fate. The chief player in the research that led to a cure for pernicious anemia, George Minot (1885–1950), carried out his elegant work in 1925. Minot had developed severe diabetes in the autumn of 1921. Banting and Best discovered how to isolate insulin in 1922, and early in .1923 a Boston physician managed to obtain some for Minot; had he not, the account that follows might have been quite different.

The discovery of insulin illustrates how experiments on animals and a chance clinical observation have led to the control of a serious disease. In the present story the road to success was different. Observations made at the bedside of patients, simply by talking to them, paved the way to the generation of a hypothesis that, when tested, provided the cure for to a previously fatal condition. This is a genuine example of clinical research and remains one of its finest achievements.

Thomas Addison (1793–1860), a physician at Guy's Hospital, London, is customarily considered to have recognized pernicious anemia for the first time, in 1855. Whether he really described it is uncertain, but it was subsequently learned that there was a severe form of anemia, a reduction in the number of red blood cells, that affected individuals during the middle and later years of their life and that was invariably fatal. Hence "*pernicious* anemia," a term first used some twenty years later by the German physician Anton Biermer (1827–1892). Later it was noticed that some of these patients also developed progressive weakness and paralysis of the legs and gross incoordination of movements of their hands and feet, together with a sore tongue. By the end of the last century this distressing picture was well known to clinicians on both sides of the Atlantic. In 1908 the Boston physician Richard Cabot (1868–1939) reviewed twelve hundred patients and found that survival from the onset of the disease was one to three years; only six recoveries had been reported.

By the later half of the nineteenth century hematology, the study of blood, was already well developed. The cells of the blood could be examined under the microscope, and thanks to the work of the German pathologist Paul Ehrlich (1854–1915), it was possible to stain blood cells and to examine their shape and size. All our blood cells are the offspring of parent cells in the bone marrow, where blood is formed. In 1880 Ehrlich noticed that patients with pernicious anemia have large, primitive red cells in their blood, which he called megaloblasts, and he concluded correctly that these cells had escaped from the bone marrow. He also found that they were much larger than normal progenitor blood cells in the bone marrow, which he called normoblasts. Since red blood cells decrease in size as they mature, the idea evolved that this process is defective in pernicious anemia.

There followed many years of speculation about why this might happen. For a long time it was thought that it might be due to the action of a toxin that prevents the maturation of blood cells. This agent never materialized, however, and thoughts about the cause of pernicious anemia turned in other directions. Ever since 1747, when James Lind (1716–1794) showed that citrus fruits could prevent scurvy, it had been apparent that disease might result from dietary deficiency. Around the turn of the century there was a major revival of interest in this possibility. In 1897 a Dutch military doctor, Christiaan Eijkman (1858–1930), found that chickens fed a diet of polished rice, similar to that consumed by inmates of jails and hospitals, developed a paralytic disorder resembling a condition called beriberi, which was well known to occur in these instititions. Furthermore, the disease could be prevented or cured by feeding unpolished rice. The experimental basis for a general theory of dietary deficiency was soon developed in the laboratories of Gowland Hopkins (1861–1947) in Cambridge, England, and by scientists in the United States, including T. B. Osborne (1859–1929) at Yale and E. V. McCollum (1879–1967) at Wisconsin and, later, Johns Hopkins.

In 1912 Casimir Funk (1884–1967), a Polish-born biochemist working in London, reviewed the literature of this new field and invented the term "vitamine" to describe the missing factors in deficiency diseases such as beriberi and scurvy. Not surprisingly, these seminal ideas were read by workers in many different disciplines; those who were interested in blood began to wonder if some forms of anemia might be due to a deficiency of factors in the diet. Beginning in 1918 George Whipple (1878–1976), working in California and later in Rochester, New York, carried out a series of experiments on dogs that he made anemic by regular bloodletting. If the animals were then maintained on a diet of salmon and bread, very little more blood needed to be removed to maintain a severe degree of anemia. However, if their diet was supplemented with animal or vegetable foods such as liver, beef muscle, or spinach, red cell production was increased and more blood had to be withdrawn to keep them anemic. By the mid-1920s Whipple was already advising clinicians to pay attention to diet in the management of patients with anemia.

The story now moves to Boston. George Minot, an assistant professor of medicine at Harvard, had since his days as a medical student been fascinated with the subject of anemia and other blood disorders. Minot spent many hours talking to patients about their illnesses and gained the impression that those with pernicious anemia had a poor appetite and an aversion to meat. These observations seemed to be in line with Whipple's findings and suggested to Minot that pernicious anemia might result from a dietary deficiency.

At first he simply advised his patients to try to eat meat and liver, but the improvement he noted was not impressive. In 1925 Minot invited William Murphy, a member of the staff of the Peter Bent Brigham Hospital, in Boston, to join him in setting up a trial of a particularly rigorous and unpleasant diet for patients with pernicious anemia. They were fed about half a pound of lightly cooked beef liver every day. Given that this disease has a very debilitating effect, it must have taken a great deal of patience and persuasion on the part of Minot and Murphy and the nurses who helped them to induce these unfortunate persons to eat this nauseating concoction.

But eat it they did, and within a week or so they all showed a quite astonishing regeneration of their red blood cells. In a few months their anemia was cured. Indeed, a particularly cantankerous old lady in this group of patients, and the one Murphy found most difficult to feed, underwent a dramatic change of personality on about the seventh day of treatment, just before her blood began to show signs of regeneration. In retrospect this is an especially interesting anecdote; many years later the term "megaloblastic madness" was coined to describe the mental changes that sometimes accompany pernicious anemia. On May 4, 1926, Minot and Murphy reported these extraordinary results to the annual meeting of the Association of American Physicians, in Atlantic City, New Jersey. Over the next few years their findings were confirmed all over the world, and in 1934 Whipple, Minot, and Murphy received the Nobel Prize for their discovery. After many years of intensive research the missing dietary factor was isolated and named vitamin B_{12}. Its structure was worked out by Dorothy Hodgkin of the University of Oxford in 1961.

But of course the central question remained. Why do patients with pernicious anemia fail to absorb the factor from their diet unless they are fed such

FIGURE 10

William Castle (1897–1990).

enormous quantities? To find out we have to return to Boston and continue our story with the work of the main unsung hero of the pernicious anemia saga, William Castle (1897–1990) (figure 10). In 1926 Castle was a young resident physician in the Thorndyke Laboratory of the Boston City Hospital, where he became interested in the disease. He knew of the work of Whipple and Minot and was also aware that there is something unusual about the gastric juice of patients with pernicious anemia. Evidence had accumulated over the years that, unlike normal gastric juice, the stomach secretions of these patients are almost completely lacking in hydrochloric acid. Castle wondered why most of us do not get pernicious anemia and why those with the disease have to eat half a pound of liver a day to correct the apparent deficiency that leads to the disease. Is it possible, he speculated, that the stomach of normal persons can derive something from food that for them is the equivalent of eating pounds of liver?

To test this hypothesis Castle carried out a simple but brilliantly conceived series of experiments involving patients with pernicious anemia. For ten days he fed them a modest quantity of rare hamburger steak made from lean beef muscle. Their blood showed no sign of improvement. At the same time he fed normal men the same amount of steak and recovered their gastric contents through a stomach tube. The contents was incubated for a few hours until it liquefied and then fed through a stomach tube to the patients with pernicious anemia. After about six days of this treatment the patients' blood showed clear signs of regeneration. He then went on to ask whether the response was due to the stomach contents alone or due to its action on the beef muscle diet. He found that stomach juices alone were ineffective, no response occurred unless the stomach juice and the beef muscle were given together. He concluded that there must be an interaction between an unknown "intrinsic factor" in the gastric juice and an "extrinsic factor" in the beef muscle.

The extrinsic factor turned out to be vitamin B_{12}, of course, and the intrinsic factor was later found to be a substance that is secreted into normal stomachs and is involved in transporting vitamin B_{12} across the wall of the lower bowel, where it is absorbed. Pernicious anemia is caused by a deficiency of gastric intrinsic factor, and hence the vitamin B_{12} in the patients' food becomes unavailable to them unless it is present in enormous quantities.

In medicine, as in all branches of science, the solution to one question usually poses a new series of problems. Although Castle had identified the basic defect in pernicious anemia, it remained unclear why some individuals destroy the lining of their stomach and are unable to produce hydrochloric acid and intrinsic factor. Many years later scientists in Denmark and England found that, like the insulin-dependent form of diabetes, marked by self-destruction of the pancreatic islet cells, pernicious anemia occurs because of the production of antibodies against the stomach lining. We still do not know why this happens. Hence pernicious anemia joins the long list of what are called autoimmune diseases, which contains many of the unsolved mysteries of modern medicine; apart from diabetes the list includes rheumatoid arthritis, thyroid disease, other diseases of endocrine glands, some forms of anemia and other diseases of the blood, chronic diseases of the nervous system, and many more.

And there are still many other gaps in our understanding of the way in which a deficiency of vitamin B_{12} leads to the clinical picture of pernicious anemia. The vitamin is needed for some of the key steps in the production of nucleic acids, the major players in the passage of genetic information from cell to cell. Presumably this property is essential for the maturation of rapidly dividing blood cells. If so we might expect that its deficiency would lead to abnormal production of blood cells, since every twenty-four hours we replace about 1 percent of them. Similarly, a defect of this type would be mirrored in other cells that are being replaced regularly. This might explain the sore tongue and gastrointestinal symptoms. But we still do not know exactly how vitamin B_{12} deficiency damages the spinal cord and nervous system.

What is clear, however, is that a monthly injection of vitamin B_{12} will control all the symptoms of pernicious anemia and restore patients to complete health. A uniformly fatal disease has been conquered although we do not really understand how it arises.

In addition to offering one of the first examples of an advance in clinical science emanating directly from observations at the bedside, the story of how Minot and his colleagues arrived at a cure for pernicious anemia is illuminating in other ways. Minot considered it a deficiency disorder, similar to other vitamin deficiencies. When he talked to his patients and found that they had a poor appetite, this would have been a reasonable assumption. We now know that pernicious anemia leads to gastrointestinal symptoms, including a poor appetite, and that it is due not to a dietary deficiency but to an inability to absorb a crucial vitamin from the diet. Similarly, in retrospect it seems very likely that Whipple's dogs were anemic because he was depriving them of iron rather than other factors that are important for the production of blood. In medicine, as in other branches of science, it is possible to get the right answer even if the basic premises on which the work rests are wrong; successful medical research requires luck. Particularly when we consider the role of William Castle in working out the cause of pernicious anemia, it also becomes apparent that the major contributors do not always receive all the acclaim they deserve. Most important, we see how curiosity-driven research—in this case, involving Whipple's dogs—may have profound and unexpected benefits for clinical practice.

The Control of Infectious Disease

In *Airs, Waters and Places, an Essay on the Influence of Climate, Water Supply and Situations on Health,* Hippocrates wrote, "Whoever would study medicine aright must learn of the following subjects. First he must consider the effect of each of the seasons of the year and the difference between them. Secondly he must learn the warm and cold winds, both those which are common to every country and those peculiar to a particular locality. Lastly, the effect of water on the health must not be forgotten." Over subsequent centuries ideas about the impact of the environment on disease developed along two main lines. The contagion theory supposed that there are specific agents that produce particular diseases. The miasma theory held that polluted or otherwise unpleasant environments have a more generally debilitating effect. Malaria— the word means "bad air condition"—was, for example, attributed to the effects of evil emanations from swamps. The contagion theory stemmed from the observation that epidemics seemed to require contact to produce disease. Already in the sixteenth century the Veronese doctor and poet Girolamo Fracastoro (ca. 1478–1553) suggested that the seeds of disease are minute animals capable of reproducing their kind, although it was not until three hundred years later, and the discovery of bacteria, that the germ theory of disease put the idea of contagion on a firm scientific footing.

The story of the control of infectious disease and the development of bacteriology and immunology starts in the middle of the eighteenth century. The concept of vaccination was first developed by Edward Jenner, who noticed that patients who had suffered from cowpox caught by milking diseased cows seemed to be protected against human smallpox. In 1796, after many years of observing this strange phenomenon, Jenner injected material obtained from a cowpox vesicle on the hand of a young woman into the arm of a young boy called James Phipps. Some months later he injected Phipps with material from a human case of smallpox, and no disease developed.

It is often said that the eradication of smallpox that the World Health Organization announced in 1980 goes back to an experiment that a present-day ethics committee would never have sanctioned. But this is unfair on Jenner. Throughout the eighteenth century, and based on the observation that an attack of smallpox confers lifelong immunity, it was common practice to try to prevent the disease by infecting people with material from patients, in the hope of producing a mild disease that would prevent further attacks. It is not known whether this killed more people than it protected, but since Jenner himself underwent this procedure as a child, his experiment on young Phipps was not so reckless as it sounds. Vaccination, so called because "vaccinia" is the Latin name for cowpox, represented the first major triumph in the conquest of infectious disease.

At first sight the story of Edward Jenner appears to be different from that of most of the other scientists who have made signal contributions to medical knowledge. He spent what was apparently a quiet and simple existence as a retiring, slow-talking country doctor whose other researches went little further than the hibernation of the hedgehog and the antisocial habits of the cuckoo. But was he, as his recent biographer Richard Fisher has suggested, "an ordinary man struck by one colossal lightening-bolt of serendipity"? It depends on what is meant by "ordinary." Through the sponsorship of the anatomist John Hunter, Jenner's work on the cuckoo led to his election to the Royal Society, a body that, incidentally, rejected his classic paper "Inquiry into the Natural History of a Disease Known in Gloucestershire by the Name Cowpox." And even though his ideas were not accepted by many of his colleagues, some of whom denounced him as a quack, he did a great deal to pursue and develop the concept of vaccination and to devise methods for the preservation of vaccine lymph.

The careers of Robert Koch (figure 11) and Louis Pasteur (figure 12), the founders of modern microbiology, whose work was ultimately to lead to the control of infectious disease, were very different from Jenner's. After service as a medical officer in the Franco-Prussian War, Koch was appointed as a district physician in a small town in Posen in 1872. He set up his own laboratory and isolated the organism that causes anthrax. Later he moved to the Imperial Health Office in Berlin and studied the bacteriology of wound infection, and in 1882 he isolated the organism responsible for tuberculosis. On a subsequent trip to Egypt and India he identified the cause of cholera. His famous postulates are still the yardsticks against which the causal relation between an infectious disease and a particular organism are measured. Louis Pasteur, by contrast, started his academic career in physics and chemistry. He showed that fermentation in wine, beer, and milk is due to microorganisms that are not spontaneously generated but that normally abound in the atmosphere. He went on to investigate diseases of silkworms and found that two different microbial disorders were ruining the French silk industry. Even after suffering a stroke, he continued his work in microbiology and pathology and made one of his most important contributions to the future control of infectious disease.

In 1880 Pasteur was working on a fatal disease of chickens, which he could induce by injecting cultures of a cholera bacillus. He left some of these cultures standing on his laboratory shelf during a summer holiday and discovered on his return that injections of the same culture produced a mild disease. Furthermore, he found that new cultures, although usually lethal for chickens, had little or no effect on those that had received the old cultures. Thus, by following up a chance observation, Pasteur discovered how to reduce the power of bacteria to cause disease while retaining the ability to immunize

FIGURE 11

Robert Koch (1843–1910). Lithograph by an anonymous artist, 1891.

FIGURE 12

Louis Pasteur (1822–1895), aged about forty. Lithograph by Schultz after a photograph by Pierre Petit.

against it. He called this treatment vaccination, in memory of Jenner, and was soon able to produce a vaccine to protect sheep against anthrax. Pasteur even went on to develop a crude vaccine for rabies and successfully treated a child with this condition; the patient lived to become the gatekeeper of the Pasteur Institute. This remarkable achievement is commemorated today by a bronze statue outside the institute.

At about this time there was some progress toward understanding how the body defends itself against microorganisms. The Russian zoologist Elie Metchnikoff (1845–1916) found that cells such as amoebas and mammalian white blood cells are able to surround and ingest particles, including microbes. He called this process phagocytosis and realized that it must play an important role in defense against infection. Indeed, he wondered if vaccines might stimulate phagocytes to attack particular organisms. Thus began the branch of medical science later known as immunology. However, because work over the next few years suggested that natural and acquired immunity were more likely to be mediated by factors circulating in the blood, Metchnikoff's work was largely forgotten until the renaissance of cellular immunology in the 1950s.

In the 1880s the bacteria that cause diphtheria and tetanus were isolated, and these diseases were found to be caused by a toxin produced locally and released by the bacteria as they grow. In 1890 Emil von Behring (1854–1917), working in Koch's laboratory in Berlin, showed that animals become immune to lethal doses of these toxins if they have been injected with a series of minute doses. Even more remarkably, if blood was taken from these animals a few days after the injections, the serum was able to neutralize fresh samples of toxin. These sera, if injected into other, untreated animals, enabled them to survive a lethal dose of the appropriate toxin. Behring suggested that this form of immunity was due to protective substances in the blood that he and Paul Ehrlich later called antibodies. This new knowledge was used for the treatment of a human disease for the first time in 1891, when a child in Berlin was successfully treated for diphtheria by means of antitoxin.

Toward the end of the nineteenth century it was becoming apparent to Koch and others that the bacteria they were identifying might not be the only organisms capable of causing disease. In 1886 Adolf Mayer (1843–1942) found that when he filtered the sap of tobacco plants suffering from tobacco mosaic disease, the filtrate infected healthy plants. Since bacteria are too large to pass through the pores of the filters that Mayer was using, this suggested that the offending microbe must be unusually small. This work was extended by Martinus Willem Beijerinck (1851–1931), who demonstrated that this type of infectious organism can reproduce only within living cells. This discovery marked the beginning of virology, although it was many years before it was realized how many important human diseases are caused by viruses;

not until 1939 was the first picture of a virus obtained, by electron microscopy.

During the first half of this century rapid progress was made in vaccination and immunization against bacterial and viral diseases. Once viruses were identified as the cause of disease, the main problem was how to grow and attenuate them so that they could be used for vaccination. In 1931 Ernest Goodpasture (1866–1960) discovered that it was possible to use certain types of hen's eggs to grow the virus of fowl pox. Later, with the advent of potent nontoxic antibiotics, it became possible to grow human and animal cells in nutrient media. In 1949 John Enders (1897–1985) and his colleagues in Boston first grew the poliomyelitis virus in tissue culture and set the stage for the development of a poliomyelitis vaccine by Jonas Salk. Vaccines were also developed for important bacterial infections such as diphtheria. The results of vaccination depend on the nature of the disease; conditions that produce lifelong immunity after a single infection are much more likely to be prevented than those with a long and relapsing course—tuberculosis, for example.

These advances in the prevention of infection have had a major impact on the pattern of illness over the last fifty years. Effective vaccines have been developed against smallpox, measles, rubella (German measles), diphtheria, tetanus, mumps, typhoid fever, yellow fever, pertussis (whooping cough), and poliomyelitis. But attempts to immunize with purified fractions of bacteria have proved less successful. And while antisera gainst tetanus, diphtheria, and rabies confer protection if given during the incubation period of the illness, this approach does not work in the case of many other common infectious diseases caused by bacteria. In the event, it turned out that the conquest of bacterial illness required the development of a completely new approach.

The other thread in the control of infectious disease was the discovery of chemotherapeutic agents, that is, substances toxic to bacteria but not very toxic to the host. The father of this field was the great German medical scientist Paul Ehrlich. After making important contributions to the field of antibody-mediated immunology and after standardizing preparations of antitoxins, Ehrlich pursued the idea that simpler substances might inactivate bacteria. In 1910 he introduced the arsenic derivative Salvarsan, the first drug of any real value for the treatment of syphilis.

The central theme of much of Ehrlich's work was that substances that might have a chemotherapeutic effect must be fixed by the cells on which they act. Since dyes behave in this way, various dyestuffs were investigated as potential sources of drugs of this type. The first successes using this new approach resulted from the work of Gerhard Domagk (1895–1964), director of research in Bayer Laboratories. In 1932 he discovered that a dye, prontosil red, could control streptococcal infections in mice. Streptococci are organisms that produce a variety of unpleasant infections in humans, including erysipelas, tonsillitis, scarlet fever, and diverse forms of rheumatism. It was soon found that the chemotherapeutic effect of this agent was due not to the dyestuff itself but to the sulfonamide group that splits off from it in the body. Domagk successfully treated his own daughter for a streptococcal infection with this agent, and by the mid-1930s a number of simpler sulfonamide compounds were developed. They showed varying activity against bacteria; unfortunately, they were not active in the presence of pus and did not kill the organism *Staphylococcus aureus,* a major cause of virulent skin, wound, and bone infections. Over the next five years the clinical value of the sulfonamides in fighting streptococcal infections was established by many different clinical studies. Particularly impressive was their effectiveness in the treatment of puerperal fever.

The steady downward trend in deaths from infection in the second half of the nineteenth century and in the early part of this century undoubtedly resulted from improvements in social and economic conditions and the use of Listerian methods in surgical practice. But despite the introduction of various aseptic techniques into obstetric practice, maternal mortality remained on a plateau from the 1850s to the mid-1930s. Many of the maternal deaths in pregnancy were due to puerperal fever, an infection of the womb with streptococci in the period after delivery. The high maternal mortality was not confined to the United Kingdom; it remained more or less constant in the United States even though most deliveries there took place in hospitals after the 1920s. The sulfonamides became widely available for the treatment of puerperal fever in the mid-1930s and, according to the English medical historian Irvine Loudon, it seems certain that the introduction of these agents was most responsible for the dramatic fall in maternal mortality from 1937 to the early 1940s. Although later the use of penicillin, blood transfusions, and better obstetric care, both in hospitals and in the community, played an important role in further lowering maternal mortality, the role of sulfonamides in improving the disgraceful mortality figures of the mid-1930s constitutes a major success story for scientific medicine.

But although the sulfonamides made considerable inroads into the management of erysipelas, scarlet fever, and puerperal sepsis, and although later derivatives of these drugs showed some activity in the presence of pus, other bacterial diseases continue to pose a major problem. In particular, the ubiquitous organism *Staphylococcus aureus* remained a leading cause of virulent infections. In 1929 Alexander Fleming (1881–1955) described the antibacterial properties of an extract of the common mold *Penicillium notatum.* This new agent had astonishing antistaphylococcal properties and was active in the presence of pus. Penicillin was given to a patient for the first time on

February 12, 1941, in the Radcliffe Infirmary, Oxford. The story of the development of penicillin over this twelve-year period has been recounted many times, but because it illustrates the strange way in which discoveries in the basic sciences sometimes acquire practical importance, it is worth briefly reconsidering.

The tale of the mold spore that blew in through the open window of Fleming's laboratory facing Praed Street, London, and that one of Fleming's biographers, André Maurois, referred to a as "the mysterious mould from Praed Street," is well known. The skepticism of later writers who thought that it was more likely a laboratory contaminant has done little to dull the romantic aura of this episode. In any event, Fleming identified a specific mold, *Penicillium notatum,* which when grown on a particular nutrient broth confers strong antibacterial activity. In addition to being toxic to staphylococci, this agent, unlike most antibacterial substances studied up to that time, had no ill effects on animals and, in particular, did not damage white blood cells, the natural scavengers of infection. Although Fleming carried out a limited clinical trial of his mold broth, using it to treat some local infections of the eye, he did not take the purification of penicillin any further. The story of how this was achieved in Oxford some years later provides an especially good example of the unpredictable relation between basic and applied science.

In 1935 the University of Oxford, in a series of devious maneuvers characteristic of that ancient institution, appointed to the professorship of pathology an Australian called Howard Florey (1898–1968) (figure 13), who had had first come to England as a Rhodes scholar. After working at Oxford, Florey carried out important research in different aspects of cell biology in several centers in the United Kingdom before returning to take up the chair. When he came to Oxford, he was particularly interested in the properties of a substance called lysozyme. To understand why, we must return briefly to Alexander Fleming. In 1922 Fleming had noticed that while he was suffering from a cold, a drop of his own nasal secretions dissolved, or lysed, certain bacterial colonies. This was a tantalizing observation because it suggested that there might be a substance in secretions such as tears and saliva that, though not toxic to normal tissues, might have an antibacterial effect. These hopes were not to be fulfilled, however, because most bacteria that cause disease did not seem to be vulnerable to lysozyme.

In 1935 Florey investigated secretions from body surfaces, particularly mucus, and wondered whether these apparent lubricants might have other roles, such as providing protection against infection. He was not a chemist, and when he moved to Oxford he decided to invite the young German immigrant Ernst Chain (1906–1979) (figure 14), an outstanding chemist, to join him and study the chemistry and functions of lysozyme. In a brilliant series of experiments Chain and his colleagues found that lysozyme is an enzyme,

FIGURE 13

Howard Florey (1898–1968).

FIGURE 14

Ernst Chain (1906–1979).

whose action is directed specifically against complex sugars, or polysaccha-
rides, in the cell walls of sensitive organisms. Chain went on to identify the
particular polysaccharide that is involved and thus was able to account for
the antibacterial activity of lysozome.

In the late 1920s and early 1930s there was a great deal of interest in
antibacterial agents. The story of the development of the sulfonamides was
already familiar. In the United States, Selman Waksman (1888–1973), a Rus-
sian-born microbiologist, having observed an inhibition of growth around
colonies of the soil bacteria actinomyces, had also decided to undertake a
systematic study of antibiosis, the antagonistic or synergistic behavior among
microorganisms. In 1927 René Dubos (1901–1981), a former student of
Waksman's, came to the Rockefeller Institute Hospital to test the notion that
natural soils contain microbes able to attack other bacteria. In 1939 Dubos
and Rollin Hotchkiss isolated a crystalline antibiotic, which they called tyrothricin, from the culture media of the soil organism *Bacillus brevis*. Subsequently it was found that this is a mixture of two antibiotics—gramicidin and
tyrocidine. It turned out that tyrothricin was active against a range of
important bacteria but that it was too toxic to be used for the treatment of
important bacterial infections in human beings. However, these were
extremely important observations and gave a major impetus to the development of more effective antibiotics, a fact Florey later acknowledged.

His interest in lysozyme must have made Florey well aware of these developments. In 1935 he and Chain decided to carry out a literature survey of all
the antimicrobial agents described up to that time. Among those that they
thought most promising for further study was *Penicillium notatum*. Because of
an unfortunate series of disagreements with the British funding body, the
Medical Research Council, Florey and Chain had considerable difficulty in
obtaining support for this project, but in the end they received a modest
grant from the Rockefeller Foundation in the United States. They set up a
small team to isolate and purify penicillin; it included another outstanding
chemist, Norman Heatley (figure 15).

Despite the disruptions caused by the onset of the Second World War,
Heatley and Chain were able to purify sufficient penicillin that on Saturday,
May 25, 1940, as the German armies were sweeping across France, Florey
was able to inject eight mice with a virulent strain of streptococci; four mice
were set aside as controls, and the others were treated with penicillin, as it
was now called. An all-night vigil was set up, and by the time Heatley finally
left the laboratory, in the early hours of Sunday morning, all the untreated
mice were dead. By the next morning, when Florey, Heatley and Chain came
back to the laboratory, the extraordinary result of this experiment was apparent; three of the treated mice were lively and well, and the fourth was what

FIGURE 15

Norman Heatley.

Florey called "a little piano"—in other words, not too well but still in the land of the living.

Thanks mainly to the technical ingenuity of Heatley (figure 16), it was possible to produce sufficient penicillin to carry out the first clinical trials in February 1941. The first patient to receive the drug was a forty-three-year-old policeman who had been a patient in the Radcliffe Infirmary, Oxford, for many weeks, suffering from dreadful staphylococcal infections over many parts of his body. There was a striking improvement in his clinical condition, but sadly, despite the repurifying of penicillin from his urine, there was an insufficient supply to maintain his improvement, and he died on March 15, 1941. Four other patients were treated over the next few months, and it became clear that this agent was nontoxic and highly effective against staphylococcal infection (figure 17). It was subsequently found to have activity against many bacteria and against the spirochete that causes syphilis.

The next problem was how to develop the production of penicillin on a scale large enough for clinical practice and for use on the battlefields of the Second World War. The findings of Florey and his colleagues, exciting though they were, did not stimulate the war-weary British pharmaceutical industry, and in the early 1940s Florey and Heatley set off for the United States to obtain help. Heatley went to work in the U.S. Department of Agriculture's North Regional Research Laboratory, in Peoria, Illinois, which already had considerable experience in fermentation processes. There, together with his American colleagues, he was able to make major progress in developing the bulk production of penicillin. Meanwhile, Florey traveled around the United States trying to interest pharmaceutical companies in penicillin. In the end he succeeded, and several firms started to produce the drug on a massive scale. Subsequently it was learned that there is a whole family of penicillins with different properties.

The seed was now laid for a period of dramatic development in the production of antimicrobial agents. In 1944 Selman Waksman announced the discovery of an antibiotic produced by the soil microorganism *Streptomyces griseus,* which he called streptomycin. It turned out that this agent was active against tuberculosis. There followed the production of other families of antibacterial agents and the control and treatment of many infectious diseases that were both common and life-threatening at that time—gonorrhea, syphilis, tuberculosis, pneumonia, meningitis, typhoid and typhus fever, plague, leprosy, and serious infections of the heart valves and urinary tract.

The effect of the development of vaccines and antimicrobial agents on the pattern of human disease in the period immediately after the Second World Ward was remarkable. The statistics for tuberculosis are revealing. In 1900, in the United States, this disease was responsible for 110 deaths per 100,000 of the population; by 1950 it had fallen to 22; by 1960, when several different

FIGURE 16

Some of the primitive apparatus used in the early attempts to prepare penicillin.

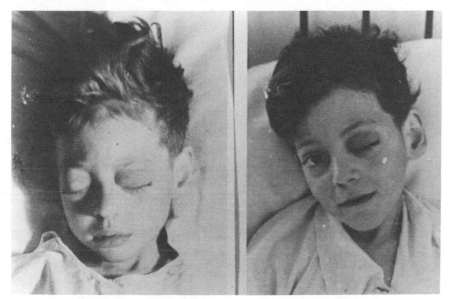

FIGURE 17

Clinical photographs of one of the first patients to be treated with penicillin, in the Radcliffe Infirmary, Oxford, in May 1941. The boy had developed intracranial infection following a sty on each eye. He was comatose on admission and developed a thrombosis of the cavernous sinus, a major vein behind the eye. He made a partial recovery after penicillin, but died eighteen days later.

antituberculous drugs were widely available, it was 5.4; in 1975 it was 1.2. Of course, those who wish to play down the role of scientific medicine simply see the action of these drugs as a final cleaning-up operation after improvements in public health, living conditions, and nutrition had caused such a dramatic decrease in the incidence of this disease in the first half of this century. In the period just after the Second World War tuberculosis was still a dreaded disease, however. Patients spent months in sanatoriums, lungs were collapsed in an attempt to deal with tuberculous cavities, and babies with tuberculous meningitis almost always died or, if they recovered, were often left with lifelong complications, including mental retardation and deafness. We will never know whether the prevalence of the disease would have continued to decline even if antituberculous drugs had not been discovered. But regardless of the numbers involved, their availability had an enormous impact on clinical practice in the 1950s. Within a few years most sanatoriums had been either dismantled or converted to "chest hospitals," in which were treated diseases that replaced tuberculosis as the bread and butter of chest physicians—chronic bronchitis and lung cancer, for example.

But perhaps the most extraordinary instance of the control of infectious disease is the eradication of smallpox. Epidemics of smallpox had decimated populations all over the world for centuries. At the Eleventh World Health Assembly, in 1958, a major global attack on the disease was proposed. A suitable freeze-dried vaccine was developed, and by 1967 the program had been established in each of the thirty countries in which the disease was endemic. The last-recorded case of smallpox occurred in Somalia in 1977.

As a medical student of the late 1950s, I find it interesting to reflect on the enormous mood of optimism that followed the development of vaccines and antibiotics. It was not to last for long, however. Antibiotic-resistant strains of organisms soon started to appear, and the ensuing years saw a remarkable battle between the world of microorganisms and the ingenuity of the pharmaceutical industry. During the 1960s, when powerful agents for the treatment of cancer were beginning to be used, a completely different set of infective agents, the names of which we had never heard, made their appearance. These were organisms that do not attack healthy people but that thrive on those who have had their immune defenses damaged by treatment for cancer or, as it turned out later, during preparation for organ transplantation. Despite extensive research in the universities and the pharmaceutical industry, there was little progress toward the development of drugs to treat virus diseases. Influenza epidemics continued to kill the young and old, the common cold was ever with us, and viral disease of the nervous system, leading to death or permanent disability, still attacked young people all over the world. Clearly, infectious disease had not been conquered, although it had been controlled to such a degree that it was no longer a major cause of death in industrialized societies.

The situation in the developing world is quite different. Against a background of poverty and malnutrition, infective diarrheal, respiratory, and parasitic illnesses remain major killers. There exist powerful agents with which to treat them, even many of the parasitic diseases, but poverty and inadequate health services combine to restrict their availability. Furthermore, extensive experiences in eradication programs for parasitic diseases such as malaria have made it quite clear that we cannot pin all our hopes on the environmental control of tropical diseases. Just as resistance to antimicrobial agents occurs because of genetic changes in bacteria, parasites and the insects that disseminate them are equally ingenious in the methods they have evolved to develop resistance to chemotherapeutic agents or insecticides.

The control of infectious disease involves a constant battle between drug manufacturers and the genetic cunning of bacteria and parasites. Added to this is the constant danger of the emergence of completely new infectious agents. The dreadful epidemic of AIDS in Africa is a clear warning that although infectious disease has been partly controlled, it has certainly not been eradicated. Indeed, as this book is being written, frightening reports are appearing from the United States and Europe about the emergence of strains of the bacteria that cause tuberculosis that are completely resistant to all known chemotherapeutic agents.

LEGACIES FROM THE MEDICAL SCIENCES OF THE FIRST HALF OF THE TWENTIETH CENTURY

In outlining how clinical and basic science combined in the work leading to the control of diabetes, the cure of pernicious anemia, and the conquest of many infectious diseases, we have highlighted three great success stories for medical science in the first half of this century. The cases of insulin and vitamin B_{12} demonstrate how good clinical observation backed up with careful animal research and the early involvement of the pharmaceutical industry can lead to major advances in medical practice. The case of diabetes shows how a major discovery that results in the partial control of a disease may produce a completely new set of clinical problems. The long-term complications of diabetes have played an important role in the genesis of modern high-technology medicine and form an important part of the work of specialists in diseases of the eye, heart, blood vessels, kidney, and nervous system.

The development of microbiology, and the realization that different fevers have specific causes, was to have profound effects on clinical practice and on the future direction of medical research. Until then studies at the bedside and autopsy room had suggested that they might have nonspecific, multiple, and unrelated causes. Now it was clear that even though individual susceptibility might vary according to nutrition, age, and so on, fevers could be related to infection with particular organisms. Several medical historians have suggested that this was the point at which modern Western medical thinking turned away from earlier Western medicine, on the one hand, and from non-Western systems of medicine, on the other. It certainly concentrated the activities of medical practice and science on the investigation and treatment of disease, perhaps at the expense of considering patients in their broader environmental contexts. And it may have led medical thought and research to seek single causes for the illnesses that replaced infection as the major killers—cancer and heart attacks, for example. As will become apparent when we consider current ideas about the nature of these diseases, the assumption of a single cause has turned out to be very far from the truth.

The story of the control of infectious disease in the period after the Second World War brings out the way in which clinical practice feeds on the basic sciences and how medical advances result from years of painstaking work, often carried out in unrelated fields and with completely different objectives. Microbiology grew out of problems in the brewing industry, and vaccination was born of the chance observations of a country practitioner. The significance of Fleming's observations, though they smack of good fortune, would have been missed had he not spent many years looking at routine bacterial plates and hence preparing himself to recognize that a contaminated culture, particularly one in which something had killed staphylococci, might be telling him something of interest. Florey and Chain's work on antibiotics was stimulated by their curiosity about the mechanisms of action of Fleming's lysozyme and the properties of bacterial cell walls. The success of the Oxford group was at least in part due to the chance association of scientists with different temperaments and skills and the pressure to produce antimicrobial agents in the shadow of the Second World War. As much as anything else this story underlines the critical role of the pharmaceutical industry in improvements in health care in the first half of this century. Its exploitation of the work on penicillin, and its subsequent development of many other chemotherapeutic agents, contributed immensely to the health of society.

A haphazard story in some respects, but one that contains all the elements underlying many of the genuine advances in medical practice: good basic science, often done for its own sake rather than with any specific practical end in view; the slow and painstaking development of the related clinical sciences that had to interpret the messages coming from fundamental research; teams of scientists with the breadth of skills to put the whole thing together; the social stimulus of a major health problem; the ability of the pharmaceutical industry to identify and follow up a promising lead; and, not least, a fair amount of good luck.

Writing about this extraordinary period in the development of medical science many years later, Rollin Hotchkiss tried to categorize the roles of the various players in the story: "We would be overcasting if we fitted these pioneers too nicely into stereotyped character roles in our historical drama: Fleming, the last and most acute of the ancient observers; Dubos, the astute explorer; Waksman, the busy prospector; Florey and Chain and our British colleagues, the enterprising developers and engineers. In truth they all had to be part observer-explorer and engineer, and their intellectual efforts and the patient, innovative work of their hands have opened new doors." In short, the development of antibiotics shows a pattern seen again and again in the history of medical advance—the coming together of a number of previously disparate clinical and laboratory sciences at a time when each was ripe for practical exploitation.

There is another, less obvious message for today from this remarkable period in the development of medical research. Robert Koch announced his discovery of the organism that causes tuberculosis, *Mycobacterium tuberculosis,* in March 1882 at a meeting of the Berlin Physiological Society. The news of this work caused enormous excitement throughout the world and was summarized by scientists in letters to the London *Times* and later in the *New York Times*. Not unreasonably, editorial writers assumed that Koch's discovery would lead to a treatment for tuberculosis, perhaps along the lines of Pasteur's vaccines. People wanted to believe that this would happen; for centuries tuberculosis had been feared as the "captain of the men of death," and in the nineteenth century it caused about one in every seven deaths. It killed young people, and despite what Louis Magner describes as the "perverted sentimentalism associated with the disease" and the misguided thinking that it heightened creativity before an early death, it was among the most dreaded diseases of the time. It is not surprising, therefore, that the news of Koch's discovery raised unrealistic hopes for its immediate control. In fact it was over sixty years before this happened.

History is repeating itself today with respect to the perceived practical outcome of major new discoveries. Rarely a week goes by without the announcement by the press of a new "breakthrough." A gene for an intractable disease is discovered, and it is assumed that a cure is just around the corner. When a few years go by and no cure is at hand, a sense of disillusionment with the medical sciences sets in. The story of the sixty-two years that elapsed between the announcement of Koch's discovery of the tubercle bacillus and Waksman's development of streptomycin should be required reading for every government official who has responsibility for funding medical science, not to mention every journalist who writes about it.

The triumphs I have just outlined rank among the most spectacular ones of medical science in the first half of this century. But, as we have seen, each of them was incomplete and raised another set of questions for scientists and clinicians and new problems for those who administer our health services. The situation with many of the common diseases of industrialized societies is even less satisfactory. While genuine progress has been made in our understanding of how to manage heart disease, stroke, rheumatism, the major psychiatric disorders, cancer, and so on, we have only reached the stage at which we can control our patients' symptoms or temporarily patch them up. Our lack of success over the last fifty years, in getting to grips with the basic causes of these diseases, combined with our increasing understanding of the pathological consequences of diseased organs, has bred modern high-technology medicine. Much of it is very sophisticated and effective at prolonging life, but it neither prevents nor cures many diseases. Granted, in the last ten years or so we have started to understand how some of them may be partly controlled by changes in our environment and lifestyles, but they all pose a major challenge to our health services.

In the next chapter we will look at what has been achieved in regard to the commonest killer in our industrialized societies—heart disease. We will see a remarkable example of how the combined ingenuity of the pharmaceutical industry and medical science has been applied to the management of a common group of diseases about which we have very little genuine understanding, a situation that prevails in much of modern medical practice.

PART III

How Much
Do We Really
Know?

A Glimpse of
High-Technology Medical
Practice

So it was all modern and scientific and well-arranged.
You could die as privately in a modern hospital as you could in the
Grand Central Station, and with much better care.
                              —*Stephen Vincent Benét (1898–1943)*

It is difficult to comprehend the extent to which medical practice has changed over the last fifty years. In 1975 Paul Beeson, a distinguished physician and former editor of one of the leading American textbooks of medicine, compared the seventh edition of his book, published in 1947, with the thirteenth, which appeared in 1971. Selecting the investigation and treatment of heart disease, because this subject encompasses both medical and surgical practice, he was able to list no fewer than twenty-one areas of major advance that had transformed this one speciality.

As a telling example of the changes in the management of heart disease over this period Beeson cites a short report that describes the last illness of President Franklin Roosevelt, written in 1970 by H. S. Bruenn, the cardiologist who attended the president almost daily from March 1944 until his death from a stroke in April 1945. As Beeson points out, a mere thirty years earlier an important world figure suffering from heart failure and high blood pressure could be treated only by the administration of crude digitalis and dietary restriction of salt. Had Roosevelt suffered from the same illness in the early 1970s, he would have had the benefit of a battery of drugs designed to control his blood pressure and remove excess fluid from his body, and hence ease the burden on his heart and reduce the likelihood of a stroke. If, as seems probable, he had coronary artery disease, he would have been investigated by means of a variety of sophisticated techniques and received powerful drugs to open up his coronary arteries. A U.S. president with a similar illness in the 1990s might have the additional benefit of having a coronary artery bypass operation or perhaps, if all else failed, a heart transplant. And—perish the thought—there is even an outside chance that he would be the recipient of one of the first artificial hearts.

Such improvements in clinical care are to be found in almost every field of medical practice. There are few diseases for which nothing can be done to relieve pain and suffering. But just how far have we progressed, and to what extent is medicine, as practiced in our industrialized countries, based on solid scientific evidence?

To try to answer this difficult question we must take a closer look at a few examples of current practice and the science that lies behind it. It will emerge that, while we have become increasingly efficient at controlling the symptoms and progression of many of the important diseases whose victims fill our hospital wards, we have made less progress in understanding their underlying cause and hence in learning how to prevent or cure them. On the other hand, our ability to patch up patients and to prolong their lives seems to be almost limitless. This curious anomaly, more than anything else, has led to the dramatic increase in the cost of medical care and, to some extent, to a dehumanizing effect on its practitioners and the hospitals in which they work.

An adequate picture of present-day medical practice would fill many volumes. The second edition of the *Oxford Textbook of Medicine,* which my colleagues and I are foolish enough to edit, runs to two volumes of over four million words, 500 authors, and 3,400 pages, and it weighs sixteen pounds. Yet even this worthy effort offers only an incomplete outline of one aspect of modern medicine. What follows is a bird's-eye view of modern cardiology, the prime example of how high-technology practice has developed, and a brief look at some other specialties that illustrate what we have achieved amid our ignorance about the causes of most of our serious diseases.

MODERN CARDIOLOGY

Cardiovascular disease—that is, disease of the heart and blood vessels—is responsible for about 40 percent of all deaths in advanced Western societies. Even though we all have to die of something, and though a quick demise from a heart attack might seem a better way to go than most, the problem is not as simple as this. Although the average life expectancy for men in a rich, industrialized country is about seventy-six years, and a little longer for women, many of us expire much earlier. Diseases of the heart and blood vessels are the most important contributors to these premature deaths. Before we look at how the diseases are being tackled, it might help nonmedical readers if we digress briefly to see why they are so important.

What Is Heart Disease?

The function of the heart is to pump blood containing oxygen to the tissues. It is a powerful muscle that encases four cavities, two collecting chambers, or atria, and two pumps, called ventricles. Blood returning to the heart from the tissues enters the right atrium, from where it passes into the right ventricle and is then pumped through the pulmonary arteries to the lungs. Oxygenated blood returns from the lungs and enters the left atrium and thence the left ventricle. It is then pumped into a large blood vessel called the aorta, and distributed around the body. Strong valves between the atria and ventricles and at the outlet of the ventricles prevent regurgitation of blood into the chambers of the heart. Its action is coordinated by a complex electrical conducting pathway that, in turn, is connected to a specialized part of the nervous system, called autonomic, because we have no control over it, which allows the heart to respond to appropriate stimuli, such as exercise or the sight of a beautiful woman (or man). The heart expends a great deal of energy as it pumps away, and receives its own oxygen and nutrients from blood carried through the coronary arteries, which emerge from the first part of the aorta and divide into a series of vessels supplying the muscular wall.

Through a complex system of wiring and chemical sensors in the brain and tissues, the heart is able to adapt itself quite remarkably to the normal needs of the body and to compensate for a variety of diseases. It can increase its rate and output in response to exercise, stress, or conditions of reduced oxygen. Much of modern cardiology is concerned with trying to repair the damage that occurs when these compensatory mechanisms break down and the heart fails.

The end result of most serious heart diseases is called heart failure, that is, an inability of the heart to pump sufficient blood to meet the oxygen needs of the tissues. This can happen in many ways. Most commonly it results from disease of the walls of the coronary arteries, which causes them to narrow. This, in turn, reduces the blood supply to the muscle of the heart. Muscle that is starved of oxygen produces abnormal chemical products that may cause pain; this is the basis of angina, a crushing central chest pain often indicative of disease of the coronary arteries. Sometimes the diseased arteries are acutely obstructed because of the formation of a blood clot on their roughened linings.

Blood clots in the coronary arteries occur either in vessels that have become very narrow as a result of disease of their walls or, quite often, on a trivial fatty bump, or plaque, which until then has caused no symptoms. The clot may be formed in response to a crack in a plaque of this kind. This may occur spontaneously or after a sudden rise in blood pressure caused by stress or unusual physical effort, such as shoveling snow.

The formation of a blood clot in a coronary artery may completely occlude it. This leads to serious damage to part of the heart muscle, which, in this case, may have no blood supply. An episode of this kind is called a heart attack, or coronary thrombosis. Another term for a heart attack, "myocardial infarction," describes what actually happens to the muscle; "infarction" means death of tissue. The heart may also fail because it must work harder than usual by having to pump blood against an increased pressure in the arteries; high blood pressure, or hypertension, is an extremely common cause of heart disease and strokes. Often, particularly in older people, the heart fails because of a combination of narrowing of the coronary arteries and high blood pressure.

There are several other ways in which increased strain can be placed on the heart. Damage to the valves may result from rheumatic fever in childhood, infection, disease of the coronary arteries, and a variety of other disorders. A narrow valve requires a stronger contraction to force blood through it and hence puts a strain on the muscle of the chamber of the heart that it protects. A floppy, or incompetent, valve allows blood to leak back after contraction, with the same result. In addition to a narrowing of the coronary arteries the heart muscle may be damaged by a variety of infections, poisons, including alcohol, and metabolic diseases, that is, disturbances of the chemistry of the heart muscle. And, finally, the beat of the heart may become inefficient because of disease of the electrical conducting system. Occasionally this may be because we are born with an abnormality of conduction; much more commonly it is due to a narrowing of the coronary arteries and a resultant failure of blood supply to critical regions of the conducting tissues.

Heart failure takes several forms. Occlusion of the coronary arteries leads to a reduced blood supply to the heart muscle. This may result in sudden death because of an abnormal rhythm of the heartbeat or simply failure of the pump as a result of extensive damage to the muscle. If there is a more gradual reduction in the output of the heart, the kidneys are not adequately supplied with blood. This results in an accumulation of fluid in the tissues, particularly the lungs, and the patient becomes breathless and waterlogged.

The heart may be the site in a variety of infectious diseases. The valves may be attacked by bacteria, either acutely or as part of a more chronic infection. The end result of infections of the heart valve, a condition called bacterial endocarditis, is destruction or deformation of the affected valve. The tough outer covering of the heart, the pericardium, may be involved in infection.

The rhythm of the heart may go haywire in many different ways. Normally the heart is activated by the coordinated passage of an impulse across the atria and then through specialized conducting tissue to the ventricles. But sometimes abnormal foci lead to disordered activation of the heart muscle. Instead of beating effectively, the atria or ventricles may go into a state of completely disordered electrical activity called fibrillation. Alternatively, if there is disease of the conducting system between the atria and ventricles, the pathway of stimulation from the atria may be disrupted, causing what is called heart block. In its most extreme form, the ventricles simply beat away on their own without any normal stimuli and at an extremely slow rate. The heart may actually stop from time to time, causing patients to black out.

Another common group of heart diseases result from abnormalities of the circulation to the lungs, or from disease of the lungs themselves. For example, blood clots may lodge in the pulmonary arteries and impede the output from the right ventricle. Or increasing scarring and stiffening of the lungs may impede their circulation, with subsequent strain to the right side of the heart. This is usually the end result of chronic bronchitis and recurrent respiratory infections.

Considering the extraordinarily complexities of the maneuvers that occur in early embryological life to convert what starts out as a single tube into a four-chambered heart, it is not surprising that they occasionally go wrong. Congenital heart disease takes many forms, ranging from the persistence of gaps between the right and the left sides of the heart to major displacements of the main blood vessels and serious defects in the valves.

Apart from the diseases that affect the heart directly, cardiovascular disease has other common and important implications. A combination of high blood pressure and diseased arteries often leads to stroke—that is, damage to the brain as a result of either the formation of a clot in an artery or the rupture of a vessel, with hemorrhage into the brain substance. Occasionally a small blood clot, or embolus, may break off from the heart and lodge in a vessel to a critical part of the brain. A narrowing of blood vessels to the limbs may reduce the blood supply to the muscles of the legs, which often causes pain in the calves during walking. Ultimately the feet may be completely starved of blood and become gangrenous. Diseased arteries also tend to develop bulges, or aneurysms. These may affect any blood vessel but are particularly common in the abdominal aorta, the main vessel leading from the heart to the lower extremeties. Rupture of an abdominal aneurysm constitutes a major catastrophe.

What Causes Heart Disease?

Clearly a great deal can go wrong with our heart and blood vessels. But by far the commonest types of heart disease and stroke are due to a thickening of and progressive damage to our arteries. It is only in the last twenty years or so that epidemiological studies of large populations have started to turn up evidence that a furring up and narrowing of our arteries may result, at least in part from our unhealthy Western diets, combined with sedentary lifestyles and bad habits such as smoking. When I was a medical student in the late 1950s, almost all the diseases of the heart and blood vessels that we have been describing were clumped together under the heading "degenerative disorders," implying that they were the natural and inevitable results of aging. It is not surprising, therefore, that the main thrust in the development of cardiology over the last fifty years has been directed at the investigation and symptomatic treatment of heart disease. The period has seen increasingly sophisticated clinical practice at the bedside and in the laboratory, together with remarkable ingenuity on the part of the pharmaceutical industry in discovering ways to patch up these conditions. But it is a story of investigating the end results of a failing organ and learning how to deal with them and *not,* for the most part, of gaining a genuine understanding of the causes of disease or ways to prevent or cure it.

Does this mean that, until recently, medical science neglected research into the cause of heart disease? Not at all. Over the last half century the laboratory and clinical sciences have made enormous efforts to try to discover why arteries become furred up and narrow as we age. And a great deal has been learned about the mechanisms of high blood pressure. But, in general, all this activity has produced little progress toward understanding the basic causes of heart disease. What it has told us, however, is that, unlike infectious disease, the thickening of our arteries or high blood pressure has no *one* cause. Rather, there are many different routes to these illnesses, and hence their underlying pathology is extremely complex. This is the central problem of modern medicine; it appears that most of our problem diseases have multiple causes, reflecting both our environment and our genes, set against the background of aging. We will return to this theme in later chapters.

How Did Modern Cardiology Evolve?

During the late nineteenth century a great deal was learned about the pathology of the heart and how to recogize heart disease. These advances, many of which started in Paris, were due mainly to Laennec's invention of the stethoscope and to a careful comparison of what could be heard at the bedside with what was seen at an autopsy examination. Physicians taught themselves how to relate unusual sounds, or murmurs, to damage to the valves, and by the turn of the century their interpreting of abnormalities in the sounds made by the heart valves, or auscultation, for the investigation of heart disease had become quite sophisticated. They also recognized that heart failure is often indicated by an increased distension in the veins of the neck, a sign that remains of inestimable value to clinicians.

By the beginning of the twentieth century much was known about diseases that affect the structure of the valves of the heart and about the infections

FIGURE 18

Sir James Mackenzie (1853–1925).

that damage them. The clinical picture of failure of one or the other ventricle was established, although the major textbooks of medicine of the time had very little to say about diseases of the coronary arteries and blood vessels.

Around the turn of the century there began a gradual change in approach to studying diseases of the heart, with increasing emphasis on understanding its function in dynamic terms. One of the most remarkable figures of this period was the Scotsman James Mackenzie (1853–1925), a family doctor in Burnley, Lancashire (figure 18). Among the first and most successful of the new breed of clinical investigators, he believed that general practice was the only place to carry out clinical research. Starting from the notion that an understanding of irregularities of the pulse might hold the key to cardiac disease, he invented a polygraph, which allowed him to record the pressure waves of the arterial and venous blood (figure 19). In this way he was able to distinguish between harmless alterations in the rhythm of the heart and the more serious ones, indicative of severe heart disease. In 1908 he published a book entitled *Diseases of the Heart,* which, like his earlier one on the pulse, was to have a great influence on the founders of modern cardiology, particularly the English physician Thomas Lewis (1881–1943).

Meanwhile the Dutch physiologist Willem Einthoven (1860–1927) revolutionized electrical studies of the heart, later called electrocardiography, by developing the string galvanometer. It enabled him precisely to record the waves that spread through the atria as they contract and to follow the electrical events that accompany ventricular contraction. He developed a way of assessing the pattern of activation of the heart by analyzing the electrical current flowing between leads attached, respectively, to the right and left arms and to the left leg. The original Einthoven equipment occupied two rooms and took five people to operate, but later more manageable instruments were designed. The different abnormalities of heart rhythm that could be identified at the bedside were correlated with specific changes in the electrocardiograph (ECG), work that Thomas Lewis (figure 20) pioneered and that he summarized in a book written in 1911, *The Mechanism of the Heart Beat.* In the years that followed, many different abnormal rhythms were identified, and it became possible to recognize increased thickness of the walls of the ventricles and diseases of the pericardium, the tough membrane covering the outside of the heart.

At this time, too, there slowly accumulated information, based on both pathological and clinical studies, about the mechanisms of anginal pain and coronary thrombosis. The Chicago physician James Herrick (1861–1954) is usually singled out as the author of the original description of coronary thrombosis and, in particular, as the first to observe characteristic changes in the ECG. It was soon determined that coronary artery disease plays a central

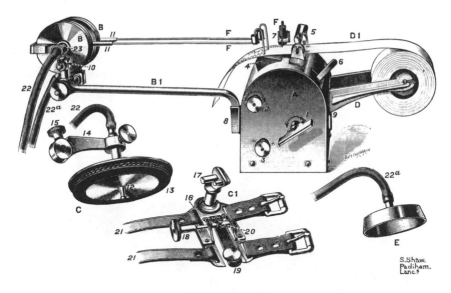

FIGURE 19

Ink polygraph. From Sir James Mackenzie, Diseases of the Heart *(Oxford: Oxford University Press, 1908).*

role in cardiac practice. Shortly after Herrick's description of the ECG following a coronary thrombosis, it was found that occlusion of the coronary arteries of dogs produces dramatic changes in the ECG. That finding led to the development of an additional electrode attached to the chest wall over the heart. Leads of this type became a vital part of the investigation of heart attacks. Soon more chest leads were used to pinpoint regions of damaged muscle, and it was discovered that a lead placed in the esophagus (gullet), which is anatomically very close to the atria, is of value in the diagnosis of complex disorders of the heart rhythm.

By the early 1920s cardiology was well developed as a speciality on both sides of the Atlantic, and the field continued to make rapid strides with this new and more dynamic approach to the study of heart disease. While these advances in bedside diagnosis and in understanding the electrical activity of the heart were important, it was clear that further progress would require better methods for visualizing the heart. In 1929 Werner Forssmann (1904–1979), a German surgeon, placed a hollow tube, or catheter, into a superficial vein in his arm and advanced it until its tip entered the right atrium of his heart, an experiment he performed with the objective of finding a way of injecting drugs directly into the circulation. Not unnaturally, he was anxious to know where the catheter had ended up, and he therefore confirmed its position by having his chest x-rayed.

FIGURE 20

Sir Thomas Lewis (1881–1945)

Over the next few years Dickinson Richards (1895–1973) and André Cournand (1895–1988) found a way to insert a catheter into the right side of the heart and the pulmonary artery, a technique later used throughout the world for studying congenital heart disease. Catheterization of the left side of the heart was introduced a year or two later by passing catheters through the septum, the thin wall that separates the two sides of the heart. This made it possible to measure the pressures in the various chambers of the heart, information of enormous value for the development of surgery for congenital heart disease and narrowed valves. Finally, catheterization was used to outline the chambers of the heart and the coronary arteries with dyes that could be visualized radiologically; angiocardiography, as this rather hair-raising pastime is called, is now used routinely in the study of coronary artery disease.

These advances were later augmented by the development of noninvasive methods of studying the heart. During the two world wars echo sounders were used in ships to determine the depth of the seabed or the presence of mines or submarines. This principle was later applied to the study of the heart, in a technique called echocardiography. The heart can be scanned by a moving beam of impulses, which are reflected by its different structures, and the resulting echos are integrated on a cathode-ray screen to build up a moving image. Increasingly sophisticated (and expensive) modifications have produced color-flow images, which allow further details of the anatomy and blood flow in the heart to be seen. This technique, which is absolutely safe, is of great value for studying the activity of the valves and for pinpointing areas of the heart muscle that are not contracting properly. It can also be used to identify some forms of heart disease in unborn babies.

Other sophisticated methods for producing detailed anatomical pictures of the organs have been developed in recent years. These include computerized axial tomography (CAT scanning) and nuclear magnetic resonance (NMR scanning). The technique of CAT scanning involves taking large numbers of x-ray pictures in the form of slices through the body and then synthesizing them electronically so that the resulting signals are integrated on a computer to give a remarkably detailed image of an organ. The advent of NMR scanning has provided even greater anatomical detail. This technique is based on the principle that water in the body contains protons, or hydrogen nuclei, which are potentially electromagnetic. If our bodies are placed in a powerful electromagnetic field (a harmless environment unless we are wearing a cardiac pacemaker!), these protons tend to line up and resonate to a specific high-frequency radio wave and at the same time absorb some of its energy. By scanning the body by a principle similar to that of x-ray scanning and integrating the signals on a computer, we can build up pictures of the various organs, depending on their water content (figure 21).

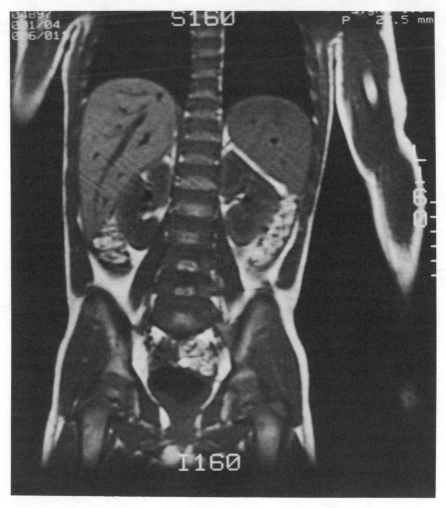

FIGURE 21

A section through the body showing various organs as seen by nuclear magnetic resonance spectroscopy.

As if all this were not enough, cardiologists have invented other highly sophisticated (and equally expensive) ways of visualizing the heart, which involve the use of radioisotopes injected into the circulation. Their passage through the blood and heart muscle is followed with an instrument called a gamma camera, which records the radioactivity produced in the tissues. Many different radioisotopes have been used for this purpose, including those of technetium, xenon, gold, nitrogen, and indium. This technique allows cardiologists to determine the amount of blood that the heart ejects with each beat and to study the function of different areas of heart muscle.

These remarkable diagnostic advances have been accompanied by equally impressive progress in managing established heart disease. For many years the only drug of any value for treating heart failure was digitalis, discovered by the English physician and botanist William Withering in the mideighteenth century, after learning from an old countrywoman that the purple foxglove was "good for the dropsy." Withering published *An Account of the Foxglove* in 1776, but although various extracts and preparations of digitalis have been used to treat heart failure ever since, we still do not fully understand how this old standby actually works.

The first major pharmacological advance in present-day cardiology was the introduction of oral diuretics, drugs able to control the accumulation of water in the tissues and invaluable for the treatment of the failing heart. Previously, the only way of managing this distressing symptom was to insert tubes into the tissues and allow the fluid to drain out. With increasing knowledge of the way in which kidneys handle salt and water, the pharmaceutical industry has produced a wide range of powerful drugs to counteract fluid retention. Similarly, although the cause of raised blood pressure is still not known, the industry has been remarkably successful in discovering drugs to lower it. Their development reflects an understanding of the way in which the autonomic nervous system controls the tone of blood vessels and great ingenuity on the part of pharmacologists in interfering with this complex process. The medical treatment of angina has been much improved by the development of drugs that dilate the coronary arteries, the latest family of which cause the vessels to open up by interfering with the movement of calcium, which is involved in the regulation of the contraction of the muscle in their walls. Recently it was found that some of these dilator agents also help improve otherwise intractable heart failure by reducing the workload on the heart.

Another important landmark in the management of heart disease was the discovery of drugs that interfere with blood clotting. The formation of blood clots in the veins of the legs and pelvis is a common event in patients who are immobilized, during surgical operations or with a serious illness, for example. If the clot stays in the veins, all may be well, but occasionally pieces break off and lodge in the lungs. If they are large enough, they may block the circulation to the lungs and cause death from acute heart failure. The development of drugs that interfere with clotting has revolutionized the prevention and treatment of this serious complication. Until recently the use of such agents in the treatment of acute blockage of the coronary arteries was controversial. However, in the last few years a new family of drugs have been developed that dissolve clots. Large-scale clinical trials have shown that they reduce mortality after heart attacks by dissolving clots in the coronary arteries.

In the light of this frightening expansion in the pharmaceutical armamentarium of cardiologists, it is reassuring to hear that a cheap and easily obtain-able old family remedy has recently returned to the scene. It has been found that aspirin helps to reduce the number of deaths after heart attacks, particu-larly if used together with clot-dissolving drugs. The formation of blood clots reflects a complex interaction between the various clotting factors in the blood and small cells called platelets, which are designed to patch up holes in our blood vessels. It turns out that half an aspirin tablet is enough to reduce the stickiness of blood platelets and hence to make them less effective in helping to generate blood clots. This discovery is an interesting example of how, by the use of large clinical trials, it is possible to reevaluate clinical practice and to make better use of drugs that are already available.

These developments in the pharmacological treatment of heart disease have been followed by astounding advances in its surgical management. Although there were isolated attempts at cardiac surgery in the early part of this century, the field really started to develop just after the Second World War. In the late 1940s surgeons in the United States and England corrected narrowed heart valves by inserting a finger through the left atrium and cut-ting them. In 1945 Alfred Blalock (1899–1964), professor of surgery at Johns Hopkins Hospital, in partnership with the cardiologist Helen Taussig (1898–1986), carried out the first surgical correction of a congenital anomaly of the heart.

In the mid-1950s heart-lung machines were developed that made it possi-ble to pump oxygenated blood around a patient's circulation while at the same time bypassing the heart so that it could be operated on at leisure. Major improvements in the design of these machines have since made "open-heart" surgery routine in hospitals throughout the world. Diseased valves can be reconstructed or replaced by prostheses, and even the most complex congenital abnormalities of the heart can be repaired. In the late 1960s tech-niques were developed to bypass obstruction of the coronary arteries by join-ing a piece of a patient's leg vein to the aorta and then to the coronary artery beyond the block. Over the next ten years it was found that coronary bypass surgery of this type could relieve anginal pain that was not responsive to drugs and that, in certain circumstances, it might prolong the lives of patients with serious disease of their coronary arteries. By the early 1980s approxi-mately 100,000 patients were undergoing coronary artery bypass surgery in the United States each year.

More recently a less invasive approach to the treatment of narrowing of the coronary arteries has been developed. It rejoices in the name "balloon angioplasty." This brand of plumbing involves passing a catheter into the coronary artery and, when it is in position, inflating a small balloon designed to stretch the walls of the diseased vessel. Although narrowing may occur again, it has a useful role in the management of coronary artery disease. It is also used instead of clot-dissolving drugs in the early stages of heart attacks.

Progress in managing abnormalities of the rhythm of the heart resulting from disease of its electrical conduction system has been no less spectacular. In one of the most serious disorders of this type, heart block, there is a complete failure of conduction, leading to a slowing of the pulse and intermittent attacks of unconsciousness. This is a common disorder in the elderly and in patients who have had a heart attack. Its management has been revolutionized by the development of artificial pacemakers. At first electrodes were introduced into the heart, either through a vein or during surgery, for temporary pacing with electronic apparatus outside the body. However, the development of microelectronic circuits made it possible to construct implantable pacemakers. Initially impulses were transmitted to the implanted receiver by means of a surface coil; later fully implantable units powered by a mercury-zinc battery were inserted under the skin of the chest wall and connected to the right side of the heart by means of an electrode inserted through a vein. Pacemakers are now even more sophisticated and come complete with a "demand" circuit so that they can sense a patient's heart rate and switch on electrical impulses only if the heart slows to a particular preset rate.

It is also possible to treat previously fatal abnormalities of the heart rhythm, especially ventricular fibrillation, a common complication of a heart attack. This dreaded condition is characterized by a state of electrical chaos in the heart muscle, acute heart failure, and death. It was noted at the end of the last century that ventricular fibrillation could be arrested in the hearts of dogs by treating them with a discharge from a capacitor or an alternating current. This Frankenstein-like act was first perpetrated on a patient in 1947, and ten years later a defibrillator was designed that involved placing electrodes on the chest and delivering a shock of five amperes of 60 cycles alternating current passed through the long axis of the heart for a period of 0.25 seconds. A few years later a direct-current capacitor discharge system was used for depolarizing the heart in the "safe" part of the cardiac cycle. The instrument employs a capacitor discharge triggered by the electrocardiogram, and hence it can also be used to correct other abnormal rhythms. These can now be combined with a pacemaker and are fully implantable, though they cost some tens of thousands of dollars.

In addition to these sophisticated aids to cardiac resuscitation, work carried out over the last thirty years has made it possible to maintain an output from a heart that has stopped temporarily or developed an abnormal rhythm, by intermittent pressure on the chest wall. This procedure, called cardiac massage, which is combined with mouth-to-mouth respiration, is now used in hospitals and by paramedical personnel throughout the world.

While these physical methods for dealing with abnormal heart rhythms were being developed, the pharmaceutical industry was busy discovering new drugs to prevent or treat them. Today's cardiologists have an ever-expanding battery of drugs at their disposal, some of proven value, some fashionable, and nearly all expensive.

The final tour de force of high-technology cardiology—a successful heart and, later, heart and lung transplant from one individual to another—was the fruit of many years of basic research in immunology. If a piece of skin is grafted from one position to another on the same person it will heal and survive, but if the graft is taken from someone else it is rapidly destroyed. The one exception to this rule is that grafts between identical twins "take" quite successfully. Stimulated by seeing the ghastly burns of a pilot who was shot down in the Second World War, Peter Medawar and his colleagues discovered that the reason we cannot accept tissue from other people is that our body mounts an immunological reaction against it because it recognizes it as "foreign."

Since it is possible to graft tissues between identical twins, it was reasoned that the body's ability to recognize foreign tissue must be genetically determined. The genes involved in this curious phenomenon were discovered by the French scientist Jean Dausset. Extremely complex, they are inherited in groups, and their products can be identified by a procedure called tissue typing, or matching. There is a one-in-four chance that a brother and sister will be compatible for a transplant; organ grafts between relatives are much more successful if they show a complete match. The first successful organ transplant was carried out in 1962, when a kidney was transplanted from one relative to another. However, even after a well-matched organ transplant the recipient may raise an immunological attack on the graft, and further progress in this field awaited the discovery of drugs that suppress the body's immune response, a variety of which have now been developed. The combination of tissue typing and immunosuppression has opened up the whole field of transplantation surgery, using organs from cadavers or, less commonly, from relatives.

The first heart transplants were carried out by the South African surgeon Christiaan Barnard in 1967. The first patient died after two weeks, but the second survived for a year. Since then improvements in immunosuppression and tissue typing, together with better surgical techniques, have led to the successful transplantation of many hundreds of hearts from cadavers to living patients with heart disease. It is even possible to transplant both the heart and the lungs into patients with severe heart disease secondary to disorders of the blood vessels to the lungs. Surgeons who indulge in this pastime tell me that their first view of a living patient with a chest completely devoid of all its organs, except, one hopes, the gullet, is one of the most daunting experiences in modern surgery. Over recent years there has been a growing interest in the development of artificial hearts, but so far the installation of these has not become an established part of clinical practice.

The Scientific Origins of Modern Cardiology

From this bird's-eye view of the growth of modern cardiology, it is clear that the field was founded and developed by a small group of clinician-scientists like Mackenzie, Lewis, Herrick, Forssmann, and others. But, as we have seen modern cardiology is a highly technical field that relies on the fruits of work in many disciplines, among them electrophysiology, physics, immunology, and pharmacology. How did all this happen? In the preceding chapter we saw how some of the major scientific advances in medicine in the early twentieth century originated in scientific discoveries not aimed at the management of human diseases. Has this trend continued, or is modern cardiology the result of applied science directed at the management of heart disease?

This question was addressed by two American scientists, Julius Comroe and Robert Dripps, in one of the few genuine studies of the role of the basic sciences in clinical medicine, published in 1976. Comroe and Dripps asked 140 specialists what they considered to be the top ten clinical advances in cardiovascular and chest medicine and surgery in the preceding thirty years. When they had arrived at a consensus, they analyzed the goals of the authors of the 529 key scientific articles that they believed had led to these achievements. Their analysis asked specifically whether the key research was "basic"—that is, not directed at a particular clinical question—or whether it set out with the express objective of tackling a medical problem. Remarkably, they found that only 60 percent of the seminal work in modern cardiology was clinically directed and that the remainder was in the basic sciences and had not been carried out with any particular clinical end in view.

By now the reader should not be surprised at the outcome of this study. We have already seen how the great successes in the control of infectious disease followed exactly the same pathway. Present-day cardiology has been similarly underpinned by curiosity-driven science, the fruits of which have been taken up by physiologists and cardiologists and applied directly to clinical practice. Although very few other aspects of current medical practice have been scrutinized in such detail, it is likely that they have followed much the same path.

The story of modern cardiology offers another interesting lesson. While its development has relied on key advances in the basic sciences, both conceptual and technical, historians of the period have pointed out that this was not enough. The field also required talented physicians able to understand enough about the scientific exploration of the functions of the heart to apply its technology in the clinic. Cardiology was one of the first subspecialties of general internal medicine, in which groups of physician-scientists broke away from the mainstream of clinical practice and concentrated their efforts in one field. For example, in Great Britain a small group of physicians interested in heart disease formed the Cardiac Club in 1922. It included some of the founding fathers of modern cardiology and epitomized the importance of a new brand of physician who was able to both converse with basic scientists and apply their tools to patient care. This pattern of specialization has underpinned advances in many branches of medical practice over the past fifty years.

How Does Cardiology Work in Practice?

How has all this new technology affected the lot of patients with heart disease in the 1990s? Coronary artery disease and its complications form a major part of the work of a modern cardiologist. So what happens to patients who develop chest pain and are suspected of having had a heart attack? By one route or another they find themselves being transported through the doors of a hospital's emergency department. Here they are greeted by a doctor who takes their history, examines them, and arranges for an electrocardiogram to be carried out. On the basis of this information, a rapid decision is made whether they are likely to have had a heart attack, a diagnosis that may have to wait for a day or two for confirmation, as changes in their blood evolve. If there is a high likelihood that they have suffered a heart attack, they are given an injection to ease their pain, together with aspirin and a clot-dissolving drug, and whisked straight into a coronary care unit. These specialized wards were first developed in the 1960s. Their existence is based on the notion that because of the high mortality as a result of an abnormal rhythm of the heart or other complications that occur in the first few hours after a heart attack, the condition is best managed in a specialized environment in which patients can be monitored continually, at least for the first few days after the attack. This involves close clinical observation and a continuously recording electro-cardiogram, which is watched by specially trained staff. A patient's first view of a unit of this type may be extremely daunting, with its rows of human beings wired up to monitors and a variety of other unfriendly-looking devices.

The objectives of management over the next day or two are reasonably clear-cut and logical. First, if the heart develops an abnormal rhythm, it is corrected by the use of drugs or an electric shock, as described earlier. Second, every effort is made to protect the heart muscle from further damage. If a blocked coronary artery is detected after the powerful clot-dissolving agent has been given, a balloon angioplasty is considered. Drugs are given to prevent abnormal rhythms and to reduce the work of the heart muscle. If the blood pressure is raised, it is lowered with drugs, and if the heart shows any signs of failure, others are given to remove fluid from the body and to further ease the work of the heart. If the heart muscle is severely damaged and is unable to maintain an output, the resulting state of shock is treated by drugs designed to maintain blood pressure and protect vital organs such as the kidney. The pharmacology of coronary care is expensive and constantly changing, and its current list of drugs resembles the telephone directory of a modest-sized town.

It may be that there is no evidence that the heart has been damaged and that the chest pain is simply a warning of an incipient heart attack. If the pain persists—a condition called unstable angina—every effort is made to prevent further blockage of the coronary vessels and the development of a full-blown heart attack. Drugs are given to make the blood clot less rapidly, and a whole battery of pharmacology is brought into action to reduce the work of the heart. If the situation does not improve, the patient may undergo coronary angiography to see whether there is a region of narrowing of the coronary arteries. If so, a balloon may be fed into the coronary artery and blown up in an attempt to open up the narrowed vessel. Or the patient may be advised to have a bypass operation there and then.

What happens after recovery from a heart attack? This is another constantly changing scene and varies from country to country and center to center. After two or three weeks, if there is no more pain and the patients are otherwise well, they may be asked to perform a vigorous exercise test while attached to an electrocardiogram. If this test shows evidence that part of the heart muscle is being starved of oxygen, they will undergo coronary angiography and, if there is a narrowing of any of the vessels, bypass surgery. In some parts of the world coronary angiography has become almost routine after a heart attack, particularly in younger patients. Alternatively, if the exercise test is negative, the patient may be investigated no further and simply advised about a lifestyle designed to prevent further heart attacks.

Many patients with coronary artery disease report to their doctors with chest pain after exercise. Their management depends on the doctor and on local fashion. Some have their pain controlled by medication; only if this fails do they proceed to coronary angiography and, if necessary, bypass surgery. But in many centers, the investigation and the treatment of angina are more aggressive, and coronary angiography is carried out whether or not the pain is relieved by drugs.

It is easy to see why the management of coronary artery disease is producing a heavy burden on our health services. And this is only part of modern cardiology. The investigation and the surgical correction of congenital heart disease and of acquired abnormalities of heart valves are other important parts of the cardiologist's work. The care of patients with abnormal heart rhythms, particularly the elderly, who often require a pacemaker, is becoming a major industry. There are younger people whose heart muscles have been irreparably damaged, either by a mysterious disease called myocarditis, which may result from a virus infection, by overindulgence in alcohol, or by the effects of widespread coronary artery disease. When the heart muscle is damaged to this degree, and the patient is suffering the miseries of intractable heart failure, the only option is a heart transplant.

Has Modern Cardiology Improved the Lot of Patients with Heart Disease?

How much of modern cardiological practice is of proven effectiveness, and how can we tell? As we saw earlier, the application of quantitative methods to clinical practice was pioneered by Louis in Paris in the middle of the nineteenth century. But although a few clinicians on both sides of the Atlantic urged that trials of clinical outcome be adopted, and although the principles of numerically based experimental design were set out in the 1920s by the geneticist Ronald Fisher (1890–1962), the use of statistical analysis and properly designed trials to assess different types of medical treatment did not start to have a major influence until after the Second World War. Much of the credit for the new thinking belongs to Sir Austin Bradford Hill (1897–1991) and the British epidemiologists who followed him, notably Richard Doll and Archie Cochrane (1909–1988).

The idea is very simple. If a form of treatment is of uncertain value, or if there is doubt about which of two approaches is the better, patients are randomized to one or the other and the outcome assessed by statistical methods. The randomized control trial (RCT) is one of the most important developments in medical research in the last fifty years and has been applied widely and in many different guises to the study of treatments in almost every branch of clinical practice. It has been extended to compare the results of particular forms of therapy with the administration of placebos, inert substances prepared to resemble as closely as possible the drug being assessed. It was soon found that most treatments have a "placebo effect"—that is, there is some improvement even though the placebo could have had no effect on the body's function. In other words, the very act of giving any form of treatment may have some therapeutic value, though how this is mediated is still not clear. As a further refinement, trials are often carried out in which neither the doctor nor the patient knows whether a drug or an inert placebo is being administered. In these "double-blind placebo trials" the drug being assessed and the placebo may be exchanged at some point during the study. Where differences in various forms of treatment are marginal, results have been improved by the use of large numbers of patients. Quite recently a mathematical trick called meta-analysis has been devised, whereby one can combine the results of many different trials to try to give an overall picture of trends of significance.

Of course, some forms of treatment do not require randomized trials to prove that they work. There was no need to carry out a trial of vitamin B_{12} therapy for pernicious anemia, or to test the benefits of penicillin for certain bacterial infections. On the other hand, it was vital, and an early vindication of the use of clinical trials, to establish the best combinations of antibiotics for the treatment of tuberculosis. Many forms of therapy, particularly those which aim at controlling rather than curing diseases, or which may be of marginal benefit, require properly executed trials before they can be considered to be of genuine value.

The medical profession has not always been as enthusiastic as it might have been in subjecting new developments to clinical trial. There are many reasons for this. Not unreasonably, doctors like to believe that their treatment is useful, and, as we saw in chapter 2, medical practice still suffers to some degree from dogmatic authoritarianism. Trials are difficult to design and may take years to complete, and the need to do *something* for sick people is often so great as to force clinicians into actions of unproven value. Doctors may feel that the perceived value a particular form of therapy would make it unethical, or at least unnecessary, to subject it to a randomized trial. Medicine is still a field of fashions, and often new treatments become established before there has been enough time to determine whether they are of any real value. And while the pharmaceutical industry has had a good record of assessing its products by clinical trials, there is no doubt that, in a highly competitive market, practitioners may sometimes be induced to use new forms of treatment before they have been fully evaluated.

Another problem is the dissemination of information. With hundreds of clinical trials going on all over the world, how is the busy practitioner to remain aware of their results? And often the findings are at odds with each other. The design of trials with much larger numbers of patients, together with the recent development of meta-analysis and electronic technology for updating results, may help overcome some of these difficulties.

Set against this extremely complex background, how far has modern cardiological practice improved the quality and length of life of patients with heart disease? There is little doubt that the large family of drugs available for the treatment of patients with heart failure has done much to relieve their symptoms. Similarly the development of agents for lowering blood pressure has greatly improved the outlook for those with very high levels. Advances in cardiac surgery, which have made it possible to correct serious congenital heart defects or to replace diseased valves, have allowed patients with these previously intractable conditions to enjoy a normal life. And the availability of powerful antibiotics has made it possible to treat the dreaded infections that used to destroy diseased heart valves and lead to a rapid demise. Many of these advances did not require clinical trials to show that they worked.

But what of the leading killer of Western society, coronary artery disease? The development of powerful drugs has doubtless made it possible to relieve anginal pain caused by narrowing of the coronary arteries. And recent large-scale clinical trials have shown beyond all reasonable doubt that many lives can be saved by the early administration of drugs to dissolve clots and reduce the stickiness of blood platelets after an established coronary thrombosis. Similar studies have found that the use of drugs to lessen the work of the heart after a heart attack improves the chances of survival.

But what of the particularly high-technology management of heart attacks—coronary care units, detailed monitoring, angiography, and bypass surgery? Here we enter the kind of gray area that bedevils the assessment of modern medical practice. Many people die immediately after a heart attack, either because the heart enters a state of electrical chaos in which it can no longer pump blood around the body or because severe damage to its muscular wall produces a similar result. It might seem self-evident, therefore, that to train individuals in cardiac resuscitation and to monitor patients carefully in an environment where these events can be dealt with must improve the outlook after a heart attack. Curiously, it has been difficult to obtain clear-cut evidence that patients in hospitals with coronary care units do much better than those who are looked after at home or in an ordinary ward. Such trials as have been published, most of which were inadequate to answer the question, suggest that the difference in outcome is minimal. Despite this, I know very few physicians who would stay at home with a bottle of scotch whiskey and a good book and chance their fate after having a heart attack. Coronary care units have become an accepted and increasingly expensive part of the work of most major hospitals, and a great deal of money is spent in training ambulance staff and other paramedics in the resuscitation of patients after heart attacks.

The role of surgery for coronary artery disease is constantly changing. Early trials suggested that coronary artery bypass surgery had a great deal to offer patients with severe angina that could not be controlled with drugs. In Great Britain most patients are still given an adequate trial of drug treatment for angina, although, as in the United States, there is an increasing tendency to investigate them by coronary angiography and to proceed to bypass surgery or balloon angioplasty if defects in the coronary arteries are seen. One reason for this trend is that large-scale trials involving many centers that carry out cardiac surgery have suggested that if there is a severe narrowing of the main stem of the coronary arterial system, or disease in three of its main branches, the lives of patients can be prolonged by surgery. Furthermore, it has been found that the quality of life after bypass surgery may be greatly improved, mainly because it is often no longer necessary to prescribe powerful drugs to relieve anginal pain.

One of the difficulties in assessing this field is that things have moved so fast over the last ten years that there has been little time to step back and see where it is going. There are batteries of new drugs for relieving anginal pain and resting the heart muscle; at the same time coronary artery bypass surgery has become much safer and easier to perform. It seems unlikely that the type of trials that would be required to evaluate much of modern coronary artery surgery will ever be carried out. The whole field has gathered such a momentum, and fashions in surgery are changing so rapidly, that it seems certain to expand in an uncontrolled and uncontrollable fashion for the foreseeable future.

It appears, therefore, that the dramatic expansion of cardiology that has occurred over the last few years, and which is threatening to bankrupt our health services, is based on a mixture of solid science and fashionable, though unproven, practice. In 1989, 368,210 coronary bypass operations and 258,796 operations to dilate the coronary arteries with balloons were performed in the United States. Although it is now accepted that bypass surgery is an effective treatment for the symptomatic relief of angina, its impact on life expectancy remains in dispute. While it is widely believed that surgery prolongs the life of patients with certain specific patterns of coronary artery disease, the results of some trials indicate that life expectancy may not improve for the majority of those who undergo bypass surgery, although others suggest that it does. Yet other trials have suggested that better operative technique, though it has improved the surgical outcome, has not shown any benefit over medical treatment, after patients have been observed for periods of up to ten years. Trials of drugs to prevent abnormal rhythms of the heart have not shown any reduction in patient mortality. On the other hand, a recent meta-analysis involving 27,000 patients in twenty-seven trials suggests that a modest reduction in mortality is achieved if drugs to relieve the work of the heart are given after a heart attack. Even less is known about the longer-term results of balloon angioplasty, though a 30 percent restenosis (narrowing) rate may limit its capacity to improve life expectancy.

It is clear from these examples that, overall, modern cardiology is improving the lot of patients with coronary artery disease, although just how much is very difficult to assess. And we must always bear in mind that patients who are treated in trials are often monitored much more carefully and frequently than those in the real world of community medical practice. On the other hand, we should not be skeptical about what are apparently the marginal benefits disclosed by studies of very large numbers of patients in trials. After all, a 10 percent reduction in mortality for a very common disease is worth striving for.

Our ability to control very high blood pressure levels, which lead to death from heart or kidney failure, or stroke, has no doubt saved many lives. But what of the milder degrees of high blood pressure that affect so many of us as we get older? Recent meta-analyses indicate that a moderate elevation in blood pressure increases our risk of having a stroke and, probably, a heart attack. Since blood pressure levels of this type affect a high proportion of older people, should we be making efforts to treat this increasingly large proportion of our population? To address this question, several large clinical trials have been carried out on both sides of the Atlantic. Until a short while ago the results were ambiguous. It was not entirely clear how effective control of the blood pressure really is, and there was considerable concern about the side effects of the agents given to treat mildly elevated levels. Recent meta-analyses, however, seem to have provided better evidence that treatment of moderately elevated blood pressure levels can reduce the number of strokes. It has also been found that this reduction is not related to age; the treatment is effective in individuals sixty-five years of age and over. Since a stroke is one of the most debilitating of all ailments, these findings have important implications for the provision of health care and for population screening for high blood pressure.

Science and Art in Modern Cardiology

Even though we still do not know much about the causes of many of the heart diseases that we encounter, readers might assume that the remarkable advances in their investigation and symptomatic control that we have just reviewed should have removed much of the uncertainty from cardiological practice. But just how far has modern cardiology evolved from an art to a science?

How good are we at managing heart attacks? The diagnosis is suggested by the patient's story of chest pain. It may be confirmed by changes on the electrocardiogram that are present when the patient first appears. A few days after the episode the products of the damaged heart muscle are shed into the bloodstream, where they can be easily measured. Thus at least the diagnosis of a heart attack should be reliable, based as it is on well-defined principles of electrophysiology and biochemistry.

Unfortunately, things are not as simple as this. First, there is no one, infallible test to tell a doctor that a patient has had a heart attack, certainly not if seen shortly after its onset. While the patient may complain of crushing central chest pain, very often the pain is in a different place or has the wrong characteristics, or both. It may, for example, be localized to the abdomen, the back, the shoulder, or the left side of the chest. Even for the most experienced cardiologist, the distinction between genuine cardiac pain and pain arising from other causes may be extremely difficult or impossible to make. Some patients are unreasonable enough to have heart attacks with no pain whatever. And quite often the changes on the electrocardiogram take hours or even days to evolve. In these cases the doctor may have to decide to treat a patient for a heart attack on the clinical evidence alone. In the days when there was no definitive treatment this did not matter, but now that it is routine practice to administer powerful clot-dissolving drugs, which have side effects, the decision is much more onerous.

The difficulty of determining the cause of chest pain is not confined to the diagnosis of an established heart attack; it is even commoner when one tries to decide whether a patient with this symptom is suffering from angina, a frequent reason for referral of patients to a hospital. The pain may be quite unlike that of angina and the electrocardiogram may be normal, even after exercise; yet it may still be indicative of coronary artery disease. Some patients appear to have genuine anginal pain, which is due to starvation of the heart muscle from oxygen; yet the coronary arteries are normal, even when examined radiologically after the injection of dye. It is believed that this type of pain reflects a spasm of the coronary arteries, although it is not known why this occurs. On the other hand, there are many other causes of pain in the chest, all of which have to be excluded. The same problems arise in determining the significance of breathlessness, another important symptom of heart disease. Breathlessness is highly subjective, can result from many different causes, and is not amenable to many scientifically based investigations. Assessing its significance, like evaluating that of chest pain, relies very much on doctors' skills at talking to patients and on their experience and judgment.

The treatment of heart disease is no easier. As we have seen, clinical trials have given us valuable information about the outcome of different ways of going about it. But trials involve large numbers of people, and when it comes to the individual patient things may not be so straightforward. There is enormous variation in how different people respond to drugs; an agent that corrects an abnormal heart rhythm or that relieves fluid retention in one person may not work for another. And clinical trials are usually directed at the analysis of one particular form of treatment—the use of clot-dissolving drugs or drugs to rest the heart, for example. While each of these approaches may have been shown to be effective alone, sometimes there is very little information about how to use them in combination. The treatment of an individual patient is often a matter of trial and error, backed up by experience, fashion, guesswork, and common sense.

These few examples of the doubts and uncertainties in the management of common forms of heart disease reflect the pattern of practice right across modern medicine. We have a certain amount of knowledge and a few facts that are based on good scientific evidence. But, because of the complexity of the diseases that we are treating, and the many different ways in which patients display them and react to them, day-to-day practice remains very much a mixture of experience and empiricism, set against a background of slow progress in the application of adequate statistical evaluation of what we are doing.

Only within the last decade have genuine efforts been made to evaluate our clinical skills on a more scientific basis. Some forms of clinical examination and laboratory tests are being subjected to analysis by probability theory, and attempts are being made to construct algorithms for more logical forms of management, a series of preset alternative pathways in which the answer to one question automatically determines the next. And computer programs are being designed in an attempt to improve diagnostic accuracy for complex symptoms like chest pain. While it is too early to determine whether these new approaches will make a major impact on clinical practice, it seems likely that they will. They have already shown us that some of our most cherished methods of examining patients have a level of precision such as to render them valueless.

MUCH OF MODERN HOSPITAL MEDICINE FOLLOWS A SIMILAR PATTERN

Modern cardiological practice provides an excellent example of the complex mixture of success, failure, and uncertainty that typifies the era of high-technology medical practice. Until recently the field was concerned mainly with the detection and management of advanced disease for which the cause was largely unknown. The pharmaceutical industry, together with remarkable advances in our understanding of the physiological consequences of organ failure, supportive care, and modern surgery, has allowed us to control the symptoms of most forms of heart disease and to prolong life and improve its quality. The same theme runs through most other branches of medicine.

Most specialties have changed beyond all recognition, thanks to technical and pharmacological advances over the last fifty years. Physicians and surgeons who specialize in diseases of the gastrointestinal tract and liver now have an amazing battery of instruments for visualizing much of the bowel and the various ducts that run into it, and for treating many of their disorders without resort to surgery. The management of ulcers that involve the stomach or duodenum has been revolutionized by drugs that reduce the amount of acid secreted by the stomach, thus greatly decreasing the need for surgery. Bleeding from ulcers or from other disorders of the upper intestinal tract is still a serious problem. By pinpointing the site of the bleeding by means of endoscopes (the medical equivalent of a telescope) inserted into the stomach, together with early operative intervention in the elderly, doctors have been able to improve the lot of patients with this serious complication, although its mortality remains considerable. Yet we still have much to learn about why stomach ulcers occur and how to prevent them. The common inflammatory diseases of the lower bowel remain a mystery, although their symptomatic treatment has improved. And the other major puzzle of this specialty, the irritable bowel syndrome, remains intractable and affects many of us during our lifetime.

Chest medicine has also changed dramatically. Up to the end of the Second World War chest physicians spent most of their time treating pulmonary tuberculosis and its complications. Today they deal mainly with asthma, the cause of which continues to be a mystery, and with the end results of smoking-related diseases, such as chronic bronchitis, respiratory failure due to chronic lung damage, and cancer of the lung. Surprisingly, despite the advent of powerful antibiotics, pneumonia persists as a major problem, particularly in the elderly. It still claims an unacceptably high mortality in Western societies, and despite new sophisticated methods for obtaining specimens from deep down in the lungs by intubation or biopsy, the causative organism is not always found. Pneumonia is especially common in those whose immune systems are depressed by alcohol, by AIDS, or by the increasingly widespread use of powerful agents to treat cancer that also damage the immune system. Chest physicians now spend more and more of their time managing patients with pneumonia caused by the activities of their colleagues in other specialties; one of the unfortunate by-products of modern medicine is iatrogenic disease, that is, illness caused by the side effects of treatment.

The diagnostic work of neurologists, physicians who care for patients with diseases of the brain and spinal cord, has also been transformed by high-technology diagnostic aids. We have already met the new CAT- and MRI-scanning techniques; the latter show quite remarkable anatomical detail of the nervous system (figures 22 and 23). And now there is PET (positron emission tomography), a development that provides functional information about different parts of the brain. Yet the management of the commonest conditions looked after by neurologists, strokes caused by a blockage or rupture of blood vessels to the brain, has progressed little, if at all. With modern scanning technology it is possible to pinpoint the site of damage to the brain, but, with a few exceptions, surgery has little to offer; the best that can be done is to attempt to make optimal use of what function remains, with physiotherapy, speech therapy, and other forms of rehabilitation. Similarly the causes of many of the chronic degenerative diseases of the nervous system— multiple sclerosis and Parkinson's disease, for example—are still not known.

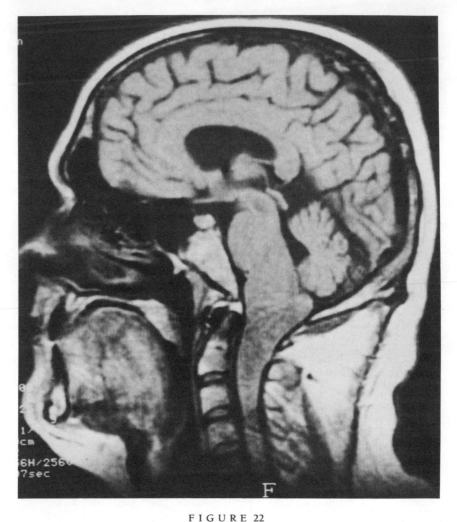

FIGURE 22

Magnetic resonance imaging to examine the brain and spinal cord.

However, there has been some progress in understanding the chemical abnormalities of the brain in different forms of Parkinson's disease, and hence its pharmacological management has improved. And much can be done to help those with progressive, incurable diseases of the brain by new methods of rehabilitation.

Much of modern surgery also deals with highly sophisticated patch-up operations. Furred-up arteries to the legs are grafted; distensions of the main artery from the heart to the abdomen and legs, aortic aneurysms, are removed and replaced with grafts; and narrowed arteries to the kidneys or brain are

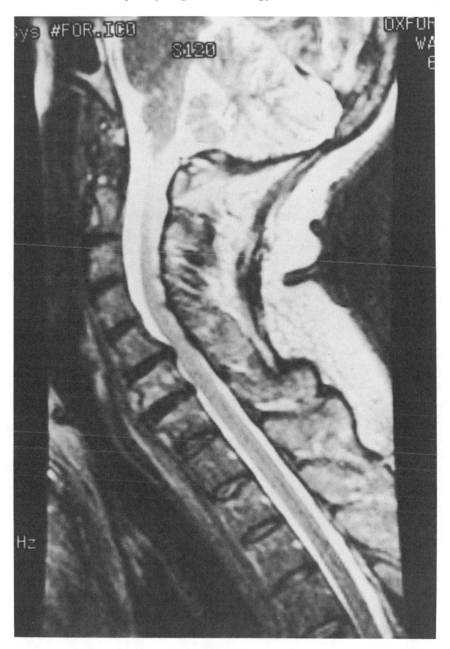

FIGURE 23

Magnetic resonance imaging showing spinal cord compression by the protrusion of a vertebral disk.

repaired. These operations are often carried out on old people with severe heart disease and with a generalized silting up of their arteries. Yet some remarkable results are obtained; the long-term survival of patients who have had aortic aneurysms removed ranks as a major success story for vascular surgery. And every branch of surgery is being revolutionized by developments in instrumentation that allow operations to be carried out through thin tubes without the necessity of large incisions and trauma to the tissues. Much of modern orthopedic surgery, which is also benefiting from noninvasive technology, is concerned with the replacement of hips or knees damaged by osteoarthritis and the partial correction of the ravages of rheumatoid arthritis. Ophthalmic surgery has been simplified by the use of laser treatment for a variety of disorders of the retina and by developments in local anesthesia that allow common and distressing conditions such as cataract to be managed, as can many other forms of surgery, on a visit to the hospital lasting less than a day. The list of technological advances in the surgical specialties is endless, yet the number of surgical conditions for which the cause is known, or which can be prevented, has changed very little in recent years.

Cancer is the second-commonest killer in Western societies. Although there are many ways of attacking cancerous growths, the main successes have resulted from surgical excision and radiotherapy or chemotherapy—that is, the use of x-rays or drugs that destroy a malignant growth but do not, one hopes, also kill the patient. Some cancers are curable. The early successes of chemotherapy, particularly for some forms of leukemia, exemplify the mixture of science and empiricism, backed up by the use of well-conducted clinical trials, that typifies some of the best practice of the high-technology era.

Our bone marrow normally contains immature white blood cells, the progeny of which are critically important for the body's defense mechanisms against infection. In leukemia these early forms proliferate and divide in a totally disordered way, do not mature, and infiltrate the bone marrow and other organs. This results in a reduced output of the other cells of the blood, particularly red cells required for oxygen transport and platelets needed as the first-line defense mechanism against bleeding after damage to small blood vessels. Patients usually succumb to the combined effects of anemia, bleeding, and infection. Virtually nothing is known about the cause of leukemia or its prevention.

Until the early 1960s very little could be done for children or adults with acute leukemia, which proved uniformly fatal within a few months of diagnosis. The first hint of any hope came from the observation that some forms of childhood leukemia might undergo a short period of arrest, or remission, after the administration of cortisone-like drugs. At about the same time a few agents were discovered that seemed to have specific action in destroying leukemic cells, though it was not clear how they worked. It was found that the use of these drugs in combination with corticosteroids could prolong remissions. Over the next twenty years various combinations of drugs were tested in a series of clinical trials in the United States and Europe. The best mixture in one study became the standard against which other combinations were tested in subsequent trials. Gradually the remission times increased, but it was then found that in certain parts of the body, or havens, leukemia cells seemed to be protected from antileukemic agents—notably in the brain, testes, and ovaries. Methods for eradicating leukemic cells from these organs were explored and further trials carried out, including radiation of the brain and spinal cord. Ultimately it was possible to cure a high proportion of childhood leukemias.

The partial conquest of some forms of childhood leukemia marks one of the genuine medical successes of the postwar period. It is a great credit to basic science and the pharmaceutical industry and a fine example of how, by using properly controlled clinical trials, one can compare the therapeutic efficacy of drug regimes that do not have obvious differences in outcome to recommend them.

Acute leukemia in adults turned out to be much more difficult to eradicate. This setback led to the use of bone marrow transplantation after irradiation of the patient's own marrow, a daunting procedure but one that offers at least some chance of a cure. Because the drugs and radiation used to prepare patients for marrow transplantation cause damage to their healthy bone marrow and other tissues, this type of treatment has had to be backed up with the administration of blood products and the management of infections by powerful antibiotics. In most centers it is carried out in sterile containment systems in an effort to prevent infection. The treatment of adult leukemia is extremely expensive, traumatic for patients and their relatives, and dependent on a large team of doctors, nurses, and technicians. Although not a common disease, it puts a considerable strain on resources for health care.

Treatment of other forms of cancer has followed much the same line. Careful clinical trials comparing surgery with drugs have been carried out to determine which method carries the longest remission or, in some cases, leads to cure. Considerable success has been achieved for many cancers, including those of the breast, bowel, and lymphatic system. On the other hand, the outlook for patients with cancer of the lung, pancreas, and brain has improved very little.

Finally, it should not be forgotten that as the era of high-technology medical practice has advanced, so has the age of our population. A large proportion of the patients cared for in our hospitals today have pathology involving more than one system. As we get older, many of us seem to live on a physiological knife edge; when a disease hits one organ, a chain reaction follows that causes failure of several others. And old people take much longer to recover from illness than the young do. Furthermore, an increasing proportion of the very old have varying degrees of mental impairment, or frank dementia. Modern medicine has thus had to cope not only with rapidly changing technologies but also with applying them to a changing population, an increasing number of whose members suffer from the ill-understood problems of aging.

THE MODERN HOSPITAL

Modern, high-technology hospitals are complex institutions. Increasingly their staff and facilities are broken up into specialized units that deal with just one piece of the patient. Since sick people to not always have the goodness to confine their diseases to one organ system, particularly when they are old, this is not always in their best interest. These units require the backup of batteries of diagnostic services. Modern radiology is augmented by the many sophisticated scanning devices, the biochemical analysis of blood has been revolutionized by automation, and most hospitals have diagnostic radioactive isotope departments for organ scanning and assessing the function of different tissues. Every clinical specialty has access to a part of the hospital called an intensive care unit, where patients can be maintained by artificial respiration for an indefinite period (figure 24). Their work includes the care of patients after open-heart surgery or complex vascular procedures, resuscitation after overdoses from a wide variety of self-administered drugs, tiding over those who have had severe and overwhelming infections—in short, the management of any condition involving a failure of one or more organs. The ability to maintain life in this way raises new ethical problems, including extremely difficult decisions about when to stop treatment and allow patients to die with dignity.

Not surprisingly, therefore, our hospitals are viewed as rather terrifying and dehumanizing institutions. Their outpatient departments consist of row upon row of patients waiting to be processed, the wards are often the sites of frenetic activity in which the needs of patients as individuals may be forgotten, and the coronary care and intensive care units, as depicted in figure 24, are so full of frightening machinery and monitors that it is often difficult to find the patient at all. Because of the plethora of information arriving from the laboratory and other diagnostic departments, more time is spent perusing the records of the patients than in talking to them and hearing about their fears and concerns. In the worst scenario, they may leave the hospital without the faintest idea what has been done or what is the matter with them.

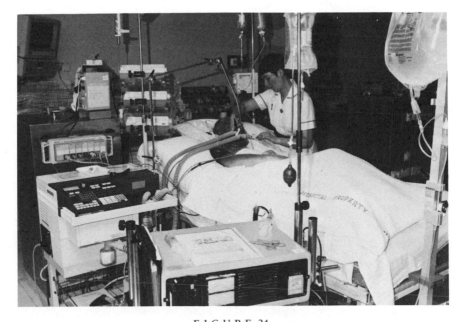

FIGURE 24

The intensive care unit of a modern hospital.

WHERE IS HIGH-TECHNOLOGY MEDICINE GOING?

High-technology medicine reflects a period of steady progress since the beginning of this century. For almost all diseases something can now be done, and a handful can be prevented or cured. But after the extraordinary advances in the prevention and treatment of infectious disease that followed the Second World War, we seem to have reached an impasse in our understanding of the major killers of Western society, particularly heart and vascular disease, cancer, and the chronic illnesses whose victims fill our hospitals. Although we have learned more and more about the minutiae of how these diseases make patients sick, we have made little headway in determining why they arise in the first place. And much of the progress in their symptomatic management has has been achieved by the development of increasingly expensive patch-up procedures. Given the ingenuity of the medical sciences and the pharmaceutical industry, there is no reason why we should not go on improving our ability to repair our patients and extend their lives. But is this the appropriate goal for scientific medicine as we move into the next century? And can we afford to go on in this way?

In recent years there has occurred some serious rethinking of these questions. Much of this has been generated by the simple reality that not even the richest of industrial societies can find a way of coping with the spiraling costs of health care. Furthermore, the work of epidemiologists brings increasing evidence that some of our commonest intractable diseases, rather than the inevitable results of aging, stem from our indulgent lifestyles and the new environments we have created. This has raised the expectation that they should, at least to some degree, be preventable. Thinking along these lines 'has led to a drastic change of attitude toward both medical research and medical practice. Almost overnight, the accent has switched from thinking about the future development of hospital practice to the prevention and management of disease in the community.

The remainder of this book examines the scientific basis on which this change of direction rests, and it tries to anticipate the future role of medical research against the background of this new way of thinking about patient care. Before tackling these difficult questions, however, we must digress briefly to try to assess what we have achieved already, not just in the richer countries but also in the developing world. For although the health problems of the poorer countries are different from those that we have been considering so far, this may not hold true for long.

How Much Has
Been Achieved?

We're all of us ill in one way or another:
We call it health when we find no symptom
Of illness. Health is a relative term.
                                    —T. S. Eliot (1888–1965)

What is health? In 1958 the World Health Organization defined it as "a state of complete physical, mental and social well-being." The problem with subjective definitions of this kind was crystallized by René Dubos in his provocative book *Mirage of Health,* first published in 1959:

> For several centuries the Western world has pretended to find a unifying concept of health in the Greek ideal of a proper balance between body and mind. But in reality this ideal is more and more difficult to convert into practice. Poets, philosophers, and creative scientists are rarely found among Olympic laureates. It is not easy to discover a formula of health broad enough to fit Voltaire and Jack Dempsey, to encompass the requirements a stevedore, a New York City bus driver, and a contemplative monk.

For the want of anything better, therefore, health is usually assessed by estimating mortality rates or the prevalence of disease in populations. While this custom is not entirely satisfactory, we will follow it as we try to evaluate how far medical practice has advanced and the role for the scientific exploration of disease in the future.

In this bird's-eye view of the current state of our health, I have split the world into two compartments—the rich industrialized nations, and the developing world. Although the pattern of disease differs widely between them, it is, as we will see, an increasingly artificial division. Even more important, it is one on which it would be unwise to base future strategies for

the development of medical research and practice. The diseases of wealthy
countries will become much more common in the developing world as standards of living improve, if by "improvement" we mean the duplication of our
self-indulgent Western lifestyles. And because of the increasing numbers of
travelers and mass movements of refugees, some of the major killers of the
Third World are already becoming part of day-to-day medical practice in
the richer industrialized countries.

● I should preface this brief digression on the current state of the health of
the nations with an apology to those readers who, like me, find pages of facts
and figures both numbing and indigestible. Andrew Long once accused a
writer of using statistics the way a drunken man uses lampposts—for support
rather than for illumination. But I hope what follows will illuminate a little,
for if we are to explore the role of science in medical care in the future, we
need to understand the problems facing as we move into the next millennium.

THE HEALTH OF RICH INDUSTRIALIZED COUNTRIES

Major Killers and Causes of Disability

In England a century ago, four out of ten babies did not survive to adult life,
the life expectancy at birth was only forty-four years for boys and forty-seven
for girls; as recently as the 1930s, 2,500 women a year died during pregnancy
or childbirth. Today life expectancy at birth is about seventy-three years for
boys and seventy-eight for girls. Things are much the same in most developed
countries (figure 25). The Japanese manage to live slightly longer than the
rest of us, an observation that may surprise those who have spent any time
in present-day Tokyo. As countries try to improve their health care facilities,
many of their efforts are directed at babies and young children. Their success
is reflected in infant mortality rates, that is, the number of deaths in the first
year of life per 1,000 live births. Here again, industrialized countries seem to
be doing well (figure 26). Infant mortality rates have slowly declined and
currently range from just under five in Japan to just over thirteen in Portugal;
Britain and the United States, with rates of about 7 to 9 per 1,000, seem to
have some way to go.

Since the early part of this century there has occurred a major change in
the causes of death in industrialized countries. The proportion due to infection and respiratory disease has declined; as figure 27 shows, the main cul-
prits are now diseases of the circulation, particularly coronary artery disease
and stroke, and cancer. Other important killers include traffic accidents, sui-
cide, violent death, including accidents in the home, and alcohol-related dis-

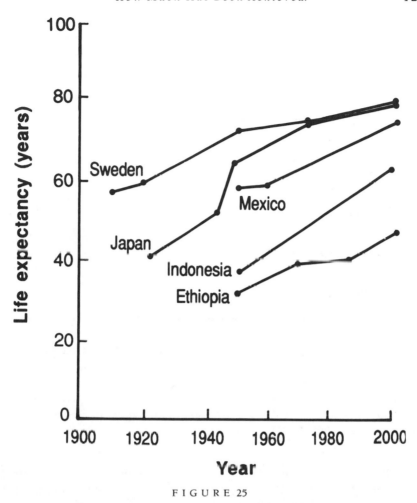

FIGURE 25

The life expectancy for different populations. Data from the Commission on Health Research and Development, Essential Link to Equity in Development *(Oxford: Oxford Univ. Press, 1990).*

eases. Although most of us in Western societies live well into out seventies, some of us are less fortunate. The causes of premature death vary in different age groups: in infancy the commonest are prematurity, congenital malformation, and infection; in childhood the pattern changes, and accidents, cancer, and the effects of genetic diseases are the major killers; and death in middle life is most commonly due to vascular disease and cancer. In the United States and Europe about 12 percent of all deaths occur before the age of fifty years; most of these are in middle life.

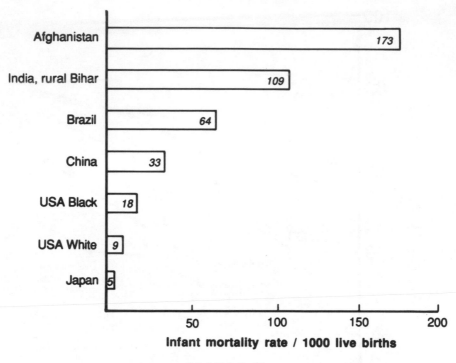

FIGURE 26

Infant mortality rates for different countries. Data from the Commission on Health Research and Development, Essential Link to Equity in Development *(Oxford: Oxford Univ. Press, 1990).*

Vascular disease, mainly coronary disease and stroke, accounts for about 40 percent of all deaths. It is estimated that in Great Britain coronary artery disease or related illnesses fill about 5,000 government health service beds every day, account for about 2.5 percent of the total expenditure on health, and result in some 35 million lost workdays each year. Strokes account for about 12 percent of all deaths and, because they do not always kill, are a major cause of chronic disability, particularly among elderly people. In England stroke-related illnesses fill about 16,000 hospital beds every day and account for about 7.7 million lost workdays each year. In many countries death rates from heart disease and stroke have been falling for the last ten to fifteen years, but they remain the most important killers.

After heart disease, cancer is the commonest cause of mortality in most industrialized societies. About one in three of us has some kind of cancer by the age of seventy-five. In men the most common site is the lung, followed by the prostate, stomach, and large intestine. Recent years have seen a steady

Industrial Countries 1990

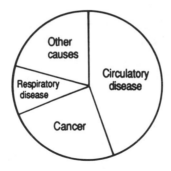

Industrial Countries 1930

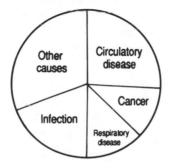

Developing Countries 1990

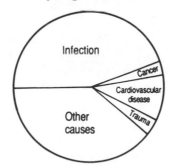

FIGURE 27

The causes of death in industrialized countries in 1930 and 1990, and in the developing world in 1990.

fall in the rate of lung and stomach cancer, mirrored by an increase in prostate cancer. In women the commonest site is the breast, followed by the lung,
ovary, and stomach. The death rate from breast cancer has risen slowly
among women in the United Kingdom, although it may have reached a plateau, but there has been a sharp increase in deaths from lung cancer. Overall,
tumors of the lung, female breast, and stomach account for about 40 percent
of all cancer deaths.

Another way of looking at the burden of serious disease is to see how we
use our hospital beds and the kinds of illnesses that take us to the hospital.
In Great Britain in the mid-1980s approximately 29 percent of all beds were
filled with patients with mental disease and 10 percent with those who had
had strokes or other vascular disease of the brain; respiratory disease, cancer,
fractures and other injuries, and heart disease each accounted for about 7
percent of the beds occupied. Over the same period the commonest causes
of admission to a hospital were accidents and diseases of the digestive system,
followed closely by cancer, heart disease, respiratory disease, and diseases of
the circulation.

If we exclude all these unpleasant illnesses, most of which we can do something about but cannot prevent or cure, just how healthy are we? Again it is
very difficult to obtain information about how we feel and, even when we
try, to interpret the findings, given individual differences in our attitudes to
illness and health. While some accept aches and pains, minor depression,
and other symptoms as part of the ups and downs of life, others demand a
feeling of constant well-being and run to their doctor with the slightest com-
plaint. Thus information collected from primary care sources is only a very
crude indicator of our health.

In Great Britain almost all the people see their general practitioner as their
first-line call on medical care. By far the commonest minor illnesses are upper
respiratory infections, skin disorders, emotional problems, and minor gastrointestinal complaints. Among the more chronic conditions necessitating frequent visits, high blood pressure, chronic rheumatism, and chronic
psychiatric disorders top the list. To put things in perspective, an average
British general practitioner might expect to see only one new case of breast
cancer and perhaps ten heart attacks in a year.

Many surveys have tried to estimate the amount of chronic physical
impairment in the community. It appears that rheumatic disorders, particularly arthritis, head the list, followed by circulatory and heart conditions,
disorders of the nervous system, and diseases of the sense organs, especially
blindness. Surveys of institutions for the chronic sick tell a similar story,
though in this case psychiatric disease, stroke, and arthritis top the list. In
recent years, as thoughts have moved toward trying to reduce the number of

premature deaths, we have tried to assess the health of our population by cross-sectional surveys of men and women of different ages. For example, in 1991 over 3,000 British men and women were subjected to a questionnaire, a clinical examination, and blood sampling. The results, some of which were compared with a dietary survey in 1986, were not reassuring. The proportion of men who are genuinely obese had increased from 7 percent to 13 percent, and the proportion of obese women from 12 percent to 15 percent. It was also found that 70 percent of all men and women had a blood cholesterol concentration above what is now considered to be a "desirable" figure. It also turned out that about 30 percent of all men and women are still smoking cigarettes and that a high proportion of Britons are drinking too much alcohol and taking too little physical exercise. Judged by these criteria, it appears that the British are both dissolute and unhealthy, and are getting worse.

Disease Is Not Distributed Uniformly within Countries

The pattern of illness in industrialized societies varies considerably among different social classes and occupations. For example, in England and Wales the professional classes have life expectancies significantly higher, and perinatal mortality rates lower, than those of unskilled workers. The strongest class association is for respiratory disease in both men and women, circulatory diseases in women, and accidents and diseases of the nervous system in men. The data in figure 26, which show a difference between infant mortality rates in the United States for white and those for black populations, presumably reflect a similar trend. A study carried out in the United States in 1986 suggested that the death rates for persons who earned less than $9,000 a year were three to seven times higher than those for persons who earned $25,000 or more.

The reasons for the inequality of health between different social classes are not entirely clear. Studies of the spouses of unskilled workers suggest that occupation may play a role, but that is not the whole story; an unfavorable environment, limited access to medical care, and lack of education are undoubtedly also very important. Whatever the cause, these discrepancies are a sad reflection of the unequal distribution of wealth and amenities in Western democracies. It has been calculated that, if the mortality rate of professional workers in Britain had applied to the lower classes during 1970–72, the lives of 74,000 people under seventy-five would have been saved. And, even worse, this estimate includes nearly 10,000 children and 32,000 men of working age.

Infectious Disease Is Still with Us

Although we like to think we have controlled infection, infectious diseases are a constant danger and new ones are appearing all the time. For all the successes of immunization and antibiotics, it was always clear that the body's natural defense mechanisms would remain an important factor in determining our response to infection. This was exemplified in the 1960s when powerful drugs were first used to treat patients with cancer or to damp down immune responses in preparation for organ transplantation. In those whose immune systems had been suppressed in this way, a new pattern of infection started to emerge because of the activity of microorganisms that do not bother healthy people.

Between October 1980 and May 1981 five young homosexual men were treated for unusual infections in hospitals in Los Angeles. The strange thing was that these patients had infections usually associated with an ineffective immune system, and yet there was no obvious reason why they should have been affected in this way. About the same time a curious form of skin cancer was being noticed with increasing frequency in young men in New York City and in California. These observations heralded the emergence in the United States of what was a "new" disease, subsequently named acquired immunodeficiency syndrome, or AIDS.

During the early 1980s it became clear that AIDS is transmitted not only by a sexual route but also by blood transfusion or blood products and, even more worryingly, during the perinatal period; babies of infected mothers might be affected. These observations implicated blood as a route of transmission and confirmed the suspicion that an infectious agent was involved. In a remarkably short time the culprit, a specialized type of virus called a retrovirus, was isolated by research teams in France and the United States.

By October 1986, 33,217 cases of AIDS had been reported to the WHO from 101 countries, from all continents. The largest number, over 28,000, were from the Americas; Europe had 3,245, Africa 1,008, and Asia only 55. During the second half of the 1980s the number of cases reported in the United States and Europe steadily increased, although possibly as the result of public education, there appeared to have been some leveling off of numbers. At the same time information started to emerge from Africa that gave some inkling of the terrible epidemic of AIDS that was rapidly spreading across in the developing world.

While the AIDS epidemic is the most newsworthy story of how a "new" infection has the potential to change the pattern of health in a developed society, it is by no means the only example of medical practice's constant battle with the ingenious microorganism. We are still plagued with major epidemics of influenza that take a large toll, particularly among the very young and very old. And other novel microorganisms appear with frightening regularity. In 1976 an outbreak of pneumonia occurred among American Legionnaires who were attending a convention in a Philadelphia hotel. There were 102 cases of what became known as Legionnaires disease and 29 deaths. Although the offending organism was soon isolated and antibiotics to which it is at least partly susceptible were discovered, it remains a cause of serious outbreaks of pneumonia in industrialized countries. Similarly in 1977 there was an outbreak of painful arthritis among children and adults in Lyme, Connecticut, which was ultimately traced to an infection with a newly identified organism transmitted by a variety of ticks. Within a few years outbreaks of Lyme disease were reported from other parts of the United States, Europe, and Australia. More recently it has become clear that this is an extremely common infection.

It appears, therefore, that we still live in a hostile world and that infectious disease is always with us. As we will see later, the AIDS epidemic is continuing to spread. Whether as the result of the emergence of a new and lethal organism of this type, which may result from genetic changes in a once innocuous strain of virus, or whether, as with Legionnaires disease, by changing our environment we have created conditions for breeding a nasty organism, it is clear that we will always be fighting a battle for survival in the constantly changing world of the microorganisms in which we coexist.

Self-inflicted Diseases

From the time of the earliest human civilizations, we have relied on some form of chemical escapism. With the possible exception of Eskimos, no populations have managed to get by without psychoactive drugs of some kind; Eskimos have had to find other ways of comforting themselves simply because they cannot grow anything. Modern Western societies are no exception. It is currently estimated that in Great Britain about 30 percent of all adults smoke cigarettes and that this habit may be responsible for up to a third of all deaths in middle age. About one in four men and one in twelve women consume more than what is considered to be a safe level of alcohol. Diseases in which excessive use of alcohol plays a role may affect as many as 20 percent of the patients who fill our medical wards, and in England and Wales alcohol consumption causes anywhere from 5,000 to 50,000 deaths a year. Recent studies have suggested that alcohol is the major factor in approximately 25 percent of all deaths in traffic accidents and in 40 percent of all deaths from falls or fires. Alcohol consumption has increased steadily in most Western societies over the last twenty years.

Although it is difficult to obtain accurate information, it seems certain that our consumption of other addictive drugs has also increased steadily over recent years. In many states in the United States the building and maintaining of prisons is the fastest-growing item in the budget. In nearly every state over 20 percent of those sentenced to prison in recent years have violated drug laws. For example, in Florida 785 people were sentenced for crimes of this type in 1980; the number was 15,000 in 1989; the figures for California were 1,063 in 1980 and 10,445 in 1988. Although part of this disturbing story may reflect a tightening up of legislation to try and curb the problem, there is no doubt that drug usage is on the increase in most industrialized countries.

The drug habits of British schoolchildren have been the subject of several recent surveys. The results are frightening. Over 70 percent of secondary school children, fifteen to sixteen years old, have been offered drugs, and over half have taken one or more. This compares with about 25 percent in the late 1980s. Cannabis usage is increasing, LSD is freely available, and "new" drugs such as crack (cocaine rocks) and ecstasy (an amphetamine derivative) are in vogue. Heroin, benzodiazepines, amyl nitrate, and a variety of organic solvents, together with alcohol, also seem to be used widely among young people. Police records suggest a 70 percent increase in cases of drug offense in young people over the last year.

The sooner we start to understand the reasons for these social changes, the better. Because of the age and wide range of social backgrounds of the young people in these surveys, it cannot all be blamed on economic factors and unemployment, or breakdowns in family life and stability. It is a serious enough problem with respect to the increase in crime and social decay that it reflects, but its potential for precipitating acts of violence and for generating serious ill health in young people is even more worrying.

The Changing Environment

Another potential risk to the health of Western societies stems from their efforts to destroy their environment because of their hunger for energy. In 1985 the world's total energy consumption reached the equivalent of 9.3 million tons of oil, causing the emission of nearly 6 billion tons of carbon into the environment. It has been estimated that by the year 2020 we will be using 75 percent more energy and that most of it will still be supplied by coal, oil, and nuclear power. Greenhouse gases derived from burning fossil fuels are accumulating in the outer atmosphere, with the result that the temperature of our planet is increasing. We appear to be slowly destroying our protective layer of ozone by continuing to pour noxious chlorofluorocarbons from aerosols and refrigerators into the atmosphere, and our motor cars are responsible for emitting, in addition to carbon dioxide, unpleasant acidic gases that produce acid rain. While nuclear energy reduces our reliance on fossil fuels, it has—witness the Chernobyl disaster—the potential to cause hazards on a massive scale. Smaller leaks of radiation have been blamed for clusters of leukemia and other radiation-related diseases in those living near nuclear installations, though much of the work that has led to these concerns is extremely controversial.

It is difficult to assess the long-term effects of these insults to the environment, some of which may have been exaggerated, but it appears that we are not behaving very sensibly. The greenhouse effect, with subsequent global warming, has the potential to raise our seas to a dangerous level. The thinning of the ozone layer could increase the incidence of sun-related skin cancers; a 10 percent reduction in total ionizable ozone, which might occur by the late twenty-first century, could result in an extra 16,000 cases of skin cancer in the United States each year. And even though we have partly controlled some of the lethal dense fogs that filled our cities, the result of domestic and industrial burning of high-sulfur coal, we are still exposed to a cocktail of pollutants emitted from motor cars, many of which are respiratory irritants and potential cancer-producing agents. Thus, although many of the gloomy prognostications of those who worry about our environment are not always based on convincing evidence, and may be overdrawn, there is little doubt that Western countries are slowly but surely damaging their environment.

An Aging Population

Industrialized societies will have to face another large drain on their health resources over the next century. In Great Britain between 1981 and 1989, the number of people aged seventy-five to eighty-four rose by 16 percent, and that of people aged eighty-five and over by 39 percent. The current population of males aged eighty-five or more is 204,000; by the year 2026 it will be 431,000; for females the figures are 603,000 and 819,000, respectively. The proportion of the very old in our population will continue to rise; the better we become at preventing the major killers of middle life, the greater this effect will be. Although our more optimistic health experts believe that if we look after ourselves properly we will grow old gracefully with all our faculties and then politely keel over, anyone who cares for patients in the hospital or the community has good reason to be less sanguine. Estimates vary but it is likely that at least 20 percent of all people over the age of eighty suffer from dementia, a loss of intellectual function sufficient to render it impossible for them to care for themselves.

There is very little information about the quality of life as we get older. The 1986 General Household Survey (GHS) carried out in the United States indicates that restricted activity per year among people over the age of sixty-five years was forty-three days in men and fifty-three days in women. A more recent study suggests that about 30 percent of the people of this age group are restricted in their activities because of rheumatic disease or visual impairment. However, information of this type, while giving some indication of the health of the elderly, is of limited value because it applies only to old people who live in private households and says nothing about those who are institutionalized. And what of the loneliness and isolation of old age? The same GHS report suggests that 84 percent of the old people in private households see relatives or friends once a week or more; no comparable information is available about the elderly in institutions. When we remember that the numbers of those who are surviving to old age is increasing so rapidly, it becomes clear that we need much more detailed information about their quality of life.

The idea that our societies of the future will be full of vigorous old people enjoying their declining years, and the fruits of their healthy lifestyles, may simply be wishful thinking on the part of the medical profession and governments alike, for they are ducking the issues of aging. Just as the developing world will have its population explosion, so industrialized countries will have an epidemic in the numbers of old people.

The Growth of Complementary Medicine

We cannot finish this sketch of the health of the richer countries without considering the significance of the increasing popularity of complementary, or alternative, forms of medicine—that is, varieties of treatment that fall outside standard Western clinical practice. A survey in the United States in 1990 showed that patients made over 425 million visits to unconventional therapists, compared with 388 million to primary care physicians. It has been estimated that Americans are now spending about twelve billion dollars each year on complementary treatments.

As medical practice has evolved over the centuries, there has emerged a gradual split between orthodox, or "establishment," medicine and rival, unorthodox systems. In England the foundation of the Royal College of Physicians in 1518 gave its president the power to fine unorthodox healers, but this had little effect on their activities. In 1858 the Medical Act, by registering conventional practitioners, placed some constraints on clinical practice. But to this day, with the exception of midwifery and dentistry, anyone can practice medicine or surgery. Considering the high social standing of cats and dogs in Britain, readers may not be surprised to hear that its veterinary laws are much stricter.

In Great Britain there are now some 30,000 practitioners of complementary medicine. Of these, just over 2,000 are medically qualified, while nearly 17,000 belong to no professional association whatever. What do they offer? There seem to be about sixty different types of complementary therapies, which fall into several different groups. The first comprises physical treatments: naturopathy and herbal medicine; different forms of manipulation, including osteopathy, chiropractic, and reflexology; oriental therapies, including acupressure; particular systems of medicine such as acupuncture, homeopathy, and anthroposophical medicine; exercise/movement therapies, including yoga and dance; and sensory therapies that involve music, art, or color. The second group is made up of different psychological therapies: psychotherapy; humanistic psychology; primal work; rebirthing; encounter; and transpersonal psychology. Rounding things off are various paranormal activities: hand healing; radionics; exorcism; and paranormal diagnosis, which includes palmistry, astrology, and iridology. And even this formidable list does not include all the individuals and organizations that claim the power of healing. Among the better-publicized are bodies like the Unification Church of Sun Myung Moon, whose followers are known as Moonies, or the Church of Scientology, established by the late L. Ron Hubbard. In Great Britain over one hundred groups are now operating in such a manner that they have come to the attention of an organization called FAIR (Family Action Information and Rescue), which was first formed in 1976 in response to requests for help from distressed relatives of young people who had come under their influence.

Many of these forms of complementary medicine have a long history. For example, homeopathy was founded by a German physician, Samuel Hahnemann (1755–1843), who set out the principle "Like is cured by like," a formula that seems to have been used earlier by Paracelsus. This form of treatment, which now has many different variations, is based on the principle of administering tiny doses of drugs, sometimes in dilutions at which the patient may not receive a single molecule. Osteopathy was founded in 1874 by an American, Andrew Taylor Still (1828–1917). This practice holds that most diseases are due to small dislocations of the spine that result in pressure on nerves or blood vessels. Hence treatment is largely manipulative. Chiropractic was also founded in the United States, in the late nineteenth century, and seems to have many features in common with osteopathy. Herbalism, yoga, and acupuncture are, of course, treatments of great antiquity (figure 28).

It appears, therefore, that the spectrum of unconventional therapy stretches from those with a serious pretense of a rational basis for their activities to what can only be described as quackery. There will, of course, always be quacks, whether they are crooks or cranks. The "Emperor of Quacks," the Scotsman James Graham (1745–1794), seems to have been a bit of both. His Temple of Health, with its awsome Magnetic Crown and Grand Celestial Bed, designed to help those who were sterile or impotent, though not a great financial success, at least suggests that he was endowed with enviable entrepreneurial skills. When this failed, he changed to a new venture, based on

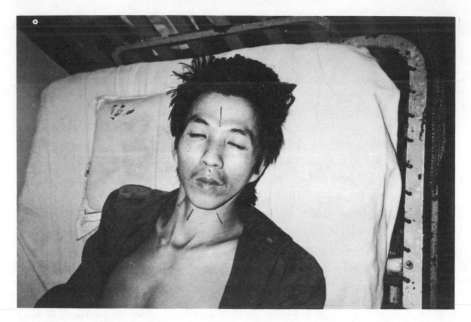

FIGURE 28

Acupuncture for the treatment of Japanese B encephalitis, a virus disorder that affects the brain. (Vietnam, 1990.)

the notion that life comes from the earth; he and a young girl removed all their clothes and were buried up to their chins, after which their hair was powdered and pomaded such that a bystander remarked that they resembled a couple of fully grown cabbages. Following the collapse of this enterprise Graham developed a religious mania and, before his death, had to be placed under restraint.

While present-day society also has its share of James Grahams, many of the practitioners of complementary medicine are serious healers and seem to provide an increasingly important service to the community. But very little of their work has been examined by properly designed clinical trials. A few studies, including treatments with acupuncture or osteopathy, have shown some benefit, but for the most part virtually nothing is known about their value or about the extent to which they rely on placebo effects. Those who are attempting to study alternative practices by modern analytical techniques claim that it is difficult to adapt them to some forms of alternative practice, particularly in a "blind" fashion. However, it is vital that work of this type continue and that, ultimately, all these practices be subjected to critical analysis.

What is this interesting phenomenon telling us about the health of Western societies? As I intimated in the introduction to this book, part of the success of complementary medicine undoubtedly reflects a society disillusioned by the inability of conventional medicine to deal with some of its major diseases—whether they are intractable chronic disorders, such as backache, or forms of cancer that have not responded to conventional medicine. In addition, there is little doubt that many Western populations feel that constant rude health is their right and that they will go to any lengths to obtain it. Forty years ago ill health, and even death, was a much more common occurrence, even among young people, and seemed to be accepted as part of the natural course of things.

But I suspect that the main reason for the success of complementary medicine is that its practitioners spend more time talking to their patients about their worries and fears, and trying to remove some of the mystery surrounding their illnesses. In the frenetic world of current high-technology medicine, doctors have far less time to spend comforting sick people and acting as counselors, friends, and confidants. Many of the minor illnesses that take patients to complementary practitioners simply mirror the pace and complexity of modern life. And those who go with more serious illnesses must undoubtedly reap the benefits of having somebody to talk to at leisure about their fears and uncertainties.

The growth of complementary medicine cannot be ignored. It may be hinting that there is a lot more chronic illness and unhappiness in our Western communities than is apparent from standard surveys of our health. And it must be telling us that conventional medicine should look at its own shortcomings, particularly with respect to its pastoral skills. But as Sir Austin Bradford Hill said nearly half a century ago, faced with uncertainty about the relative merits of different forms of care, "it may well be unethical . . . *not* to institute a proper trial."

How Well Are We Doing?

It therefore appears that, judged by crude mortality figures, the health of people in industrialized societies is still improving. Infant mortality rates are falling, and, if we survive the first few months of life, on the average we can now expect to live well into our late seventies or eighties. Although we have not come to grips with the root causes of our major killers, cardiovascular disease and cancer, we have learned how to patch up patients with these conditions. These advances, together with our inability to deal with some of the chronic disabling diseases that do not kill—rheumatism and psychiatric disease, for example—and our rapidly aging population, are the main reasons for the spiraling cost of health care, which threatens to cripple our economy. In addition, we appear to be living in increasingly violent societies and may be doing serious harm to our planet. Although we are proud of our Western democratic processes, there are still serious discrepancies in between different social classes, a state of affairs that is a sad reflection on the achievements of our governments. And the expansion of complementary medicine suggests there may be more chronic ill health in our societies than we suspect.

THE DEVELOPING WORLD

The doctors of today have some 5.57 billion people to care for. Their practice has increased fivefold in the past hundred years and will double again even faster unless there is a colossal famine or similar disaster. In its 1991 Report on World Population, the United Nations Population Fund predicts that the world's population will grow even faster than forecast in 1984 and will pass the 10 billion mark by 2050. Current estimates suggest that it may stabilize at about 11.6 billion by the year 2150. About 95 percent of this growth will occur in the less developed countries. By 2050, 84 percent of the world's population will be living in countries that generate only 15 percent of the world's GNP. About 1 billion people live in absolute poverty, and 600 million people are on the edge of starvation. Their condition is described by the World Bank as a state of life so characterized by malnutrition, illiteracy, disease, high infant mortality, and low life expectancy as to be beneath any reasonable definition of human decency.

In the light of this disturbing picture is it really worth discussing the health of the developing world in the context of the role of science in medical practice? Surely the problem is much broader, reflecting gross inequalities between the developed and developing countries and a lack of interest in the plight of the latter on the part of the governments of our more affluent societies. While this is true, it is still important to consider the diseases that result from dire poverty and to try to anticipate how far they can be controlled by economic and political measures rather than by the intervention of medical science.

Population, Malnutrition, and Natural Disaster

Before looking at the diseases of the developing world, we must review the factors that contribute to the hostile environment that makes them so common, the most important of which are overpopulation, poverty, malnutrition, and natural disasters.

Concern about the increasing size of human populations is not new. In 1798 the Reverend Thomas Robert Malthus (1766–1834) published the first edition of *An Essay on the Principles of Population*. Malthus, an economist for the East India Company, set out an extremely depressing scenario: as populations increase faster than food supplies, struggle and starvation inevitably result. He calculated that the population of the world would increase more rapidly than its means of subsistence and that the only possible checks to this process would be epidemics of disease or famine. Ferdinand Braudel, the distinguished French historian, in *The Structures of Everyday Life* (1981) describes the history of civilizations in Europe from the fifteenth to the eighteenth century. The picture he paints is not too far removed from that predicted by Malthus. The period was marked by a very high infant mortality, famine, chronic malnutrition, and a series of devastating epidemics. Famine was ever present; France suffered eleven major famines between the sixteenth and the eighteenth century. Following these episodes, or after serious epidemics such as plague, the population tended to recover and expand again. Population histories of this period show a series of cycles; as soon as the population increased, it outstripped the ability of the environment to provide subsistence and a further disaster occurred. Eventually, economic and social improvements in most European countries combined to reduce poverty and lower the birthrate, so birth and death rates reached an equilibrium, but at a lower level.

This pattern reflects the way in which populations grow. First, both birth and death rates are high and the population grows only slowly. During the second stage, living and health conditions improve and death rates fall, but birthrates remain high and the population grows rapidly. Finally, a stage of stability is achieved, as occurred in Europe. Populations with rapid and sustained growth in the second stage are in serious danger of exceeding the capacity of their local ecosystems, particularly if these are fragile, as many of them are in the tropical world. The British epidemiologist Maurice King emphasizes the instability of the second stage of transition, which must be negotiated quickly if the population is not to enter what he calls a demographic trap. In other words, if the birthrate does not fall, the death rate will ultimately rise again, so the population is trapped and finds itself in an unsustainable state, with high birth and death rates and ever-increasing pressures on its resources. Unless the population is rescued by support from outside, King believes, the only possible outcome of this unstable situation is either the death of most of the population from starvation, disease, war, or genocide, or mass movements of refugees.

Although there are many pitfalls in predicting the effects of growth of populations—and some of the predictions already appear to have been too pessimistic—it is clear that the present rate of increase cannot continue indefinitely. At a rate of growth of 1.7 percent a year, the world population would double about every forty years and increase one thousand-fold in 400 years; in 800 years there would be only one-quarter of a foot of land per person. Clearly this is nonsense. But even allowing for the likelihood that the demographic transition in the Third World will be much more rapid than it was in the industrialized countries, an increasing population is obviously the chief problem that faces us as we move into the twenty-first century.

Between 1970 and 1980 the annual population growth in the developing world was 2.6 percent, compared with 0.8 percent in industrialized countries. Why did this happen? Ignorance and lack of education must be important but cannot be the whole story. In many parts of the Third World there are legal or religious obstacles to modern contraceptive practices. Religions such as Roman Catholicism and Islam, which exert a powerful influence across Central and South America, parts of North Africa, the Middle East, the Indian subcontinent, and the Philippines, preclude the use of any form of contraception, ignoring the high prevalence of infanticide or illegal abortion in these populations. In many Third World countries large families offer apparent security, and children are seen as a vital source of support as their parents age. Surveys in Asia and Africa have suggested that high birth-rates are perceived as a defense against high rates of mortality and morbidity. But population growth varies considerably in different parts of the developing world. The reasons for this variation are not clear. Indeed, very little is known about the social and cultural factors that determine our reproductive habits.

Malnutrition is the other major problem for the developing world. That there are already many populations of starving people is particularly disturbing, since it is believed that, at least at the moment, world food supplies may be adequate. Over the last thirty years food production has increased dramatically, mainly as the result of the use of chemical fertilizers, pesticides, and disease-resistant seeds and the widespread irrigation of land. In Africa and Latin America there have been substantial additions to arable land, while in Asia increased food supplies have resulted mainly from the application of new technology developed in Japan, Korea, and Taiwan. Despite this, in 1980 one-third of the peoples in the developing world lived in countries in which food supplies were insufficient to meet the requirements of the population, even when efforts were made to distribute it according to need. And, as the economist Amartya Sen has pointed out, this is rarely done in an equitable way. Indeed, he argues that many countries that experienced famine had an adequate food supply for the population; it was simply not distributed properly.

There is nothing new in this. During the catastrophic potato famine in Ireland in the midnineteenth century, large quantities of cattle, corn, and other foodstuffs were produced and exported to England in amounts that would have been quite adequate to avert the famine if the Irish people had had enough money to buy them. The British government reduced the price of grain in 1846 by repealing the Corn Laws in an effort to make grain more easily obtainable in Ireland. But the Irish tenant farmers grew grain to pay rent to their landowners; the falling price of grain simply increased their rent and hence their poverty. In Bangladesh in the early 1970s, when famine caused the death of up to a million people, the number of ounces of food grain available per capita per day was higher than it had been for several years. However, local flooding severely reduced the opportunity for workers to earn an income. At the same time the price of food was increased in the expectation of a poor harvest in 1975, and as the result of panic buying and hoarding. It was this price increase, together with the inability of groups of workers whose land had been damaged by flooding to buy food, that led to the catastrophic famine.

Inequitable distribution of food together with inadequate production beset many parts of the world during the 1970s. Production failed to keep pace with population growth in 70 countries, out of a total of 126. The situation was especially serious in Africa, where food output increased by about a quarter and the amount available per head declined. Because of unusual weather conditions and other factors, at least 26 African countries appeared in the mid-1980s to be in serious danger of having grossly inadequate food supplies; the widespread famines of the last few years have shown that these predictions were correct. Countries with borderline food production are in a particularly precarious position because of the instability of world food prices and the wide variation in the rates of their own domestic production and in the purchasing power of individual households. Under these unstable conditions famines may easily be precipitated by wars, floods, crop failures due to unusual weather conditions, or loss of purchasing power with high food prices. As Sen has noted, lack of availability and poor distribution are not always the primary causes, and geographical factors play an important role. India, for example, enjoys an overall food surplus, yet for many years there has been serious famine in the north, mainly among the eight million people who live in very arid regions.

Taking the longer-term view, environmentalists leave us in no doubt that although there are now adequate food supplies in the world, this situation may not last for long. Air pollution, ozone depletion, land degradation, and acid rain, together with altered rainfall patterns, are already causing a reduction in world harvest. The world's grain harvest currently increases by an average of 15 million tons per year. However, this is well below the estimated 28 million ton increase required to keep pace with population growth. Environmentalists also estimate that the world's grain harvest could fall by as much as 14 million tons per year if we continue to ruin our environment. And our attempts to produce food are certainly causing serious damage to our longer-range potential for doing so. Overgrazing and other types of misuse of land have led to serious degradation, erosion, and even the creation of new deserts. Each year we manage to chop down an area of tropical forest the size of Austria. Apart from increasing the buildup of carbon dioxide, this may also spell disaster for biological diversity, given that more than half the species on earth live in the tropical rain forests.

What are the medical consequences of malnutrition? At its worst it produces a sight all too familiar to anybody who reads newspapers or watches television. Advanced malnutrition, particularly lack of protein, causes a clinical picture called kwashiorkor, from a Ghanaian word meaning a disease suffered by a child displaced from the breast. Children with this condition appear grossly wasted and have potbellies and sunken eyes. At this stage they are listless and apathetic and lose the desire to eat. In this pitiful state they are very prone to infection and, if not treated, invariably die.

Severe malnutrition has many other effects which, though less striking than kwashiorkor, are nonetheless extremely debilitating. In childhood it increases proneness to all forms of infection. And it is responsible for the widespread occurrence of anemia. This is particularly common in children and pregnant women, although in many countries adult males are also severely affected. For example, surveys of the prevalence of anemia in northern India suggest that about 50 percent of all males, and from 80 to 90 percent of all pregnant women and schoolchildren, are anemic. The major cause is a deficiency of iron due to an inadequate diet, the use of foodstuffs that inhibit the absorption of iron from food, and widespread infestation with parasites that cause bleeding into the bowel and hence loss of iron. Several studies by the World Health Organization have shown that the working efficiency of adults who are anemic is significantly reduced. In many populations anemia is exacerbated by additional deficiencies, particularly of vitamins such as folic acid and B_{12} that are essential for the production of blood. Other vitamin deficiencies are also widespread. Lack of vitamin A is especially serious because it causes damage to eyes that often leads to blindness.

What does all this mean in global terms? A recent survey from the World Bank, *World Development Report, 1993,* estimates that about 40 percent of the two-year-olds in the developing world are short for their age. The prevalence of stunting may be as high as 65 percent in India and about 40 percent in sub-Saharan Africa. Severe wasting occurs in about 11 percent of all young children. Protein-energy malnutrition, severe enough to reduce physical and mental capacity, affects about 780 million people worldwide.

In rounding off this background to the health problems of the developing world, we must not forget natural disasters. They are rare and, for the most part, cannot be prevented. During the last twenty years they have claimed the lives of about 3 million people worldwide, but their overall effects have been much greater. It has been estimated that more than 800 million people have been adversely affected, suffering homelessness, ill health, and severe economic loss leading to further famines, not to mention numerous personal tragedies. A single disaster of this type can set back economic progress in a developing country by as much as five years; about two-thirds of the world's people live in countries that bear 95 percent of all disaster casualties.

The Health of the Developing World

World Development Report, 1993 demonstrates that, despite these adversities, the developing countries have made great strides in improving the health of their populations over the last four decades. Life expectancy, overall, has risen from about forty to sixty-three years, and the number of children who die before their fifth birthday has fallen from nearly three in ten to about one in ten. On the other hand, child mortality rates remain about ten times higher than those of the rich industrialized countries, and maternal mortality rates may be up to thirty times higher.

The World Bank has introduced a new indicator of the state of our health called a DALY (disability-adjusted life year), which estimates the present value of the future years of disability-free life that are lost as the result of premature death or cases of disability occurring in a given year. This assessment, which tries to combine the effects of illness and its economic consequences, suggests that although there have been genuine improvements in the health of the developing world, there are major differences between countries. The average life expectancy may have risen by twenty-three years, but the picture in sub-Saharan Africa and India is much less encouraging. And if we consider the total burden of disease attributable to premature mortality and disability, as expressed in DALYs, we see enormous discrepancies not just between the rich industrialized countries and the developing world but also among the poorer countries themselves.

Diseases of Poverty

Given the conditions in which so many of the people of the developing world live, it is not surprising that the pattern of illness is very different from that of the industrialized countries. The major causes of death are diarrheal and parasitic illnesses, respiratory infections, and malnutrition. Together, these illnesses kill almost seven million children each year. Although the diseases of Western society, particularly cardiovascular disease and cancer, have started to become an important problem in the emerging countries, the major killers of the Third World still reflect the frightful conditions in which most of its inhabitants have to survive.

Illnesses associated with diarrhea were responsible for about a quarter of all deaths in children under the age of five in Latin America in the late 1970s, and they caused an estimated 4.6 million deaths in the developing countries, excluding China, in 1980; figures published by the World Health Organization in 1992 show little change. Eighty percent of the dead are children under the age of two years. Most of the pathogens responsible are transmitted by the contamination of water by feces. The better-known illnesses contracted in this way include cholera and typhoid fever, but many deaths of this kind are due to infections with other microorganisms. Poor water supplies and reduced immunity because of malnutrition are the major reasons for their success in the developing world. Malnutrition and poor living conditions are also responsible for the high frequency of death from respiratory infection, which is an equally serious problem for children. The pattern of infective illness in the poorer countries is in fact similar to that seen in industrialized societies before social conditions improved and before the advent of immunization and antibiotics. Thus diseases such as whooping cough, measles, tuberculosis, diphtheria, and other common killers of childhood are still rife in many of these countries (figure 29a). Recent WHO figures suggest that three million people die each year from tuberculosis and that about twenty million have the active disease.

Apart from these killers, which used to afflict industrialized societies, the developing world has the problem of common parasitic diseases that are peculiar to the tropics and of which, with the exception of malaria, most of us have never heard. Malaria constitutes a major public health problem in many parts of the tropical world (figure 29b), responsible for the death of some two million children each year. It is caused by parasites that invade red blood cells and that are transmitted from person to person by certain types of mosquito. The story of the attempts to eradicate malaria provides an interesting insight into how much we may expect to achieve in our attempts to control serious tropical infections by modifying the environment or attacking an organism directly. The WHO estimate for 1992 is that there were about 267 million people affected with malaria. In 1955 the World Health Assembly announced a program for its global eradication. The notion was to attack the mosquito and its environment, and the malarial parasite itself. First the mosquito. Here the idea was to drain swampy breeding grounds and to spray it with the insecticide DDT. Immediately a problem arose because the mosquito started to develop resistance to DDT and to other insecticides. However, it was hoped that malaria would not reappear if transmission could be interrupted and all cases treated for three years; eradication programs, it was hoped, would need to be maintained for a limited period. It sounded like good sense.

By 1970 malaria had been almost eradicated from most of Europe and the USSR, North America, several Middle Eastern countries, and parts of the Mediterranean region. It was a fine public health achievement but, as it

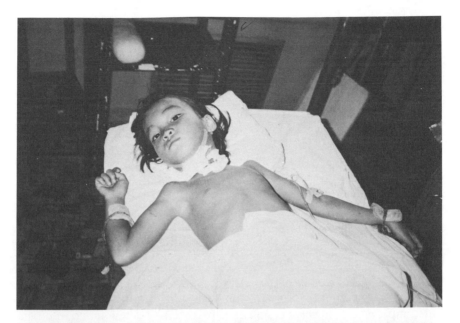

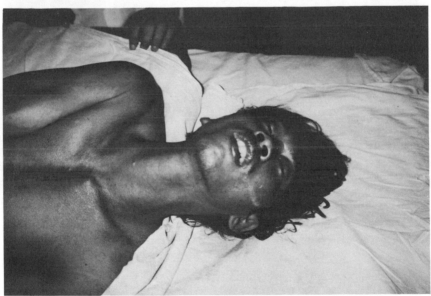

FIGURE 29

The spectrum of infectious disease in the developing world: (a) a Vietnamese child with diphtheria, which necessitated a tracheotomy to allow the patient to breathe; (b) a Thai patient comatose as a result of cerebral malaria, a condition being seen more frequently as the malarial parasite becomes more resistant to drug treatment.

turned out, not enough. Seasonal or endemic transmission continued in many tropical countries. There were difficulties in implementing eradication programs, particularly those that involved attacking the mosquito or its breeding grounds. Even more worrying, the parasite started to become resistant to drugs. Resistance to safe and cheap drugs like chloroquine started in Southeast Asia, and by the 1980s most malarial parasites in this region were completely resistant. Ten years later chloroquine resistance was being seen throughout Africa. It happens that the parasite can change its genetic makeup more rapidly than the pharmceutical industry can produce new drugs; in Southeast Asia it is even becoming insensitive to quinine, the oldest of anti-malarials. Furthermore, it has been difficult to implement malaria control programs because, with the population explosion, the numbers "at risk" are increasing dramatically. Furthermore, air travel has provided an efficient way of carrying infected mosquitos around the globe. Between 1973 and 1977 the number of cases of malaria reported to the WHO multiplied no less than 2.5 times.

Currently the number of cases of malaria in the world is changing very little; improvement in some countries is balanced by deterioration in others. It is extremely disturbing to reflect that in 1992 malaria is affecting more people than it did in 1960, and there is no prospect of a substantial improvement. On the contrary, the increasing resistance of the parasite to conventional drugs, together with the rising world population, suggests that the magnitude of this lethal disease may increase over the next ten years.

And now we have to add AIDS to the list of major infective killers in the developing world. AIDS was first identified in the United States. In the early 1980s nobody had any idea of the frightening situation that was developing, or had already developed, in parts of Africa and in Third World countries elsewhere. At first, information about the prevalence of AIDS in developing countries was slow to surface; some governments were loath to disclose the information that they had a frightful epidemic on their hands. However, the picture of the rapid dissemination of this disease in Africa and elsewhere is now well known. By 1990 it was quite clear that the industrialized nations' notion that AIDS is something that is largely confined to populations of homosexuals and intravenous drug users, and that it can be held in abeyance by adequate education, is not relevant to many parts of the world.

According to the WHO, by the year 2000, some 26 million men, women, and children will be infected worldwide, and deaths from AIDS will reach about 1.8 million. There will be about 12 million infected people in sub-Saharan Africa and about 9 million in Asia. Most of those who develop AIDS will be young men and women in their twenties and thirties, whose deaths in large numbers will devastate the population. Its main route of transmission in the developing countries is clearly by heterosexual intercourse. For example, recent studies in Thailand have tracked its origin to the country's 800,000 prostitutes; one survey suggests that 75 percent of all Thai men have had sex with prostitutes. The same picture is emerging from India. In Bombay the infection rate for AIDS jumped from 1 percent in 1987 to 30 percent in 1990 among the hundred thousand or so prostitutes. These women are averaging six contacts per night, and with a transmission rate of 0.1 percent it is estimated that 6,000 men are now being infected every month in Bombay alone. Although many of these predictions may have to be revised, and we have much to learn about the transmission, distribution, and natural history of AIDS, there seems little doubt that it will remain a grave problem for the foreseeable future.

In addition to the major killers of the developing world, we should not forget the diseases that produce chronic disability. The pattern is completely different from that in the developed world. Like malnutrition, chronic parasitic infections play a central role. For example, over 900 million people are infected with hookworms, nasty creatures that live in the bowel and cause chronic blood loss and profound anemia. In Africa malaria not only kills millions of children but also causes many more to live in chronic ill health, mainly as a result of anaemia. Another parasitic infection, schistosomiasis, affects vast populations in Africa and the Americas, with an "at risk" population in the region of 600 million people. There are approximately 45 million people with poor vision that would categorize them as "economically blind." Of these at least 28 million have eyesight that would classify them as socially blind—that is, they can move around only with assistance. The commonest cause of blindness is cataract, whose high prevalence in the underdeveloped world may be caused by sunlight and damage and dehydration of the lens. But it has been estimated that there are 100 million people with a disease called trachoma, caused by infection with an organism that attacks both the eye and the urogenital system. Of these about 7 million are blind. Some 800,000 children are blinded each year as a consequence of vitamin A deficiency. I will spare the reader any further statistics. The amount of chronic disability in the developing world poses a challenge equal to that of the high death rates from preventable causes.

In this brief survey of the health problems of the developing world, we have concentrated mainly on malnutrition and infection. Although "Western" illnesses such as vascular disease and cancer are starting to be seen more frequently, it should not be forgotten that the developing world already has its own varieties of cancer. In richer countries liver cancer is quite rare, whereas in the developing world 250,000 to one million deaths are each year attributable to this devastating disease. In recent years evidence has been accumulating that this is due to persistent infection with the virus responsible for one form of hepatitis, hepatitis B. It turns out that in many of these countries the high frequency of the carrier state for this disease is caused by the transmission of the organism from mother to baby. The virus is also passed from infected carriers, especially in conditions of overcrowding or poor sanitation. The discovery of the association of a particular infectious organism with a form of cancer marks an important advance because the development of a vaccine against hepatitis B virus offers the promise of preventing this unpleasant and common malignant disease.

Other types of cancer are also widespread in the developing countries, notably those of the nose and pharynx in Southeast Asia, which seem to be related to infection with the same virus that causes glandular fever (infectious mononucleosis) in industrialized societies. This virus has also been implicated in the high frequency of tumors of the lymphatic glands in African children. These important observations suggest that the spectrum of diseases in the world population that are due directly or indirectly to infectious organisms may be very much wider than was hitherto realized.

If we, living in our cozy Western societies, think about the health of the developing world at all, it is usually with a mixture of horror and sympathy, but we do not regard it as something that has any relevance for us. This is no longer the case. With increasingly frequent international travel, vast movements of populations of refugees, and potential changes in our climate, the diseases of the Third World are of increasing importance for all of us. It is estimated that, as a result of increasing overpopulation and civil war, over 100 million people have become international migrants. These people carry with them their own diseases, both inherited and acquired. We have only to consider the distinct possibility that the AIDS virus started out life in a tropical clime to grasp that it would be foolish for industrialized societies to believe that the medical problems of the Third World can be conveniently forgotten.

The Pattern of Disease Changes As
Countries Become Richer

In what are sometimes called the emerging countries, in which improvements in nutrition, health care, and hygiene have already been made, there has been a change in the pattern of disease. Infant mortality rates have fallen, as have death rates from infectious diseases, particularly diarrheal illnesses or parasitic infestations. But new diseases are appearing. In childhood, for example, these include a very high frequency of inherited anemias. These diseases were masked by the very high infant mortalities caused by malnutrition and infection, but once these were controlled it became clear that large numbers of children are affected.

In Africa there is a serious inherited disease called sickle-cell anemia. It is due to an inherited abnormality of hemoglobin, the red pigment that fills blood cells and carries oxygen from the lungs to the tissues. Hundreds of thousands of children die from this condition in tropical Africa each year. In the Mediterranean countries, the Middle East, the Indian subcontinent, and Southeast Asia, there is an even commoner inherited anemia, called thalassemia. The name comes from the Greek work meaning "the sea," because children in whom it was first described were from a Mediterranean background. It is also an inherited disorder of hemoglobin. In Thailand alone as many as 750,000 children have this condition. In Cyprus, where it was recognized only after the Second World War, when a malarial protection program and better nutrition had improved the health of the population, it was estimated that if all the children with thalassemia were treated, it would double the health bill of the island in about fifteen years. This is so because nearly 1 percent of all babies born in the island were affected and required regular blood transfusion to survive.

Another change in the pattern of disease that occurs when developing countries become more wealthy and start to improve the environment is the appearance of heart attacks, strokes, diabetes, and so on. Presumably this is partly because people start to live long enough to contract them, but it also reflects the acquisition of the bad habits of Western culture. In China, for example, cigarette consumption increased from 500 billion in 1978 to 1,700 billion in 1992. If this trend is similar in other developing countries, 10 percent of the 120 million who reach adult life over the next decade will die prematurely. And, as we will see later, other "Western" diseases, particularly diabetes, are also assuming epidemic proportions. Remarkably the recent report of the World Bank to which we referred earlier suggests that, in economic terms, heart disease is now the leading world health problem.

WHERE TO NEXT?

This brief survey of our current world health problems should have given some inkling of why in recent years there has been a major change in the way in which international health agencies and governments are thinking about the future directions of health care. If, the argument goes, we are to have any hope of being able to pay for a healthier world, we will have to change the emphasis of modern medicine, and its preoccupation with established disease, and concentrate its efforts more toward prevention and the better delivery of health care. In the next chapter we will explore the background to these new ideas, and some of the scientific evidence on which they are based.

PART IV

The Origins of
Our Intractable
Diseases

S I X

New Ways of Thinking
about Disease

It can be said that each civilization has a pattern of disease particular to it. The pattern of disease is an expression of the response of
man to his total environment (physical, biological, and social); this
response is, therefore, determined by anything that affects man
himself or his environment.

—René Dubos (1901–1982)

Over recent years the whole ethos of medical practice and research has been
questioned. It is argued that by taking an increasingly reductionist approach
doctors have become obsessed with the mechanisms and treatment of established illness rather than with its causes and prevention. The balance of medical education has moved too far toward the study of disease at the expense
of both patients and their environments, which, it is held, are responsible for
most of their ills. In short, medical science has become so wedded to the
Cartesian concept of man as a machine that it is failing to treat patients as
sick people and is interested only in their diseased organs; Mrs. Jones is a
case of lung cancer, not, Mrs. Jones is a frightened housewife with three
young children and an out-of-work husband who now has to cope with a
disease that, with better social circumstances and a different lifestyle, might
have been prevented.

Some critics have gone even further than this. They have analyzed the
factors that have improved our health over the last few centuries, and concluded that scientific advances have made little difference to our well-being.
Rather, almost the entire improvement in our lot can be ascribed to social
change, improved hygiene, better nutrition, and the other accompaniments
of higher standards of living. Would it not be better to concentrate our efforts
on improving our environments and lifestyles and to stop pouring resources
into high-technology practice and laboratory research?

This new way of thinking about health care is already having a major
effect on medical education, the funding of research, and the way in which

governments are planning and financing the delivery of medical services. Historians of the future may well look back on it as the most profound change of direction in medical practice and research in the second half of the twentieth century. In this chapter, after examining its origins and the new fields of medical science on which it is based, we will try to assess the evidence for the view that most of our problem diseases are environmental and therefore should be preventable.

The Origins of Doubts about
Modern Medical Research and Practice

In 1971 the distinguished Nobel laureate Macfarlane Burnet, in his book *Genes, Dreams and Realities,* described the emergence of the laboratory sciences and the way they have helped to control infectious disease; as one of the founders of modern immunology, he was well qualified to deal with these topics. He then discussed the development of the "new" biological sciences, particularly molecular and cell biology, and concluded that they have little or no potential for dealing with intrinsic disease. The major health issues of the future, he decided, would be influenced not by laboratory medicine and basic science but rather by observational and social studies. In summarizing his views on the current scene, he suggested that "future historians may speak of an age of scientific discovery that started with Galileo in 1586 and ended something less than 400 years later."

Of course, medical science has always had its detractors. The great French essayist Montaigne, not without justification, systematically took apart the activities of the medical profession of his time. In this century George Bernard Shaw did the same thing in his brilliant preface to *The Doctor's Dilemma.* Shaw was hilariously skeptical about current research on infection. His cry "stimulate the phagocytes" came straight from the laboratory of Almoth Wright at St. Mary's Hospital Medical School. But these works were written with humor and style and, especially Shaw's, seemed to be concerned most with debunking the pomposity of the medical profession. This was not true of a book called *Medical Nemesis,* which first appeared in 1975 and was written by the philosopher and theologian Ivan Illich.

Medical Nemesis is a passionate attack on modern medicine and its practitioners. A distinguished British neurologist, Henry Miller, was so incensed by this aggressive and not always balanced outburst that he was moved to rechristen Illich "Ivan the Terrible." Using a mass of statistics, Illich set out to show that modern medical practice has had no effect whatever on the health of society, that common infections such as tuberculosis and poliomyelitis were disappearing long before the advent of antibiotics and vaccines, and that modern medicine poses a threat to society as well as to individual patients. In essence, his thesis was that modern medicine does more harm than good. It generates demands for its services and encourages aspects of behavior that lead to more ill health and reduce our ability to cope with illness and to face suffering and death. Hence the medical profession, at least in its present form, should be disbanded. Although it seems unlikely that we should take such an emotional and completely one-sided attack too seriously, there is no doubt that Illich reflects a train of thought that has been building over the last twenty years.

A much more thoughtful critique of modern medicine was first published in Great Britain in 1976 as one of the prestigious Rock Carling Lectures and followed up in 1979 in a book, *The Role of Medicine*. The author, Thomas McKeown, seems to have developed his rather jaundiced views during his days as a medical student in a London teaching hospital. He was struck by two things: "One was the absence of any real interest among clinical teachers in the origin of disease, apart from its pathological and clinical manifestations; the other was that whether the prescribed treatment was of any value to the patient was often hardly noticed. . . ."

McKeown points out that diseases are due either to intrinsic (inherited) mechanisms or to the action of environmental factors. He argues that since most serious genetic diseases will have been lost by natural selection, the bulk of the killers of Western society must be due to external environmental agents, many of which should be controllable. Like Illich, he belittles the effects of advances in scientific medicine and further refines the argument that the decline in infectious disease was largely the result of improvements in nutrition, hygiene, and social conditions. While the development of immunization, chemotherapy, and antibiotics was a useful adjuvant, and undoubtedly speeded up the control of some infectious killers, the overall impact of these scientific advances was not great. Furthermore, they tended to concentrate the attention of doctors on diseases rather than on their patients. McKeown finishes with a plea for a complete reappraisal of medical education and practice, with much more emphasis on public health and preventive medicine.

In a more recent book, *The Origins of Human Disease* (1988), he restates and develops many of these arguments. Here he traces the origins of disease at different stages of our evolutionary past. During the period of hunting and gathering, man was unable effectively to control the environment or limit reproduction, but because natural selection had left him well adapted to conditions of life, noncommunicable diseases were rare. The short life expectancy of hunter-gatherers was, he suggests, due to malnutrition combined with a hostile environment and infectious disease. The arrival of agriculture saw the beginning of control of the environment and an increase in food supplies, which led to a decline in mortality and an increase in the population size. However, because reproduction was not restricted effectively, the population size increased to the point that the supply of food again set limits to growth. Furthermore, living together in large numbers in unhygienic conditions promoted the transmission and propogation of many infectious diseases, which became the predominant causes of sickness and death. Finally, industrialization led to even better control of the environment, increased food supplies, and, with time, improved hygiene, all of which led to a decline in infection. Improvements in health were not reversed by rising numbers, as occurred after the agricultural revolution, because nutritional and environmental improvements were accompanied by a sharp reduction in the rate of reproduction. However, industrialization created conditions of life far removed from those in earlier periods of human evolution, and the noncommunicable diseases that resulted from this new environment displaced infection as the major causes of sickness and death.

McKeown believes that, apart from diseases caused by genetic factors or unknown environmental agents acting during intrauterine life, the bulk of the important conditions that affect Western societies are due to the action of a hostile environment, or behavioral perversities that drive us to smoke, eat too much of the wrong kind of food, or kill ourselves in motor cars. While he grants that there are still many chronic disorders for which there is no evidence for the role of environmental agents, he believes that these will ultimately be found. In essence, the major killers of Western society are "new" diseases. Modern industrialized countries, with their complete change in diet and environment, have existed for only a few hundred years, an extremely short time in man's history. Our genetic makeup, which has been adapted over thousands of years for a totally different environment, is unable to cope with the modern world. Heart attacks, high blood pressure, and cancer are not diseases of primitive societies, but are simply the price that we are paying for our self-indulgent lives of affluent immobility.

McKeown concludes by pointing out that scientific progress has often been retarded by psychological blocks that lead us to ignore solutions to health problems that are staring us in the face. He cites as an example how it took more than a thousand years before knowledge of optics was applied for the benefit of vision. He argues that the dominance of the mechanistic approach to the problems of disease, an approach that started in the seventeenth century, has caused us to overlook important messages that the patterns of disease origins in the past have left for us, and has led us to underestimate their potential value for the development of health care in the future. If, for example, we had been thinking of disease origins rather than mechanisms, would it have taken us so long to suspect the importance of smoking or lack of exercise in the genesis of diseases of the heart?

Similar thought-provoking arguments have been raised by many other observers over recent years. The picture that they draw is undoubtedly flawed in parts. Many of those who have written along these lines have not been practicing clinicians; the world of mortality statistics is quite different from that of the doctor who has to deal with sick people. Such figures tell nothing of the thousands of children with frightful deformities caused by poliomyelitis or sanatoriums full of patients with tuberculosis, both of which were major features of medical practice early in this century. We have no way of knowing whether the falling mortality from infectious disease, which is so central to the arguments of Illich, McKeown, and others, would have continued indefinitely without the introduction of immunization and antibiotics. It is difficult to imagine, for example, that the use of antitoxins and immunization had no effect on mortality from diphtheria, a fact Mckeown acknowledges, though he still manages to find some regions in the United States where these measures were not applied and yet mortality rates still appeared to fall. As we saw in chapter 3, the large fall in maternal mortality that occurred in the 1940s, after mortality had not changed for over half a century, was undoubtedly due to the introduction of sulfonamides. And although the mortality from tuberculosis was falling, the introduction of antituberculous drugs surely speeded up the process and changed the face of medical practice in the 1950s.

Another major weakness of these claims is their almost complete disregard for the role that intrinsic biological factors such as aging and our genetic makeup play in the genesis of our current killers. As we explore what is known about the causes of the illnesses that fill our hospitals, it will become apparent that many of them reflect a complex interplay between nature and nurture, set against the background of important changes in our body chemistry that are the inevitable result of our surviving to an advanced age.

But even if, with some justification, we do not accept all the arguments of McKeown and others, there is no doubt that changing social conditions together with improvements in nutrition and the introduction of public health measures did much to reduce mortality from infectious disease in the early part of this century. We can leave medical historians to sort out which of these factors was most important. But can we use the same arguments when we consider the control of the current killers of Western society? Do we, as McKeown suggests, already know enough about the environmental agents responsible for heart disease and cancer to prevent them? Are we, in effect, wasting time and resources in our efforts to understand the underlying mechanisms of these diseases and how they arise? Or, as is already clear from the story of the partial control of infectious disease, are we more likely to succeed by adopting both approaches?

TO WHAT EXTENT ARE ENVIRONMENT AND LIFESTYLE THE CAUSES OF THE MAJOR KILLERS IN THE WEST?

Before we tackle this complex issue, it is worth digressing briefly to look at the origins and development of the new fields of medical science that have helped to shape our thinking about the environmental causes of disease. Earlier, when we touched on the historical background to the control of infectious disease, we saw that the relation between environment and health has been the subject of debate and controversy since the time of Hippocrates, and that the contagion theory of disease stemmed from the observation that epidemics seemed to arise from person-to-person contact. Thus epidemiology, the branch of medical science concerned with describing and interpreting the patterns of disease in large populations, evolved from studies of the behavior of infections in communities.

The collection of information about outbreaks of infectious diseases and the investigation of epidemics continued through the seventeenth and eighteenth centuries, and in 1836 a registration act provided for detailed annual reporting of all deaths in England and Wales. The London Epidemiological Society was formed in 1850 by a group of prominent physicians interested in the control of infection. Their methods for identifying the causes and for evaluating attempts at controlling epidemics were quantitative and objective, based on the statistical approaches to medical practice that had been pioneered by Louis and the Paris school. The influence of Louis was also felt in America; Kerr L. White, formerly professor of health care organization at Johns Hopkins University, estimates that about seven hundred American physicians visited Paris between 1820 and 1861. Among them were two who had a major influence on the development of epidemiology of infectious disease in the United States, Oliver Wendell Holmes (1809–1894) and Henry Bowditch (1808–1892). But toward the end of the nineteenth century and during the first half of the twentieth observational and numerical studies of disease became less popular and were largely replaced by the laboratory sciences, particularly bacteriology.

Modern epidemiology was to come into its own when it was first applied to noninfectious diseases after the Second World War, notably by Austin Bradford Hill, Donald Reid, and Richard Doll in Great Britain. These pioneers, who, as we saw in chapter 4, were also responsible for the development of clinical trials, devised ways of analyzing the patterns of disease in large populations. Using reliable statistical methods, they sought to relate the prevalence of common diseases like cancer and cardiovascular disease to different environmental factors and changes. There is no doubt that the emergence of clinical epidemiology marks one of the major successes of the medical sciences in the last half century.

It is difficult for anybody who has not practiced medicine over the last thirty years to appreciate how modern epidemiology has changed the way in which we think about our leading killers and chronic diseases. In the 1950s conditions such as heart attacks, stroke, cancer, and arthritis were still bundled together as "degenerative disorders." This blanket term implied that they might be the natural result of wear and tear and the inevitable consequence of aging. However, as information about their frequency and distribution, and their occurrence in people who had moved from one country to another, was analyzed, it became clear that the environment must play a major role in their genesis.

Perhaps the most convincing evidence for the importance of the environment in causing some of our problem diseases was the observation that their prevalence had changed dramatically over short time spans. While there are all sorts of dangers in interpreting information of this kind, particularly vagaries of reporting, it became clear that, in the case of cancer and some forms of heart disease, there had been quite drastic changes in death rates over periods as short as twenty to thirty years. These studies ruled out the effects of age, at least to some degree, because differences in mortality appeared to have occurred *within* the same age range. Furthermore, they had been observed over a period in which there had been no major alterations in the diagnostic criteria for the particular illnesses. For example, death certification rates for cancers of the stomach and lung rose so sharply between 1950 and 1973 that there must have been major environmental factors at work generating these diseases in different populations.

Similar, though less solid, conclusions were reached from studies of groups of people who migrated to new environments and from data on the frequency of cancer and heart disease in primitive populations. There are many difficulties in interpreting large population studies of this type, and many discrepancies between them, but the overall picture indicates that there is an important environmental component to at least some of our intractable diseases.

Cancer

It has been realized for only about the last forty years that our environment and lifestyle may play a causal role in some common cancers. Although it had been suspected since the eighteenth century that certain rare cancers of the lip and tongue result from pipe smoking, not until the late 1940s was serious thought given to whether cigarette smoking might have a deleterious effect on health. Clinicians in both the United States and Europe, writing in the first half of this century, expressed the opinion, based on their day-to-day practice, that cigarette smoking might cause cancer of the lung. Just before the Second World War it was found that cancers could be produced in the skin of rabbits by treatment with a tobacco tar. These observations were either ignored or discarded. But in 1938 Raymond Pearl (1878–1940), a statistician at the Metropolitan Life Assurance Company, noticed the ill effects of smoking from studies of family history records taken from the Johns Hopkins School of Hygiene and Public Health. Pearl made no specific mention of lung cancer, but he anticipated what was to follow when he wrote that "the smoking of tobacco was statistically associated with an impairment of life duration and the amount or degree of this impairment increased as the habitual amount of smoking increased." Again this finding was universally ignored or dismissed.

In 1947 the Ministry of Health in Great Britain wrote a letter to the Medical Research Council (MRC), urging the need to investigate the reasons for the alarming increase in deaths from lung cancer. A conference was called by the Secretary of the MRC, and possible causes were discussed. The obvious candidate was atmospheric pollution, but since smoke pollution had decreased in Britain and was almost nonexistent in Switzerland, where a similar increase in mortality had been observed, another was sought. It was suggested that tobacco, particularly when smoked in the form of cigarettes, might be the culprit.

The first study to address this problem was carried out by two of the founders of British epidemiology, Austin Bradford Hill and Richard Doll (figure 30). Although, Doll tells me, they started out on this work with no great enthusiasm for the tobacco hypothesis, their first analysis of 649 male and 60 female patients with lung cancer, published in 1950, showed that, in their group of patients with lung cancer, 26 percent were heavy smokers while in their control population without cancer only 13 percent were smokers; in the lung cancer group, only 0.3 percent were nonsmokers, while in the control group without lung cancer 4.2 percent were nonsmokers. Similar results were obtained by workers in the United States at about the same time. In 1976 Doll and his colleague Richard Peto published the results of a questionnaire on 34,440 British doctors whom they had followed for over twenty years and asked about their smoking habits over this time. The results confirmed the earlier findings and left no doubt that cigarette smoking is a major factor in the cause of lung cancer. Similar conclusions emerged from two large-scale studies organized by the American Cancer Society. Recently Richard Doll celebrated both his eightieth birthday and a forty-year follow-up of his long-suffering doctors, at least those 30 percent that were left. The results leave little doubt about the role of cigarette smoking in the generation of cancer, and not just lung cancer.

In 1981 Doll and Peto were commissioned to write a report for the Office of Technology Assessment, of the U.S. Congress, to provide background

FIGURE 30

Richard Doll.

material for an overall assessment of cancer risks from the environment. By collating all the published data up to the end of the late 1970s, they were able to produce evidence for the link between cigarette smoking and a variety of cancers, including those of the lung, mouth, larynx, esophagus, bladder, pancreas, and kidney. The strongest correlations were with cancers of the lung, mouth, throat, and esophagus; the association with other cancers were significant but weaker. These results have stood the test of time and are confirmed in the forty-year follow-up study of doctors. The authors concluded that cigarette smoking accounts for the annual deaths of about 122,000 males and 27,000 females in the United States, or 30 percent and 15 percent, respectively, of all cancer deaths.

As information of this type has accumulated over the last twenty years, it has become apparent that other environmental factors may be involved in causing of cancer. Although alcohol by itself does not greatly increase the risk, the combination of alcohol and tobacco has an additive effect in the generation of cancers of the mouth, pharynx, and esophagus, although probably not the lung. Since these cancers are relatively uncommon, as are liver cancers, in which alcohol may also play a role, Doll and Peto suggest that alcohol consumption may account for about 7 percent of all cancer deaths in men and for 3 percent in women.

Although evidence for a causal association between air pollution and cancer is very limited, substances with carcinogenic potential no doubt occur both in occupational and in natural environments. Outdoors, potential risks arise from fumes and particles from manufacturing, lead from motor exhausts, wood smoke, and asbestos. Indoor air may carry risks through cigarette smoking on the part of others, radon products, and vinyl chloride. However, the attributable risk for lung cancer from air pollution remains small, despite evidence for a slightly increased risk of lung cancer among nonsmoking women who are married to men who smoke heavily.

The role of diet has been much more difficult to assess. Foodstuffs could be involved in many ways. Cancer-producing agents, carcinogens, could be present in natural food or additives, or produced by cooking or by microorganisms in stored food. Our diet could encourage the production of carcinogens in the body or lead to their defective excretion. And it is also possible that dietary deficiency or even overeating play a role.

There are hints—but no more than that—that things we ingest may have some influence on the development of cancer. Thoughts along these lines were stimulated by experiments in mice that were carried out by Tannenbaum in 1940. He found that by severely restricting the intake of food, but without modifying the proportion of its constituents, he could halve the incidence of spontaneous tumors of the breast and lung and of a variety of cancers that can be produced experimentally by carcinogens. The underfed mice grew to about half the size of those who were fully fed, but otherwise they seemed to be active and healthy and, even more interestingly, lived on average longer than their more obese cage mates.

In 1979 the results of a study that monitored the lifestyles of nearly three-quarters of a million American men and women for thirteen years were published. The findings, though not particularly impressive, suggested that obesity might be associated with a slight increase in the frequency of certain tumors, particularly those of the uterus, gall bladder, stomach, and bowel. Correlations have also been found between the national consumption of fat and meat per head of the population and the frequency of cancers of the bowel and breast, although the effects are again small. On the basis of the original observations of Denis Burkitt in Africa, several studies seem to demonstrate a correlation between the lack of fiber in the average Western diet and cancer of the bowel.

Observers of the current medical scene might be excused for wondering whether, in our obsessional need to try to discover relations between diet and cancer, we may be wasting a great deal of time and public money. For example, an extensive study involving 89,494 women between the ages of thirty-four and fifty-nine found no correlation whatever between dietary fat intake and breast cancer. And a different long-term observational study of 35,000 women reached the same conclusion. Yet the U.S. National Institutes of Health now plan to conduct a clinical trial, which will cost over ten million dollars, to go over the same ground again. It is difficult to believe that such a vast expenditure of money, directed at this well-worn problem, can be based solely on scientific criteria; it smacks more of a political response to pressure groups, or of wishful thinking.

A great deal of work has gone into exploring the possibility that food preservatives or additives contain carcinogens; although they no doubt do, evidence that they are major players in the production of cancer in humans is still lacking. It has also been suggested that deficiency states—of vitamin A, for example—might predispose toward cancer, but, with the exception of some populations in China, there is no clear evidence that this occurs in humans. And although some experimental work suggests that high concentrations of beta carotene, which is found in high concentrations in carrots and certain leafy vegetables, or vitamins C and E might be protective, there is only limited evidence that they may affect the incidence of cancer. A recent study in China has hinted that augmenting vitamin E in the diet might reduce cancer rates in the population, but this needs confirmation. Undoubtedly some food additives, such as saccharin, are carcinogenic, but the effect has been demonstrated only in rats fed enormous doses; population studies have shown no association between the use of artificial sweeteners and human cancer.

There are some well-established occupational causes of cancer—tumors of the bladder in dye manufacturers, and of the lung and pleura in asbestos miners, for example. But occupation and industrial pollution probably play a minor role in the overall incidence of cancer.

Recent years have seen considerable interest in the relation between hormones and the development of cancer. For example, it is well known that the cells of the uterus and breast are influenced by female sex hormones called estrogens and progestogens. Because deficiency of hormones is thought to be a major factor in the development of thinning of the bones, which occurs in many women after the menopause, hormone replacement therapy has become popular in many countries. Furthermore, certain oral contraceptives combine estrogens and progestogens. Some evidence suggests that uterine cancer is related to cumulative exposure to estrogens in the absence of progestogens, while breast cancer is also related to exposure to estrogens, an effect enhanced by progestogen. Overall, combination-type oral contraceptives that contain both estrogen and progestogen seem to be associated with little increase in the risk of developing uterine cancer. It appears that there may be a slightly higher risk of developing of breast cancer after the prolonged use of postmenopausal estrogen replacement therapy.

What is this epidemiological evidence telling us about the causes of cancer? It is clear that tobacco, which may be responsible for 25 to 40 percent of all cancer deaths, is the one agent for which there is absolutely incontrovertable causal evidence. Doll and Peto, who have compiled so much of this information, have suggested that the proportion of cancers due to diet could range between 10 and 70 percent, a statement that simply highlights our ignorance about this important possibility. Apart from the small role played by alcohol, there is very little evidence to incriminate other environmental factors. Of course, the real proof that an environmental agent is the cause of a particular form of cancer is the demonstration that its frequency is reduced if exposure ceases. Evidence of this kind has come from long-term studies of smokers who have given up the habit; their risk of developing lung cancer is reduced, although it never seems to reach the level of those who have never smoked.

In short, while the relation between cigarette smoking and cancer is solid, with the exception of certain specific hazards such as asbestos exposure, there is little hard evidence that any other environmental factors are of major importance in the genesis of cancer. This does not mean that they may not exist, of course.

Cardiovascular Disease

What little is known about the factors that damage the arteries of the inhabitants of Western societies has come from two different sources. Pathologists have given us a very detailed description of what these diseased vessels look like under the microscope. But the question is how they got like that in the first place, and here the only clues have come from studies of the environments and lifestyles of patients with heart disease and, to a lesser extent, from investigations of genetic factors that may be involved in producing diseased arteries.

The arteries of babies and young children have smooth linings. As we get older, the linings of our arteries become uneven and studded with hard plaques, which result from the deposition of fatty material, which often becomes calcified, a condition called atheroma, from a Greek word meaning "gruel." Theories about the cause of atheroma have proliferated over the years, but none has been entirely satisfactory. It appears that the plaques result from damage to the lining of blood vessels followed by seepage of fats and other products and their gradual accumulation and calcification in the vessel wall. Ultimately the plaques ulcerate, and the smooth lining of the vessel wall is lost. The body's normal defense mechanism against a disruption of a blood vessel is then activated. We seal off small holes in our vessel walls with plugs of platelets, small cells that circulate in the blood for this purpose. In addition, there may be an activation of a complex series of events that leads to the formation of a clot, or thrombus. If these mechanisms are activated on an ulcerated plaque, a clot forms that obliterates the lumen of the vessel. Something like this may happen when we have a coronary thrombosis or some forms of stroke.

Why, therefore, do some of us have severe atheroma and others not? And, even more puzzling, why, since most of us develop atheroma as we get older, do only some of us have a coronary thrombosis or a stroke? Three possibilities come to mind: luck, inheritance, and exposure to different environmental agents. Large-scale studies of different populations, or of the frequency of heart attacks or strokes in identical twins reared apart in different environments, have gone some way to clarifying these issues, but have not yet provided any clear answers about the cause of vascular disease. However, because they have involved such large numbers, they have at least reduced the role of chance, except on an individual basis, as the main factor in the development of vascular disease.

Clues that our environment may play a role in the development of heart disease first came from studies of the rates of heart attacks in different populations. For example, in Finland the occurrence rate in 1970 was approximately 198 per 10,000 of the population, whereas in Japan the figure was only 15 per 10,000. This observation could reflect a genetic difference between Finns and Japanese in their susceptibility to develop heart disease, or it could mean that something in the environment or lifestyle of the two populations determines whether they develop heart disease. However, when the Japanese emigrate to countries like the United States, their rate of heart attacks rises toward that of the local population. This suggests that the environment is a significant factor in the generation of heart disease.

Several long-term studies of large populations have monitored diet, lifestyle, and the occurrence of other diseases in order to see if there are any correlations with the development of heart disease. One of the most thorough involved the follow-up of over 50,000 males in Framingham, Massachusetts, for many years. From this painstaking work emerged a number of factors— or "risk factors," as they were called later—that might make people more likely to develop heart disease. Among them, smoking, diet, especially the intake of animal fats, blood cholesterol levels, obesity, lack of exercise, and raised blood pressure seemed to be of particular relevance.

The most important result to come from many different studies of this type is the clear-cut relation between cigarette smoking and both heart disease and strokes. For example, in the study of British doctors by Doll and Peto that we mentioned earlier, it was quite clear that the mortality from coronary artery disease increased progressively, from nonsmokers through light and moderate smokers to heavy cigarette smokers, at all ages under sixty-five years. The relative risk in heavy cigarette smokers compared with that in nonsmokers was 15:1 at ages under forty-five, 3:1 at forty-five to fifty-four, and 2:1 at fifty-five to sixty-four.

Similarly, many large population studies have shown that the level of cholesterol in the blood is related to the occurrence of vascular disease. For example, the Framingham study found that the risk of heart disease varies over a fivefold range in relation to cholesterol levels found in an average American population; there is no critical value above which heart disease occurs, and the risk tends to increase throughout the range. More recent studies have confirmed these observations. If the population is divided into groups on the basis of a single blood cholesterol measurement, a steady increase in mortality from heart disease is apparent with increasing cholesterol levels. These relations hold even in Chinese populations, where the average cholesterol levels are much lower than they are in Europe and the United States and where heart disease is less common. It is important to remember, however, that correlations of this kind do not prove that there is a causal relation between heart disease and cholesterol levels in the blood.

Epidemiological studies have also shown associations between other dietary factors and heart disease in several countries. The foods that have been implicated include saturated fat, sucrose, animal protein, and, in some but not all (thank goodness) studies, coffee. Several surveys have shown a good correlation between the concentration of cholesterol in the blood and the percentage of total calories obtained from saturated fat. Long-term analyses carried out by insurance companies have suggested that obesity is associated with cardiovascular disease, although whether being overweight alone is enough, or whether other factors such as the type of diet or the level of blood pressure are involved, remains to be seen. There is clear evidence that all forms of diabetes are associated with an increased proneness to disease of the heart and blood vessels.

Since some of the fat components in our blood seem to be more liable to cause heart disease when they have undergone chemically mediated damage called oxidation, there has been much recent interest in the role of factors in our diets called antioxidants, which may inhibit the process. Among the naturally occurring substances that have this property, vitamins C and E are of particular importance. Studies of the relation between vitamin C intake and coronary artery disease have given conflicting results, but recent work appears to have shown a beneficial effect over a relatively short period for the administration of comparatively high doses of vitamin E. Although prospective studies of this type do not prove a cause-and-effect relation, they hint that vitamin E supplements may reduce the risk of coronary artery disease.

The relation between physical inactivity, until recently the standard way of life in Western industrialized societies, and heart disease is less easy to assess. One of the most quoted studies was carried out by the British epidemiologist Jerry Morris. He sent a letter to 18,000 middle-aged British civil servants that was designed to arrive on their desk on Monday morning, without previous warning. They were asked to record how they had spent each five minutes during the previous Friday and Saturday. British civil servants are honest and long-suffering folk, though to be faced with such an embarrassing question on a Monday morning must have stretched even their highly developed sense of duty to country and Queen. Nevertheless, surveillance of this group over the next eight and a half years suggested that the incidence of fatal and nonfatal heart disease was about twice as great in those who recorded no vigorous exercise in their returns. A similar study of some 6,000 San Francisco longshoremen, who, one imagines, are very different from British civil servants, also suggested that those who led more physically active working lives were protected against heart disease. These observations have been confirmed over recent years. It appears that vigorous, habitual, and continuing aerobic exercise helps protect against death from heart disease and, possibly, from some of our other major killers.

The effect of stress on the frequency of heart disease is more controversial. Many studies of this type have been attempted, including the analysis of stressful life events with respect to the frequency of heart attacks. Attempts to categorize particular types who are more or less likely to succumb in this way customarily divide us into type A or type B personalities. A type A individual is an aggressive, driving, ambitious, restless person who is excessively concerned with time and deadlines. Type B persons, on the other hand, are most easily described as smiling, easy-going cabbages. A number of studies, particularly in the United States, have suggested that type A individuals have more marked changes in their blood vessel walls than type B, and they have led to the notion that these psychosocial characteristics, as they are called, are causally linked to the development of coronary artery disease. The reason for this association has never been determined.

It also appears that there is some correlation between coronary artery disease and alcohol intake. Alcoholics or heavy drinkers seem to have an excess mortality from heart disease, while those who indulge in a moderate amount of alcohol each day may be slightly protected. At least one major international study on this emotive topic suggests that the "protective" effect of alcohol is confined to wine.

There is clear-cut evidence from major population studies, such as those in Framingham, that the risk factors for the development of vascular disease are additive; the heavy smoker who also is overweight and has a raised blood pressure is at greater risk than a smoker who has no other factors that might make him or her more prone to heart disease.

Given this plethora of epidemiological evidence, how much progress has been made in reducing the amount of heart disease by changing our lifestyles or by modifying risk factors? We know from the work of Doll, Hill, and Peto on British doctors that stopping cigarette smoking tends to reduce deaths from coronary artery disease. By comparing trends in mortality with time it, they made it quite clear that a change in smoking habits was associated with a decline in the standardized mortality ratios for coronary artery disease. Although it is difficult to interpret studies of this type, because highly motivated people like doctors may have changed many other habits during the period of study, persons who stop smoking can clearly reduce their risk of premature death from heart disease.

The results of trying to modify other risk factors have been less clear-cut. Several enormously expensive trials have been carried out to determine whether lowering cholesterol, either by diet or the use of drugs, alters the prevalence of heart disease or reduces deaths from heart attacks. The results are among the most hotly debated topics in current clinical practice. While all of them have shown some effect on the prevalence of heart disease, none has proved that reducing cholesterol levels decreases mortality. Furthermore, and possibly because of a quirk resulting from the numbers involved, some strange effects have been observed. For example, two trials found a moderate reduction in nonfatal heart attacks, but this was counterbalanced by a slight increase in mortality from other causes in patients receiving cholesterol-lowering agents; in one trial those receiving the drugs seemed to have an increased likelihood of death from traffic accidents.

Where does all this leave us with regard to cholesterol and heart disease?

Observational studies have shown a linear relation between coronary artery disease and plasma cholesterol levels. As we mentioned earlier, the recently developed statistical technique called meta-analysis makes it possible to combine the results of different trials. Where this has been done, there is fairly convincing evidence for a reduction in the prevalence of coronary artery disease as cholesterol levels fall. It takes a long time to silt up our coronary arteries, and many of the trials examining the effect of lowering cholesterol have been based on a short period of observation. Overall, even in trials lasting for about four years, a 10 percent reduction in coronary artery disease has been observed, whereas in trials lasting up to seven years the figure is about 20 percent. But even when the results of all these trials are combined, it has not been possible to demonstrate a significant reduction in mortality from coronary artery disease. The problem is that the numbers of recorded deaths were simply too small to show any effect of cholesterol lowering.

There is another problem. The use of meta-analysis for the comparison of trials that have attempted to lower cholesterol in different ways—by diet or drugs, for example—suggests that the use of cholesterol-lowering agents might have deleterious effects that could counterbalance their effect in protecting us from coronary artery disease. Thus, while it might be justifiable to employ these agents in people with very high blood cholesterol levels, or in those who have had a heart attack, it remains unclear whether it is appropriate to administer them to healthy people who happen to have a modestly elevated cholesterol level.

Similar difficulties are being encountered in assessing the value of antioxidants like vitamin E in the prevention of heart disease. While recent studies have suggested that the administration of large doses of vitamin E reduces the number of heart attacks in middle-aged women, this does not prove a causal relation. It is always possible, for example, that the ladies who religiously swallowed their vitamin pills also got more exercise and generally took better care of themselves. Furthermore, before persuading everyone to swallow vitamin E tablets every day, we must find out whether it is safe. Could this very high level of antioxidant have deleterious side effects? To answer this question large-scale trials will have to be carried out over many years. Daniel Steinberg, writing in the *New England Journal of Medicine,* summed up the position recently, "I think we must play by the rules and insist on large, long-term, double-blind clinical trials. Until they are done, please, let's hold the vitamin E."

It seems important, therefore, to set up larger and more effective trials to try to answer, once and for all, the question of whether cholesterol lowering or the administration of antioxidants saves lives. Many doctors are already convinced that the evidence is good enough and are altering their patients' lifestyles or placing them on cholesterol-lowering drugs. However, many doubts remain. This question has important implications for the future of medical care. Are we to attempt to alter the diets of all industrialized countries and to place an increasing proportion of our populations on cholesterol-lowering drugs?

Large-scale population studies have left little doubt that an elevated blood pressure is an important risk factor for cardiovascular disease, particularly stroke. For example, in the Framingham studies, 30 percent of the men and 50 percent of the women who died of cardiovascular disease had previously been noted to have an elevated blood pressure on at least three occasions.

The problem with blood pressure is that, like height and weight, it shows wide variation in the population; many of those who are said to have an "elevated" blood pressure are at one end of the distribution curve. This pattern of distribution of a biological variable suggests that both genetic and environmental factors are involved. Twin studies have shown a higher concordance rate in identical twins than in spouses, implying that genetic factors play an important role. It has been very difficult to obtain any clear evidence for the involvement of environmental factors. Dietary salt has been studied with great enthusiasm, and some controlled trials have noted a fall in blood pressure with moderate salt restriction; others have not. There are hints that the level of calcium or lead in the diet may have some effect on blood pressure, but again the results are inconclusive. Coffee drinking seems to elevate the blood pressure temporarily, and this effect is more marked if the coffee is drunk at the same time as cigarettes are smoked. Studies of the effects of cigarette smoking on blood pressure have shown slightly lower levels in smokers, although this appears to be related largely to weight. In many Western populations there is an increase in blood pressure with age, although this is not seen in a few primitive communities that have been observed over a long period. The movement of peoples from these environments into towns is sometimes associated with a rise in blood pressure. Physical exercise carried out over a moderate length of time seems to lead to a modest lowering of blood pressure. And so on.

All in all, it is clear that we know very little about why hypertension is so widespread. It seems likely that it results from the interaction of genetic susceptibility with so far unidentified factors in our environment, diet, or lifestyles, a pattern common to many important diseases of unknown cause.

Recently the relation between coronary artery disease and stroke was reassessed by an analysis of the results of nine major prospective observational studies, involving nearly half a million individuals, who between them had suffered 843 strokes and 4,856 episodes of coronary artery disease. The results showed a steady increase in the likelihood that one of these events would occur with increasing blood pressure, stretching from the normal range to one that would be considered moderately elevated. There was no evidence of any threshold below which lower levels of blood pressure were not associated with a reduction of strokes or heart disease. The results of reducing the blood pressure by treatment suggest that stroke was reduced by 42 percent and coronary artery disease by 14 percent. While this looks very encouraging, it should be remembered that the overall effect on the prevention of coronary artery disease is still quite small and that a life on drugs that reduce blood pressure is not without its disadvantages, since nearly all of them have some side effects.

Could Heart Disease and Stroke Be Influenced by Our Environment before We Are Born?

Over recent years the English epidemiologist David Barker has carried out two extensive surveys that suggest there may be a relation between nutrition in fetal life and the likelihood of developing heart disease or high blood pressure in middle age. The first, involving 1,586 men born in a maternity hospital in Sheffield during 1907–25, suggested that death rates from cardiovascular disease fell progressively with increasing weight, head circumference, and other measures of increased development at birth. The second, which involved 5,654 men born in Hertfordshire during 1911–30, found death rates from coronary artery disease to be almost three times higher among those who weighed 18 pounds (8.2 kg) or less at the age of one year than among those who weighed 27 pounds (12.3 kg) or more.

In a further series of studies Barker and his colleagues have found that low growth rates up to the age of one year seem to be associated with an increased prevalence of different risk factors for cardiovascular disease—including blood pressure, the concentration of glucose, insulin, fibrinogen, and other clotting factors in the blood—and with death rates from vascular disease. Similar findings have been reported in racial groups other than northern Europeans. There is some preliminary evidence that the particular patterns of ill health in middle life may be related to the specific period of undernutrition in pregnancy or the neonatal period.

These are intriguing epidemiological findings. Taken at their face value, they suggest that our nutritional state before we are born may, in some way, permanently modify our body chemistry. This, in turn, may make us more or less likely to succumb to some of the important diseases of middle life and old age. Clearly this important work must be repeated and substantiated. And, as we will see in the next chapter, at least some of its findings must be interpreted with caution, because of the complex inherited variability in many of the characteristics that seem to correlate between fetal and adult life.

This is a new and neglected field. I suspect we will hear a great deal more from it over the next few years. For the moment it serves to underline the enormous complexity of the ways in which our environment may shape our susceptibility to common diseases.

The Overall Implications of Tobacco and Other Risk Factors for Common Diseases

From a review of the complex and often conflicting epidemiological evidence regarding environmental factors in the causation of cancer and heart disease, it is clear that by far the most solid information relates to the effects of cigarette smoking. In October 1989 the World Health Organization convened a consultative group on tobacco-related mortality, and some of its frightening conclusions were published in the spring of 1992. It appears that the annual deaths from smoking in the developed countries alone numbered about 0.9 million in 1965 and are predicted to reach approximately 2.1 million in 1995 and about 21 million in the decade 1990–99. These figures include 5 million to 6 million deaths in the European Community, a similar number in the United States, 5 million in the former USSR, 3 million in eastern Europe, and 2 million elsewhere. It is further estimated that half these deaths will be of persons in the age range of thirty-five to sixty-nine.

At present approximately one-third of all premature deaths in developed countries can be attributed to tobacco. Ultimately about a quarter of a billion individuals, out of a current total population of just under one and a quarter billion, may be killed by tobacco. The authors of this report are cautious about the accuracy of these figures, but even if they are off by a few zeros, they provide some indication of the problem that faces the rich countries. And, as they point out, although this epidemic is at an earlier stage in the developing world, recent large increases in cigarette use in countries such as China mean that tobacco will, in a few decades, become one of the most important causes of premature deaths in less developed countries. It is vital that by education and public health methods we try to prove these predictions wrong.

What of the other risk factors that we have considered. Summarizing them recently, the British epidemiologist Klim McPherson has attempted to assess their role in contributing to the eight commonest causes of premature death. He writes, "Cholesterol level is responsible for about 23% of all deaths from these eight causes, including 43% of coronary artery disease; hypertension, 22%; obesity, 24%; insufficient exercise about 23%; alcohol a very *small* amount. . . ." Modern epidemiology is nothing if not self-confident! While the evidence we have outlined supports a role for these factors in the genesis of some of our problem diseases, it is doubtful whether it supports quite this level of precision.

Environmental Contributions to Other
Important Diseases

While our brief survey of cancer and cardiovascular disease leaves little doubt that our environment and lifestyle play an important role in their generation, with few exceptions, evidence for environmental causes of the other important diseases of Western society is less impressive. The strongest case is for diseases of the chest.

We have already considered cigarette smoking and lung cancer. There is equally strong epidemiological evidence relating bronchitis to both cigarette smoking and atmospheric pollution. Bronchitis is not serious in itself, although it is a frequent cause of morbidity; it was estimated to be responsible for over thirty million lost workdays in 1973–74 in the United Kingdom. The problem is that some patients with chronic bronchitis develop more serious damage to the lungs that leads to respiratory failure, a condition called chronic obstructive airways disease. In effect, many of them become respiratory cripples. It is not clear why this does not happen to all lifelong cigarette smokers; perhaps there is an inherited susceptibility to the development of chronic lung damage.

In many countries the effects of pollution are becoming less important, largely because of clean-air acts and the control of domestic heating and the pattern of smoke emission. Work in other industrialized societies suggests that cigarette smoking is far more important than the atmospheric environment in causing bronchitis. There is little doubt that if cigarette smoking is reduced, this group of crippling diseases will become much less common.

The environment plays an important role in causing asthma, a spasm of the airways leading to wheezing and shortness of breath. Even allowing for considerable underdiagnosis in the past, there is no doubt that asthma is becoming much more common in Western societies. A particularly worrying feature is the increasing frequency in babies. Population studies in Australia suggest that up to 20 percent of all children and 14 percent of all adults suffer some degree of wheezing, although serious asthma resulting in hospital admission is less common. It has been estimated that there are about 2 million asthma sufferers in the United Kingdom, of whom more than 700,000 are children and adolescents under the age of sixteen. This disease accounts for up to two thousand deaths a year in Britain, and, even more worrying, about 40 percent of them occur before the age of sixteen.

Asthma results from an inherited oversensitivity of the airways to a variety of environmental triggers. A remarkably high proportion of most populations are prone to sensitivity reactions of this type, a condition called atopy, which may manifest itself through hayfever, asthma, or eczema. Recent research suggests that one of the commonest sensitizing agents responsible for asthmatic attacks is the domestic mite. Minute fecal pellets produced by these creatures are particularly potent triggers of attacks. Even in the best-kept homes, in which bedding is changed several times a week, the bed sheets are covered with the excreta of house mites. Since these beasts thrive best in centrally heated, poorly ventilated buildings it is not surprising that the frequency of asthma has increased. And recent work suggests that a particularly vulnerable period for sensitization is the first three to six months of life. It is clear that by converting our homes into overheated, stuffy cocoons we are creating just the environment to sensitize our children to agents that will induce asthma as they get older. Chemical pollutants from car exhausts or industry are also common triggers for attacks of asthma.

The role of the environment in other common disorders is less clear. About 4 percent of us are likely to develop diabetes at one time or another during our lives. The insulin-dependent type that affects young people has an inherited component but may well be precipitated by environmental agents, the nature of which is unknown. The result of a genetically susceptible person's encountering such an agent is the production of antibodies that destroy the insulin-producing cells of the pancreas. Again, we have no idea how or why. This kind of "autoimmunity"—that is, self-destruction of tissues—is thought to be the basis for many other chronic disorders, including rheumatoid arthritis, thyroid disease, and other disorders of endocrine glands, some forms of anemia and bleeding disorders, inflammatory diseases of the bowel, several types of kidney disease, and even some chronic diseases of the nervous system such as multiple sclerosis. So far epidemiological studies have drawn a complete blank as to the environmental agents involved, but there is solid evidence that our genetic makeup is an important factor in rendering us prone to whatever they are.

Though less important as causes of premature mortality, the chronic degenerative disorders that affect the joints of all of us as we get older are a major cause of chronic ill health. The wearing out of our joints may be an inevitable consequence of aging, but it is not as simple as this. Degenerative disease of the hips and the knees quite often affects young people, and although those who have spent periods of their youth in violent athletic activity seem to be particularly prone, the relation is not clear-cut. In short, we know nothing about the causes of these important conditions; neither epidemiology nor laboratory science has given us any useful leads.

Similarly, there is little evidence that the environment plays a key role in the chronic diseases of the nervous system, which, though not major killers, cause much distress to their sufferers and their families. As we have just seen, some evidence points to an autoimmune basis for multiple sclerosis, and some population data suggest that locally operative environmental factors may be involved. But for the bulk of neurological disorders, including Parkinson's disease, motor neuron disease, and many other "degenerative" disorders of the nervous system, the cause is unknown. Similarly, for the distressing forms of dementia, particularly Alzheimer's disease, which is frighteningly common in the elderly, none of the numerous environmental agents that have been implicated has stood the test of time; recent reports implicating aluminum toxicity will need careful pathological and epidemiological verification.

Very little is known about the causes of congenital malformations, major causes of mortality and morbidity in early childhood. A few have been related to specific maternal infections or to vitamin deficiencies, so it is clear that the environment can play a role. But maternal age undoubtedly plays a part, and it is also clear that genetic factors are involved. It is estimated that 0.3 to 0.4 percent of the childhood population of Great Britain has moderate to severe mental retardation and that 2 to 3 percent has a milder disability. There are many different causes. Over half of the cases can be explained by specific chromosomal or other genetic defects. Only about 10 percent seem to result from definable environmental factors such as maternal infection, birth damage, or postnatal infection. In at least a third of the cases no cause can be found.

Psychiatric disorders make up an increasingly large part of the work of both primary care physicians and hospitals. There are two kinds: those that cause serious mental illness, insanity, or psychosis, and milder disturbances that affect many of us at one time during our lives and make it difficult for us to adapt to our environments. We know very little about the causes of serious psychotic illnesses, such as schizophrenia, or about the causes the bipolar affective diseases that are characterized by bouts of depression or periods of hyperactivity and totally irrational behavior called mania. Although it is possible that the stresses and strains of modern life play a role, there is scant evidence to support this notion. Unfortunately we have very little information about the pattern or frequency of psychiatric illness in the past. When making a plea for the epidemiological approach to the study of the psychoses, McKeown assumes that they are diseases of modern society. There is no evidence that this is so. And we will probably never know; even today it is difficult enough to define precise criteria for schizophrenia or bipolar affective disorders. The many years of speculation about the nature of the psychiatric illness that changed the personality of King George III is a useful reminder of the futility of the retrospective diagnosis of psychiatric illness, absorbing though it is.

To try to clarify some of these difficult issues, the World Health Organization mounted a major transcultural study in the 1960s to determine the patterns and frequency of schizophrenia in different populations. Given the difficulties of language and of establishing criteria for a psychiatric diagnosis in peoples of different cultural backgrounds, it was still possible to draw some tentative conclusions from this work. Overall, it appears that schizophrenia occurs in approximately the same numbers and with similar symptoms in many populations of the world. This suggests that either the environment plays only a small role or that, if it is important, the factors involved are widely distributed and relatively insensitive to cultural variation. Certainly the general pattern of psychiatric illness in immigrant populations is similar to that of indigenous Western populations.

These difficulties are magnified further in the case of the less serious forms of psychiatric illness. It is often hard to define what is meant by a "psychiatric illness" in a particular community. Cultural differences abound. The use of magic is still widespread, even in parts of southern Europe. In Greece, if a woman complains that the neighbors are using black magic to inconvenience her, we may not be justified in jumping to the conclusion that she is expressing a delusion. An even more exotic example of this kind of problem is a condition called koro, which is seen throughout Southeast Asia and parts of Africa. Males develop the strong belief that their penis is shrinking, and to prevent this from happening they tie it to a rock or persuade their relatives to hold on to it in relays. If witnessed in an Oxford college, this might pass for psychotic behavior, but in countries in which koro is common it is widely believed that ghosts have no genitals; perhaps it is not so irrational to hang on to one's only visible evidence of continued viability. In some African populations the experiencing of visual or auditory hallucinations is a valuable status symbol. And in Florida a willingness to admit to hallucinations seems to vary with religious affiliation.

These rather quixotic examples of the problems of transcultural psychiatry are offered not to trivialize the subject but simply to highlight the extreme complexities of studying psychiatry by epidemiological methods. Even in advanced Western societies very little progress has been made, and the central question remains whether the bulk of psychoses are due to genetic factors that alter the chemistry of the brain or whether they are unusual reactions to the stresses and strains of life or other environmental factors. It is assumed that the latter underlie many minor psychoneurotic illnesses, but even here we know very little about the reasons for differences in human behavior; the field remains wide open.

ARE THE CRITICS RIGHT?

In this brief account of some of the ways in which thinking about the origins of common disease has changed over the last twenty years, we have seen how, in the period after the Second World War, it appeared that medical practice had become obsessed with the end results of illnesses rather than with the patients who suffer from them and with the environments in which they live. Although many of the criticisms made by thoughtful writers like McKeown may have been exaggerated, they surely contained more than an element of truth. Modern epidemiology has told us that our environments and lifestyles play an important part in generating some of our intractable diseases, particularly vascular disease and cancer. However, the evidence that the environment is the main factor in many of the other chronic illnesses that make up much of modern medical practice is much less firm. We cannot assume that because we have been so successful in determining some of the important environmental agents responsible for heart attacks and lung cancer, we will, if we look at the environment for long enough, discover the causes of our other major ills.

But these observations, limited though some of them are, already have important implications for preventive medicine. We can do a great deal to improve the health of industrialized societies by dealing with environmental agents like tobacco, alcohol, addictive drugs, and reckless driving. This message is being taken up by many governments in such societies, and major efforts are being made to modify our lifestyles. Most countries have reasonably vigorous antismoking campaigns, although for a variety of political reasons governments are unwilling to ban tobacco advertising and to take other measures that might reduce the use of tobacco. In 1992 the European Community health ministers met to discuss a ban on the advertising of tobacco. In the event they postponed their decision. The number of opponents to a ban is falling; Germany will reconsider the evidence, but Great Britain seems to wish to stand alone in continuing to advertise tobacco, a decision completely incomprehensible to anybody who has weighed the evidence. In addition, efforts are being made in most countries to encourage a healthier lifestyle by a modification of diet and an increase in the amount of exercise. All these measures are sensible. However, there is a danger that because the case against tobacco has been proven beyond any reasonable doubt and because we have some evidence relating diet and lifestyle to a few of our common killers, we are lulled into a state of false security and delude ourselves into believing that we know enough already to control them.

The problem with this kind of thinking, which is based in part on a lack of appreciation of the limitations of epidemiological evidence, is well exemplified by what happens these days when one conducts a ward round on the coronary care unit with medical students and young doctors. It is not unusual to meet a lean, forty-year-old nonsmoker who spends his or her life indulging in vigorous physical exercise. Great surprise is always expressed that anybody who has no risk factors can end up with a heart attack. But epidemiology deals with trends in large populations. It tells nothing about the hundreds of people who smoke and overeat and whose only exercise is the walk from the front door to the garage where their car is housed, yet who never have heart attacks, or why many people who lead blameless lives succumb to them.

When we open our morning newspapers and read that "coffee is linked to cancer of the pancreas," our immediate reaction is to destroy the percolator. But a statistical correlation of this type does not necessarily reflect a causal relation. To prove the point requires that the removal of the environmental agent, ideally without altering any other variables, leads to a genuine reduction in the frequency of a particular illness. While these criteria have been met in the cases of cigarette smoking and lung cancer and heart disease, they have not for almost all the other epidemiological clues that have been unearthed about possible causes of our problem diseases. All this is not to denigrate the role of epidemiology in shaping our thinking about medical care in the future. The way it has made us change the emphasis of how we look at disease has been of inestimable value, and we must continue to explore the origins of our problem diseases by means of the methods it has evolved.

But the central question remains. To what extent can we hope to control our major killers and sources of chronic ill health in the future by modifying our surroundings and lifestyles? Before we address this very difficult problem, we need to look further at the origins of our current killers and the relative roles of nature, nurture, and aging in their genesis. And we must reexamine some of the major diseases of the developing world, and ask the same questions.

Nature, Nurture, and Aging:
A Closer Look at the Origins of
Our Current Killers

Nothing in biology makes sense except in the light of evolution.
—*Theodosius Dobzhansky (1900–1975)*

The plea for a change in emphasis in modern medicine, away from the study of disease mechanisms and toward the control of our lifestyle and environment, urged so strongly by writers like McKeown, is being echoed by international health agencies, by governments of industrialized countries, and by many doctors themselves. Even in the major teaching centers in the United States, and certainly in Europe, an uneasy feeling prevails among practicing clinicians that although the explosion of knowledge in the basic biological and medical sciences is intellectually satisfying, it seems to be doing very little to alter day-to-day clinical practice. Perhaps Macfarlane Burnet and others like him were right and laboratory research has had its day.

Why did Burnet, who had done so much to develop modern immunology and to lay the ground for some of the real successes of scientific medicine, finish his career with such a gloomy prognosis for the future of laboratory science? Peter Medawar, in a rather self-revealing attempt to solve this conundrum, has written, "The reason he took this view, I believe, is that Macfarlane Burnet was formerly, as I was, the head of a large medical research institute devoted to 'basic' medical research and that he was dismayed as I was at the fact that so many members of his staff were more intent upon enlarging their own reputations as 'pure scientists' than in engaging directly upon the study of medical problems."

I doubt if Burnet's reasons for predicting the demise of laboratory science in medical research were as simple as this. It seems unlikely that either Burnet or Medawar spent most of his marvelously productive years in search of projects that always had an immediate relevance to medical practice. Rather, I suspect that Burnet was disappointed when, toward the end of his career, and after the control of infectious disease and the development of tissue transplantation, he saw little further immediate practical benefit from immunology and other basic sciences. A few years ago I was fortunate enough to care for Nikolaas Tinbergen, the distinguished scientist who shared the Nobel Prize in 1973 for his work on animal behavior. Tinbergen was similarly disheartened about the future of his field and bitterly regretted its lack of impact on human biology and medicine. Arguments that the human behavioral sciences are very much in their infancy, and that it might be many years before lessons learned from observational studies on birds would have significant implications for understanding abnormal human behavior, were of no avail. I suspect that Burnet's rather jaundiced view of the future role of the basic sciences in medicine stemmed from similar feelings of frustration and impatience. Not surprisingly, it is often difficult to accept that our work, while of no immediate practical use, might be of great value in a hundred years' time. Yet, as we have already seen so often in this book, this is the way that medical science has evolved over the centuries.

How have the champions of the laboratory and basic medical sciences responded to their critics? In 1980 Peter Medawar wrote a long and detailed review of McKeown's *The Role of Medicine* for the *New York Review of Books*. Although it is clear that he was not enamored of McKeown's thesis, and found some of the arguments lacking in factual basis or logic, he did not denigrate the importance of the control of environment in relieving disease and suffering. But Medawar did not believe that the neat division of diseases into those that are part of our genetic makeup and those that are due mainly to our environment or lifestyle is valid. And he thought that McKeown had very much underplayed the role of laboratory science in the control of infection. He ended his review by predicting that the basic biological sciences would, within the next ten years, provide remedies for multiple sclerosis, juvenile diabetes, and several forms of cancer that, in 1980, were all intractable. There is little to be learned from the fact that they are still intractable in 1993, except perhaps to reflect that it is reassuring for ordinary mortals that even Peter Medawar could get it wrong occasionally.

Lewis Thomas, another writer who has dwelt at length and with great elegance on modern scientific medicine, is also critical of the writings of Illich and McKeown. Like Medawar, he has no quarrel with the epidemiologists' notion of improving environmental conditions in order to control some diseases, and he does not deny that infectious diseases were becoming much less common before the appearance of vaccines and antibiotics. But he paints a vivid picture of the limitations of medical practice before the Second World War and how things were completely changed by these advances. Like Meda-

war, too, he leaves us in little doubt about the importance of the basic biological sciences and clinical science in the developments that led up to the
antibiotic era. In summarizing his views on the future way forward for medical science, Thomas again draws on the example of immunization and concludes, "The deeper our understanding of a disease mechanism, the greater
are our chances of devising direct and decisive measures to prevent disease,
or to turn it around before it is too late."

Convincing evidence for the value of clinical science in the early part of
this century has been collated by the American physician Paul Beeson, whom
we met in chapter 4. In 1980 he published an analysis of the changes that
had occurred in the management of important diseases between the years
1927 and 1975. He based his study on a comparison of methods for the
treatment of common diseases in the first and fourteenth editions of a leading
American textbook of medicine. Some of his findings are summarized in
figure 31. He found that, of 181 conditions for which there had been little
effective prevention or treatment in 1927, about 50 could be managed satis-
factorily by 1975. Most of these advances seemed to have stemmed from the
fruits of scientific and clinical research directed at the understanding of disease mechanisms. What pleased Beeson particularly was that treatment he
had scored as useless or harmful in 1927 had been eliminated in no fewer
than 74 instances. Lord Butterfield, formerly Regius Professor of Physic at
Cambridge University, has also vigorously defended modern laboratory-
based scientific medicine, in this case by analyzing the practical applications
of the work of those who won the Nobel Prize in physiology and medicine
between 1923 and 1971. The list of discoveries is impressive: insulin; vitamins; a cure for pernicious anemia; penicillin; the hormones of the adrenal
gland that are used for treating a wide variety of previously fatal diseases;
streptomycin for the control of tuberculosis; and so on.

We could go on arguing about the importance of a changing environment
compared with that of scientific advances in the laboratory and clinic in
improving our health over the last fifty years. But the issues are extremely
complex, and even those who have played a central role in the debate are not
always consistent. Macfarlane Burnet, for example, after seeing off laboratory
research as a way of controlling common diseases in 1972 in his book *Genes,
Dreams and Realities,* made a very strong case for the role of basic science in
the pursuit of knowledge for its own sake and for the betterment of mankind,
in his William S. Paley Lecture on Science and Society, given at the Cornell
Medical Center, New York, in 1980.

However, as we saw in the preceding chapter, the debate has been
extremely valuable because it has raised a very important question for the
future. Given limited resources, and what we have already learned from the
epidemiologists, should we be concentrating our efforts, both in teaching

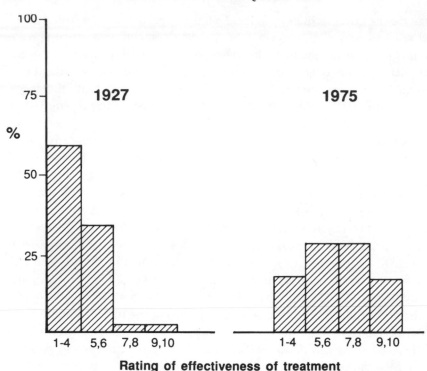

Rating of effectiveness of treatment

FIGURE 31

An attempt to rate the effectiveness of treatment between 1927 and 1975. Scales 1–4 indicate completely useless therapy, while 9 and 10 represent either complete prevention or cure of an illness. Modified from Beeson (1980).

medical students and in medical research, more on controlling our environment and lifestyle and less on laboratory studies that try to understand the causes and mechanisms of our common killers? To address this question we must look more closely at our problem diseases.

WE HAVE NOT YET CONQUERED INFECTIOUS DISEASE

As we explore the origins of our current diseases, we will often be looking at them in their evolutionary context. The genetic makeup of all living things is constantly changing as the result of alterations in its structure, or mutations. This is the mechanism by which Darwinian evolution works. In essence, mutations that provide an organism with an advantage in particular environmental conditions make it more likely to survive to produce offspring that are similarly advantaged.

This pattern of evolution is a major factor in our difficulties in controlling infectious diseases. We now have many powerful antimicrobial drugs that can attack bacteria in a variety of different ways, most of which involve interfering with particular chemical pathways. But once antibiotics are introduced into a population, they provide a strong selective pressure for the emergence of genetically variant strains of bacteria that have evolved subtle differences in their chemistry that enable them to bypass the damaging effect of the drug. Although microorganisms develop antibiotic resistance at widely varying rates, sooner or later they always do. This is why serious infections are monitored by testing the organism at regular intervals for its spectrum of antibiotic sensitivity. Sometimes the problem can be overcome by resort to mixtures of antibiotics that attack different chemical pathways in the bacteria. However, because of this constant pattern of evolutionary change in response to the selective pressures of antibiotics, or the emergence of virulent strains of microorganisms as a result of new mutations, we will never completely conquer our infectious killers.

Our attempts to control infectious disease are also hindered by changes in populations' resistance to attack by microorganisms. This may reflect poverty and overcrowding, or it may result from reduced immunity due to new forms of medical treatment or the emergence of diseases like AIDS.

In effect, therefore, we are faced with the resurgence of some "old" infections and the constant threat of the emergence of new infective agents. It appears that North America and other countries are now encountering a form of tuberculosis that is resistant to all our current drugs. This may be the result of poverty and the improper use of antituberculous drugs, together with highly susceptible populations of people with AIDS. Whatever the cause, it is a clear reminder that we can never be confident about our ability to treat an infectious disease even if we possess a whole battery of drugs to which it is usually sensitive. The emergence of drug-resistant strains of malaria offers a similar story. And recently the news has broken that there is a virulent, fast-spreading new strain of the organism that causes cholera, the "Bengal variant." In 1993 it was responsible for 110,000 cases of cholera in Dhaka, mainly among adults. The World Health Organization is already convinced that it will be controlled only if a vaccine is produced.

Over the last few years a new chapter in our understanding of infections that attack us has begun. It turns out that at least a third of the world's human population is infected by an organism called *Helicobacter pylori*. This organism infects us early in life and, by some amazing adaptive arrangements, manages to live in the linings of our stomachs for many years. This is no mean feat when one remembers that the bacterium spends its existence bathed in hydrochloric acid. Many of us manage to live in reasonable harmony with *H. pylori,* but it appears that some of us are not so fortunate. In many affected persons there is a low-grade inflammation of the stomach that may lead to what is called atrophic gastritis, that is, a slow destruction of the lining of the stomach. In a small subset of those affected in this way cancers of the stomach develop. Another, much more common complication of this infection is the development of ulcers of either the stomach or the duodenum. And, even more remarkably, other rare types of tumor may form in the stomach.

These new findings raise all sorts of important questions. Why is it, for example, that while so many people are affected with *H. pylori* only a small proportion of them develop stomach ulcers or cancer of the stomach? Although it seems likely that both genetic and other environmental factors are involved, as yet there is very little information about what they might be. And of course there is a broader issue. If chronic diseases like ulceration or cancer of the stomach can be caused, even if indirectly, by a long-standing infection by an organism that has learned to live with its host over many years, how many other chronic diseases for which the cause is unknown are produced in a similar fashion? Perhaps persistent infection with *H. pylori* is a paradigm for many other "slow" microbial diseases in human beings.

The recent production of a successful vaccine for a serious form of hepatitis, hepatitis B, which causes a major public health problem in many countries and which plays an important role in the genesis of primary liver cancer, serves as a good example of how modern science can be invaluable for the control of infectious disease. It seems unlikely that this disease would ever have been controlled by public health measures. Similarly, the spread of the Epstein-Barr virus, which causes an innocuous disease in Western societies but which is an important player in the generation of cancers of the nasal passages and lymph glands in Southeast Asia and Africa, will certainly not be controlled by public health measures. Again we are close to developing a satisfactory vaccine that could play a key role in the prevention of cancer in high-risk populations.

We still have very few effective antiviral agents and no cure for the common cold or influenza. In some parts of the world viruses that attack the eye or the brain pose a major public health problem. And, like bacteria and parasites, viruses are constantly changing their genetic makeup and keep appearing in new and more virulent forms. It is difficult to imagine that these diseases can be controlled by public health measures, and we will thus need to discover better antiviral agents or vaccines if we are to make any progress toward their control. AIDS is a particularly good example of a disease that will undoubtedly require a combined attack by education and by the production of a vaccine or an effective drug, or both.

Perhaps by the time medical historians look back on the second half of this century, it will have become apparent that the one good thing that arose from the scourges of the AIDS epidemic was that we became less blasé about our ability to control infectious disease. At least our medical schools and research-funding bodies have been jolted into the realization that infectious disease is still a subject of major importance and that, particularly with our newfound ability to change people's resistance to infection and their environments almost at will, we will always be vulnerable to attack by microorganisms. The argument for pursuing a vigorous program in the basic sciences directed toward understanding the arcane world of microorganisms and the defense systems of those they attack is overwhelmingly strong.

THE EVOLUTION AND NATURE OF OUR NONINFECTIOUS KILLERS

It is when we start to think about the noninfectious killers of Western society that are now spreading to the developing world that the case against laboratory sciences starts to look even leaner. One of the weakest parts of the arguments of McKeown and others has been in their perception of the *relative* roles of nature and nurture—that is, inborn genetic factors and the environment—in the cause of these diseases. The attractive idea that they can be easily compartmentalized into those which are intrinsic—that is, either part of what we inherit or due to some chance accident during development—and those which result from environmental factors is a misleading oversimplification. To explain why, we must digress briefly and outline the nature of the relation between inheritance and disease, a topic we will look at in more detail in later chapters.

At the moment of conception we receive equal amounts of genetic information from each of our parents. This is carried on the twenty-three pairs of chromosomes that contain our genes and, hence, determine our complete genetic makeup; one of the pair is derived from each parent. Genes are the units of heredity. Their main function is to determine the structure of proteins. The wide diversity of living things reflects the existence of numerous varieties of proteins. The tough proteins of hair and skin and the hemoglobin that fills our red cells all consist of the same building blocks called amino acids; they differ in their properties only because the amino acids are arranged in different sequences in different proteins. Our genes contain the information that ensures that the amino acids for a particular protein are always in the right order. Human beings have approximately 50,000 to 100,000 genes. In essence, they are the blueprints that determine what we are and that control every one of the thousands of chemical reactions that allow us to function. There are over four thousand diseases that result from single defective genes. Their pattern of transmission follows the laws of inheritance discovered by Gregor Mendel in the middle of the nineteenth century.

His terms are still used; a dominantly transmitted disease requires only a single defective gene to produce an illness, while recessive disorders need two defective genes, one from each parent. In the latter case the parents who carry the single defective gene do not show any features of the disease and are as a rule completely healthy.

Most single-gene disorders are rare, but because there are so many of them, they constitute a considerable burden of disease. About 1 percent of all newborn babies have some kind of genetic defect, and inherited disorders are the third-commonest cause of death in childhood. Furthermore, it is not uncommon for an accident to occur during development that leads to a major abnormality of one of the pairs of chromosomes. Since this involves many genes, these chromosomal abnormalities often result in widespread developmental abnormalities. The best-known example of this kind of disease is mongolism. Other congenital malformations have a strong genetic component or may result from environmental damage during early development. Thus single-gene disorders, major chromosomal abnormalities, and congenital malformations are responsible for a great deal of illness and mortality in early life, and for a high proportion of disability, including mental retardation, in adults.

Because many serious single-gene diseases prevent reproduction, they are not passed on to future generations and hence tend to be selected out of populations. Thus, individually, they are not a major cause of mortality or morbidity. There are exceptions to this rule, however. Some recessive diseases, particularly the anemias that occur in some tropical populations, seem to provide symptomless carriers with an advantage over the rest of us. Inherited diseases of this type become a serious problem in developing countries once social conditions start to improve and the high mortality rates due to infection and malnutrition are controlled.

Clearly, inherited diseases form an extremely important part of medical practice in our industrialized societies and are starting to become problems in parts of the developing world. Since the environment has a very limited role in their genesis, we will have to attack them by the application of modern scientific medicine, another theme to which I will return later.

But the importance of our genetic makeup does not, as McKeown and others suggest, stop at the single-gene disorders. The role of nature as well as nurture in the genesis of common disease was recognized by Hippocrates and has been the subject of much speculation over the centuries. It was discussed in 1932 in a brilliant but neglected book by the English physician Archibald Garrod (1857–1936) (figure 32). This work, *The Inborn Susceptibility to Disease,* is less well known than Garrod's first book, *Inborn Errors of Metabolism,* which established the idea that many single-gene disorders are due to defects in the enzymes involved in critical biochemical reactions. In *Inborn Suscepti-*

FIGURE 32

Sir Archibald Garrod at the time of his retirement as Regius Professor of Medicine in Oxford in 1927.

bility, published just before he died, Garrod made a very strong case for the importance of genetic susceptibility and resistance to many of the common diseases of Western society, particularly infection and "degenerative" disorders.

Those who followed Garrod have attempted to dissect the relative roles of inheritance and environment in the genesis of these diseases. One of the most rewarding approaches is to study the prevalence of disease in identical twins, particularly if they were reared apart, a circumstance that overcomes the problem of a common environment. If both of a set of twins show the same genetic trait, they are said to be concordant. It has been found that concordance rates vary widely. It appears, for example, that the especially common form of diabetes that occurs in middle or later life, maturity-onset diabetes, has a very strong genetic component, while heart disease, high blood pressure, and others seem to be the result of both genetic and environmental factors.

Although the replication of our genetic material from generation to generation is remarkably accurate, changes, or mutations, occur from time to time, and our genes alter their structure. Furthermore, one of the less obvious advantages of sexual reproduction is that during the formation of gametes— that is, sperm or eggs—the chromosomes we inherit from our parents become closely opposed to each other, and hence our genes can move from one to the other, a process called recombination. Thus, over many generations, populations change. Just as is the case in microorganisms, these are the genetic mechanisms that underlie Darwinian evolution in humans; by mutation, or gene shuffling, fitter individuals appear now and then. In this sense "fitness" means a greater ability to adapt to their environment and, hence, to survive and disseminate their more advantageous genes.

Why should our genetic makeup render us more or less susceptible to many of the common diseases of Western society? It is undoubtedly a reflection of our immense genetic diversity and the complex pattern of our evolutionary history. As we just mentioned, our genes exist in pairs, one from each parent. Genes that reside at the same place, or locus, on a pair of chromosomes are called alleles. If the genes at these positions are identical we are said to be homozygous, whereas if one of these genes differs from the other we are called heterozygotes. It turns out that on the average we are heterozygous at about 6.7 percent of our gene loci. At first this does not sound like much genetic variability. However, since we have up to 100,000 genes, it follows that 6,700 of them have different alleles. That means there are 2^{6700}, or in base ten arithmetic about 10^{2000}, potential new combinations of genetic material that we can pass on to our children. This figure is so large that it is impossible to comprehend. John Gribben, the popular science author, has tried to put it into perspective for us. He points out that the astronomer

Arthur Eddington calculated that there are about 10^{80} protons and neutrons in the universe. We therefore have a veritable astronomic propensity for mixing our genes.

The timescale of our evolutionary history is given in figure 33.We spent at least 900,000 years, or approximately 50,000 generations, wandering around in nomadic bands of hunter-gatherers. Agriculture and the formation of settled villages started about 10,000 years ago, and towns and irrigated agriculture have existed for about 5,000 years. We have been exposed to our current sedentary existence, smoke-filled rooms, and high-energy diets for a mere few hundred years. During each of these phases of evolution we were exposed to completely different environmental hazards. The low life expectancy of the hunter-gatherers may have been partly the result of infectious disease, but such evidence as there is suggests that accidental and traumatic death, together with the activity of animal predators and social mortality in the form of infanticide, cannibalism, and warfare, may also have been important contributors. The growth of agriculture is thought to have led to the emergence of many of the infectious diseases still with us today, including parasitic, viral, and bacterial killers. Once we started to huddle together in insanitary and overcrowded towns, conditions became rife for major epidemics such as plague and smallpox that devastated entire populations. During this period the fact that many of the natural hazards, periods of starvation, and infections killed people before they reached reproductive age must have led to the selection of a variety of traits that were advantageous in environments completely different from those in which we live today.

In the light of this long evolutionary history it is not surprising that modern humans have not had time to adapt to the completely new environments in which they have found themselves in the last few hundred years. Our sedentary existence, increased levels of animal fat in our diet, exposure to new hazards such as tobacco and other social drugs, and an altogether new set of potential cancer-producing agents in the environment, taken together with a remarkable facility for controlling and changing the environment almost at will over a short period of time, may well be incompatible with a genetic makeup that was adapted over a million years to utterly different conditions.

In discussing the genetic components of disease in his book *The Origins of Human Disease,* McKeown quotes Macfarlane Burnet as follows: "The average genetic constitution of present-day persons is not greatly different from what it was a hundred thousand years ago, well before the advent of any form of pastoral or agricultural activity." Developing Burnet's argument, McKeown suggests that there would not have been sufficient chance in the two hundred or so generations from the beginning of our period as agriculturalists to the time when we started to gather in cities for many genetic changes to have occurred through mutation, though he points out that we do not know what

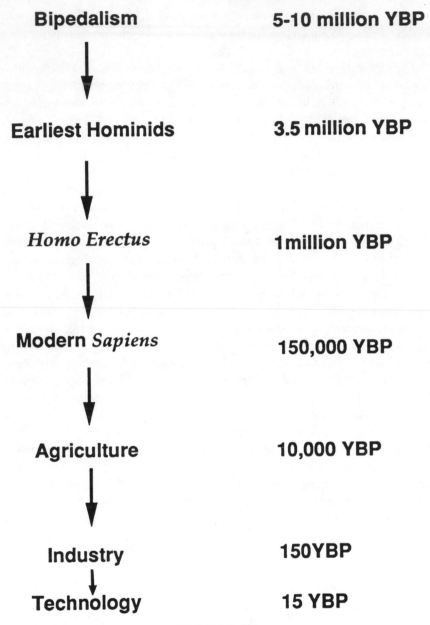

Bipedalism   **5-10 million YBP**

Earliest Hominids   **3.5 million YBP**

Homo Erectus   **1 million YBP**

Modern *Sapiens*   **150,000 YBP**

Agriculture   **10,000 YBP**

Industry   **150 YBP**

Technology   **15 YBP**

FIGURE 33

The timescale of human evolution.
(YBP—Years Before Present)

may have occurred through selection. He concludes his argument by noting that high mortalities in infancy and childhood may have led to genetically based resistance to certain infectious diseases but that it is unlikely that any other characteristics have been significantly affected. He concludes by quoting Burnet again: "As far as general bodily health and patterns of behaviour are concerned, one can feel reasonably confident that the genetic aspects of those qualities have not changed significantly since the days of Cro-Magnon man." The word "Cro-Magnon" is derived from the French sites where the remains resembling modern man were discovered. Cro-Magnons probably appeared in Europe some 40,000 years ago. If, McKeown concludes, there are reservations about the possibility of genetic change since the beginning of agriculture, they are certainly not required for the short period of industrialization, covering at the most 200–300 years and eight to twelve generations. In this time there can have been no significant variation in our genetic constitution, except possibly in relation to some infections.

No one would dispute the fact that the two or three hundred years in which we have been exposed to our increasingly artificial industrialized environments have given us little time for genetic adaptation. However, it is difficult to be so confident that, apart from some degree of susceptibility or resistance to infectious disease, the changes that have occurred in our genetic makeup in response to our early environments will be of no importance with respect to the hazards of our industrialized societies.

Apart from the removal of lethal or serious single-gene disorders, we have no idea what might or might not have been selected for during earlier periods of human evolution. It is likely, for example, that the traits that were particularly useful during our existence as hunter-gatherers are not the ideal ones for our gradual evolution to the sedentary coziness of Western civilization.

In our long evolution we have been exposed to extraordinarily diverse environmental hazards. During our period as hunter-gatherers any genetic alteration that improved our ability to cope with injury or infections prevalent at the time would have come under selection. There were periods of intense differences in temperature and times of extreme deprivation of food. When we changed our environments by the development of agriculture and, later, when we were crammed together in towns, we were subjected to a wide range of infectious killers. Over the last few thousand years numerous epidemics have wiped out entire populations. These periods of intense exposure to environmental hazards, whether physical or infectious, must have had a significant effect on our genetic makeup. This impact would certainly not have been restricted to variable susceptibility to infection but have encompassed every aspect of our ability to adapt to our environment. In the light of these diverse influences on our current genetic makeup, the question arises how far those traits that were selected for in the past have made us more or less susceptible to the hazards of the new environment we have created for ourselves over the last few hundred years.

GENETIC VARIABILITY THAT REFLECTS SELECTION BY DISEASE IN OUR EVOLUTIONARY PAST

Although the study of human population genetics and evolution is not a prominent item in the curricula of many Western medical schools, it no doubt has much to tell us about the origins of disease. Until recently one of the main difficulties in pursuing this field was that, when we attempted to interpret the frequency of different genes in populations in the light of changes in the environment that had happened hundreds or even thousands of years previously, many of the conclusions were based, at best, on inspired guesswork. However, our evolutionary history is written in our genes, and we now have the tools at our disposal to try to answer some of these questions. Indeed, several clear-cut examples already suggest that the patterns of genetic traits and inherited diseases in our present-day populations have been much modified by previous environmental hazards.

We saw in chapter 5 that some tropical countries in which childhood mortality from infection and malnutrition is declining are experiencing a high prevalence of inherited anemias in the population, particularly sickle-cell anemia and thalassemia. These diseases are part of a family of inherited disorders of hemoglobin, the pigment that fills our red blood cells and is responsible for transporting oxygen from the lungs to the tissues. Remarkably, over four hundred different inherited structural variants of hemoglobin have been discovered. At first these were named after the letters of the alphabet, but later, when no more letters were left, they were named after the place from which the first persons in which they were found had originated. There are only three very common variants, however: hemoglobins S, E, and C; normal hemoglobin is called hemoglobin A. Sickle-cell anemia results from the inheritance of the gene for hemoglobin S from both parents. The single amino acid difference between sickle-cell and normal hemoglobin causes red blood cells to form sickled shapes in parts of the body where the oxygen tension of the blood is low. This in turn causes the cells to be prematurely destroyed and to clog up small blood vessels, with consequent death of tissue. Thalassemia, on the other hand, results from a reduced rate of production of part of the hemoglobin molecule. Since normal human hemoglobin consists of two different pairs of protein chains, called alpha and beta, there are two main classes of thalassemia, alpha and beta thalassemia. Children who receive beta thalassemia genes from both parents are profoundly anemic from early in life and, without blood transfusions, die within the first few years. Babies that receive the alpha thalassemia gene from both parents are usually stillborn.

These inherited anemias are extremely common; the World Health Organization believes that by the year 2000 approximately 7 percent of the world's population will be carriers for inherited disorders of hemoglobin. Hundreds of thousands of children die of sickle-cell anemia in Africa each year (figure 34), and in Thailand alone there are between 500,000 and 750,000 children severely afflicted with thalassemia. Thalassemia occurs widely throughout the Mediterranean region, the Middle East, the Indian subcontinent, and Southeast Asia (figure 35). Hemoglobin E is confined to eastern India, Burma, and Southeast Asia. Conservative estimates suggest that about 30 million people in Southeast Asia are carriers. Homozygotes, persons who receive the gene from both parents, are mildly anemic, whereas carriers are completely normal. However, children who inherit beta thalassemia from one parent and hemoglobin E from the other are often profoundly anemic. The other common hemoglobin variant, hemoglobin C, occurs widely throughout West Africa and the Mediterranean. Like those for hemoglobin E, carriers for hemoglobin C are symptomless, while homozygotes are moderately anemic.

But if serious forms of sickle-cell anemia and thalassemia kill young children before the age of reproduction, why are they so common? Surely they should have been eliminated by natural selection. The British geneticist J. B. S. Haldane (1892–1964) in 1948 proposed an answer to this conundrum. As we have seen, all these diseases are inherited in recessive fashion, that is, they cause serious symptoms only if inherited from both parents; those who have a single dose, carriers or heterozygotes, are completely healthy. Haldane reasoned that these diseases might have flourished because symptomless carriers are protected against a major environmental hazard. Since the world distribution of the hemoglobin disorders mirrors severe forms of malaria (figure 36), Haldane made the remarkably astute suggestion that malaria might be the selective factor that had maintained these diseases at a high frequency. This turned out to be an elegant example of what geneticists call a state of "balanced polymorphism." Those who receive two doses of a defective gene (homozygotes), and hence have sickle-cell anemia or severe forms of thalassemia, die in early life. On the other hand, carriers, with only a single gene for these conditions, are partly protected against malaria. Thus a state will be reached in which deaths of homozygotes are balanced by increased numbers of heterozygotes in the population as a result of their protection against malaria; the frequency of the disease will be determined by a balance between deaths of homozygotes and the degree of protection afforded to heterozygotes.

Although it took many years to collect the experimental evidence to validate Haldane's hypothesis, there now seems little doubt that he was right.

HbE ▨ HbS

FIGURE 34

The world distribution of the genes for sickle-cell anemia and hemoglobin E.

Studies carried out in Africa forty years ago provided some evidence that the sickle-cell trait protects against malaria, but more recently it has been found that carriers for this condition may be favored by up to 90 percent protection against life-threatening malaria or against the serious anemia that is one of its consequences. Carriers (heterozygotes) for hemoglobin E or some forms of thalassemia are also protected, though not to the same degree as sickle-cell gene carriers.

How is such protection mediated? Malaria probably became a serious problem after the change to an agricultural way of life. There are several types, each due to different parasites. The most serious form in man, caused by a parasite called *Plasmodium falciparum,* follows a bite by an infected mosquito. The parasites enter the liver, multiply, and then are shed into the bloodstream, where they invade red blood cells. After further development the infected cells burst, and the parasites invade more red blood cells.

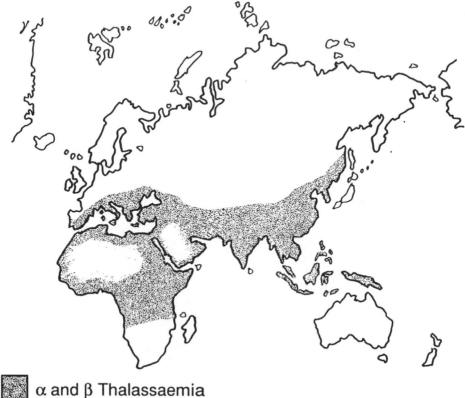

α and β Thalassaemia

FIGURE 35

The world distribution of the common forms of thalassemia.

Infected red cells adhere to the linings of blood vessels, particularly in the brain, and hence the disease is characterized by severe anemia due to destruction of red cells and by defects in the function of many organs, especially the brain and kidneys. For reasons still not clear, carriers for the sickle-cell or thalassemia genes are partly protected from the action of the parasite. Because the parasite does most of its mischief in the red blood cell, the structure of which is greatly modified in sickle-cell anemia and thalassemia, it is likely that the protection is mediated during the parasite's life cycle in the blood.

Remarkably, inherited variations in hemoglobin are not the only way in which genetic variability has been brought into action to protect against malarial infection. The red blood cell is a small dynamo that requires energy to maintain itself in the unfavorable environment of the circulation, particularly against noxious chemicals that tend to damage its hemoglobin or the membrane that surrounds it. A variety of chemical pathways have evolved to

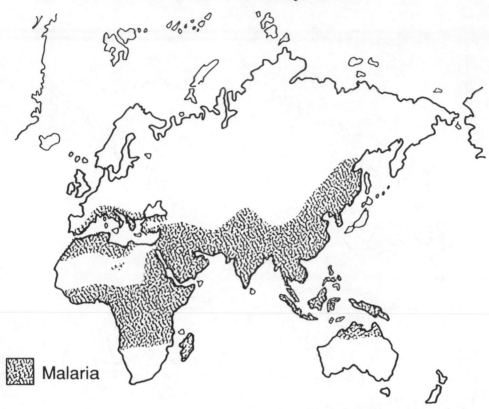

Malaria

FIGURE 36

Malaria in the Old World.

overcome these hazards. In one there is a key enzyme called glucose-6-phos-
phate dehydrogenase. It turns out that millions of people throughout the
tropical world have an inherited deficiency of this enzyme. Under normal
circumstances this may not be important, but they are highly prone to
destroy their blood cells after ingesting certain drugs or foodstuffs, especially
beans. Sensitivity to broad beans has been recognized since the time of the
ancient Greeks; the mathematician Pythagoras advised people not to eat
beans during certain seasons of the year. The catastrophic destruction of
blood cells that can follow the ingestion of beans, a condition called favism,
is widespread in the Mediterranean region and results from glucose-6-phos-
phate dehydrogenase deficiency. It appears that this enzyme deficiency is
particularly common in the tropical populations because affected persons are
more resistant than the normal population to malaria caused by *Plasmodium
falciparum*.

But genetic variability that reflects natural selection working through protection against malaria does not end with these common disorders of hemo-globin or the chemistry of our red cells. For example, there are large populations living in Papua New Guinea and nearby regions that, instead of having the usual round red blood cells, have oval ones, called ovalocytes. Remarkably, it turns out that the change in the proteins of the red cell membrane that is responsible for their unusual shape makes them resistant to invasion by malarial parasites. And protection against malaria also seems to be responsible for modifying the distribution of blood groups among human populations. Differences in human blood groups are under genetic control, and the products of the particular genes involved are expressed as complex proteins, called blood group antigens, on the red cell surface. One antigen, called Duffy, is present on the red cells of many populations but is absent in those of certain parts of Africa. It happens that this protein is involved in the complex mechanism whereby parasites for milder forms of malaria gain entry into red blood cells. Its absence in these African populations results from selection; it would have been advantageous not to have the Duffy blood group antigen in areas where this form of malaria was common.

It is interesting to reflect on the extraordinary genetic diversity of the red cell that has resulted from natural selection working toward the protection of populations against malaria. First, there are hemoglobins S, C, and E. Recent studies suggest that hemoglobins S and E may have arisen more than once during our evolutionary past. And then there are over two hundred different forms of thalassemia and many mutations that cause defective production of glucose-6-phosphate dehydrogenase. Similarly, there are numerous inherited variations in the membranes or blood group antigens of red cells, at least some of which protect against malaria. The mutations that cause thalassemia in the Mediterranean are completely different from those that produce the disease in India or in Southeast Asia. Thus hundreds of different mutations involving the red blood cell alone have been magnified in different populations by selection against malaria. In short, we seem to have made use of every conceivable way of modifying the structure of our red cells in our fight against this one serious infection.

And the story of malaria protection does not stop at the red cell. Recently it has been found that selection by malaria may be responsible for major differences in the distribution of genes for critical functions of our immune systems. A particular family of genes of this type, called histocompatibility genes, play a major role in our body defenses against invading organisms. They have this unlikely name because they were first discovered as the chief players in the body's rejection of foreign tissue grafts. Clearly they were not sitting around throughout evolution waiting for transplant surgeons, and it turns out that the products of these genes play a key role in presenting foreign proteins—viruses, for example—to the cells of our immune systems that participate in their destruction. Recent studies in The Gambia have shown that certain variants of the histocompatibility genes play a major role in the protection of young children against death or severe anemia as a result of malarial infection.

The modification of the genetic makeup of human populations by present or past exposure to malaria is one of the best-understood examples of how natural selection works. It is clear that the uneasy relation between the malarial parasite and its host over no more than a few thousand years has had a profound effect on our genetic constitution and has left us with a variety of common genetic blood diseases and with modifications of our immune system, the consequences of which are not yet understood.

What we have just been considering is, of course, an extreme example of natural selection working through the red blood cell that, because of its accessibility, has been relatively easy to study. But such adaptive changes must have been occurring throughout evolution, and much of the extraordinary genetic diversity of the different races of today results from selection for adaptation to different environmental conditions. Current evolutionary theory suggests that *Homo sapiens sapiens*—that is, modern man—evolved from a population somewhere in Africa about 150,000 years ago. It seems quite likely that our present races are the progeny of a series of migrations from Africa that began about 100,000 years ago. During the most recent glacial period, about 100,000 years ago, most of the world's surface was covered by ice (figure 37). It is probable that these conditions led to the separate evolution of whites in the west, mongoloids in the east, and blacks in the south.

Why are whites and mongoloids relatively less pigmented than their African forebears? One possibility is that their pigmentation reflects adaptation to less sunshine and ultraviolet irradiation in the habitats of these two races. Ultraviolet light is necessary for the chemical reactions in the skin that generate vitamin D, which is essential for the calcification and strengthening of bones. A deficiency of this vitamin results in ricketts, one of the important manifestations of which is bone deformity, particularly involving the pelvis. This impairs normal childbirth and would have caused the death of both mothers and infants living under primitive conditions.

The importance of our ability to calcify our bones is manifest in another common example of genetic variability between populations. There is a human blood protein that can exist in different forms, called Gc^1 and Gc^2. Other varieties of this protein have been found in different parts of the world. The Gc proteins transport vitamin D in the blood. It turns out that in the aboriginal habitats of the world high frequencies of Gc^2 are found in populations that have been living for a long time in areas with a low density of sunlight. Since Gc^2 appears to be a much more effective transporter of vita-

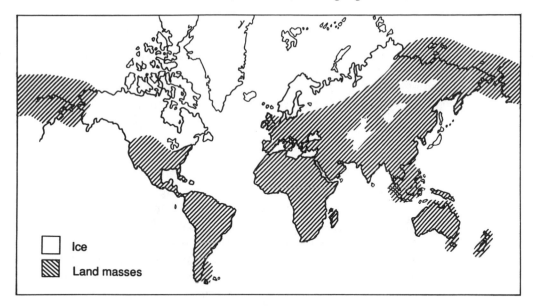

F I G U R E 37

The last ice age. Modified from Lewin (1989).

min D, this seems to be an adaptive response to a limited supply of the vitamin.

The major human blood groups, O, A, B, and AB, are distributed unevenly among world populations. For example, the Indians of Central and South America have a very high frequency of group O, which sets them apart from other populations. The pattern of other genetic markers does not suggest that these areas were once inhabited by a homogeneous population with a high group O frequency. It is possible, therefore, that blood group O reached its present frequency in this part of the world by natural selection. The population of Central and South America was completely isolated before Christopher Columbus arrived and may well have had infections not shared with the rest of the world. One candidate, thought to have been brought back to Europe, is syphilis. The German geneticist Friedrich Vogel has obtained tentative evidence that, under some circumstances, group O may be protective against syphilis. These studies are extremely difficult to carry out in these days of penicillin treatment and were based on comprehensive data on blood groups and syphilis that were collected in the 1920s, when the disease was being treated with arsenical derivatives. Although the risk of catching syphilis did not seem to be related to a particular blood group, there was a reasonable correlation between the likelihood of developing advanced syphilis and group O.

And speculations about the effects of major epidemics and the distribution of human blood groups go much further than this. For example, it is possible that the very low frequency of blood group O in the ancient plague centers of Mongolia, Turkey, and parts of Egypt reflects the susceptibility of people with this group to the organism that caused the scourges of the Black Death in these populations. Similarly, it has been suggested that outbreaks of smallpox epidemics may have been responsible for the high frequency of blood group A in some populations. And so on.

If different blood groups are related to susceptibility or resistance to infection, how might such an effect have been mediated? To address this difficult question we must digress briefly to consider the structure of blood groups. As we have seen, particular blood groups are determined by the structure of substances called antigens, which are expressed on the surface of red blood cells. A blood group is determined by the particular structure of these antigens; group A persons have A antigen, group B have B antigen, group AB have both A and B, and group O have neither. In its broader usage the word "antigen" is used to describe a protein that, when injected into an individual who has not encountered it before, produces a defense reaction characterized by the production of chemicals called antibodies. The reason that we have to have our blood matched with that of a donor before receiving a transfusion is that persons of group A have antibodies that destroy group B cells, those of group B have antibodies that destroy the red cells of group A, those of group AB have neither type of antibodies, while those of group O have both types of antibodies.

The structure of our blood group antigens is very closely related to that of substances found on the surface of many different microorganisms. Although the anti-A and anti-B antibodies that occur in the blood of people of blood groups B or A are said to be "naturally occurring," they are not present at birth, and it is believed that they develop as a response to the ubiquitous microbial antigens to which we are exposed in early life. Thus it follows that there are very close links between our blood groups and the structure of our bacterial forebears, and hence the idea that our blood groups may make us more or less prone to bacterial infection is not so wildly improbable.

A good idea is one thing, but is there any evidence in favor of these interesting speculations? In the 1950s and early 1960s central Europe encountered successive waves of infective diarrhea in infants caused by different types of an organism called *Escherichia coli*. Early studies in Austria demonstrated that the course was more severe in infants of blood group A than in those of other groups. Even more interestingly, when Friedrich Vogel and his colleagues extended this research over a longer period, they found that in some years group A infants were more frequently affected and that in other years group O carriers were in excess. They also found that antibody levels in the ABO blood group system against certain strains of *E. coli* varied considerably in individuals of different blood groups.

Although work of this type has been neglected in recent years, a similar story has emerged more recently in the case of susceptibility to bacterial meningitis, recurrent infections of the urinary tract, and skin infections due to fungi. It has been known for many years that some individuals secrete blood group antigens into their body fluids, while others do not. Whether we are "secretors" or "nonsecretors" is determined by a particular gene. It turns out that there is a major difference between secretors and nonsecretors of ABO blood group antigens with regard to their susceptibility to meningitis and infections of the urinary tract. It is believed that this is because blood group antigens in body fluids bind microorganisms and hence reduce their colonization on the surfaces of the body.

There are also hints that some of the astonishing genetic heterogeneity of the histocompatibility gene family to which we have already alluded may have been mediated through a variation in response to infection. Associations with particular genetic subtypes of this family have been found with many infectious diseases, including malaria, hepatitis B, poliomyelitis, tuberculosis, leprosy, and AIDS. As might be expected, genetic variability in response to selection by infectious disease is not confined to human beings. For example, mice, whose breeding habits make it possible to carry out extensive genetic studies, have several well-defined genes that alter susceptibility to infection, both bacterial and parasitic. And, as in humans, the histocompatibility genes seem to be major players in this respect. Since bacteria predated most other species, it seems likely that a great deal of genetic diversity in living things has arisen by selection in response to the diseases that they cause.

In recent years further evidence that our genetic makeup has been greatly modified by exposure to infection has come from several disparate sources. For example, there is a condition called Tay-Sachs disease, which is particularly common in certain Jewish populations. This disease, which, like the genetic anemias, is inherited in a recessive fashion, kills babies early in life by causing a serious neurological disability. By a remarkable piece of detective work, the high frequency of this condition has been ascribed to the protection it afforded to symptomless carriers against tuberculosis, which was rife in the ghettos of eastern Europe at the beginning of this century. Similarly, about one in twenty northern Europeans carries the gene for cystic fibrosis, another serious genetic disease, which is characterized by the production of thick, viscid mucus and chronic respiratory infection and bowel disturbance. It turns out that although there are many different mutations that can cause cystic fibrosis, in northern Europeans only one particular mutation predominates. The distribution of this gene, and its high frequency, suggests that a similar protective mechanism may have been involved, possibly carrier resistance to one of the major epidemics that decimated large European populations many years ago.

This neglected field will undoubtedly be revived in the next few years because, as we will see, much better techniques are now available to study some of these problems. However, from what little is known it is already clear that our genetic makeup has been considerably modified and adapted to our life in the world of unfriendly microorganisms. The most extreme example is the wide variety of very common single-gene disorders that have been maintained at a high frequency throughout tropical populations by selection against malaria. But this must be only the tip of an enormous iceberg of genetic variability. And of course we have considered only adaptation to infection. If selection has worked in this way, it must have been a major factor in modifying our responses to other environmental hazards, such as extremes of temperature and periods of starvation, and our ability to sustain trauma. The question, therefore, is, to what extent has our present genetic makeup, based as it is on thousands of years of adaptation to infection and a wide variety of physical hazards, left us more or less prone to the completely new insults that we have met in our few hundred years of life in modern industrialized societies?

Are We Paying the Penalty for Our Genetic Successes of the Past?

Genetic makeups that protected us against epidemics of infectious illnesses or other environmental hazards in the past may not be suited to the quite different conditions of today. Indeed, some of those selected in this way may leave us more likely to develop some of the diseases of Western society as we age. We have just seen how the distribution of human blood groups may reflect natural selection working through susceptibility or resistance to common infections. Those of us with blood group A have a 20 percent greater risk of developing cancer of the stomach than those who are blood group O. Group O individuals, on the other hand, are more prone to duodenal ulcer. It seems equally likely that traits for more effective blood clotting and so on, which may have been a distinct advantage in the bush, are just those that made us more prone to thromboses and other vascular events as we have changed our lifestyles during the last few thousand years. As we have just seen, there is increasing evidence that particular variants of the genes that regulate our immune system may have been selected by previous exposure to infection. Could these same variant genes be responsible for increased susceptibility to autoimmune diseases like diabetes and arthritis? And is it possible that a genetic makeup that was advantageous in the past in conditions in which food supplies were intermittent, and where shortages led to starvation, and relative abundance to rapid weight gain, are just those that are likely to increase the possibility of obesity, diabetes, or vascular disease in later life?

It has been known for many years that Americans of African origin are much more likely to develop high blood pressure than those of European origin. There is also evidence that the way in which the body handles salt may play a role in the control of blood pressure. It has been suggested that African Americans who retain salt more avidly than European Americans were selected for because of the limited availability of salt, either in Africa, while they were being transported to America, or during their subsequent lives as slaves. Recently, the American biologist Louis Miller has proposed quite a different explanation. He points out that the red blood cells of African Americans contain higher levels of sodium, and lower levels of potassium, than those of European Americans. He suggests that these changes, and related differences in the way in which sodium is transported in and out of red blood cells, may be genetically determined and may have offered protection against severe malaria in the past. In support of this idea Miller reminds us that malaria is not seen in animals such as dogs, cats, cows, and horses, all of which have red blood cells with relatively high levels of sodium and low levels of potassium. On the other hand, rodents, birds, and many primates, all of which have red cells with relatively high levels of potassium, are all prone to different forms of malaria. In short, genes that predispose to high blood pressure may be just those that suppress the growth of malaria parasites and hence provide a survival advantage in areas where malaria infection is common.

But how can we investigate these interesting speculations further and find out whether they have any relevance to a better understanding of our current problem diseases?

We have seen how studies that compare the prevalence of diseases in emigrant populations with their prevalence in those who stayed behind have been of great value in developing the thesis that our environment is the major player in the generation of the common killers of Western society, especially heart disease, diabetes, stroke, and cancer. While much of this epidemiological work is compatible with this notion, there are some serious inconsistencies. In particular, it is becoming apparent that the response of different populations to "Westernization" varies enormously, suggesting that genetic factors play an important role in modifying the effects of environmental insults. To illustrate these complex issues, let us briefly examine the pattern of illnesses in some population groups that have changed their environments fairly recently.

Unusually high rates of coronary artery disease were first noted in Asian populations from India, Pakistan, and Bangladesh that were living in Singapore, South Africa, and the Caribbean in the 1950s. Similar observations were first recorded in Great Britain in the 1971 census. It has been difficult to obtain reliable information about the frequency of heart disease in the countries of origin of these emigrants, but some small surveys in Indian cities have reported the prevalence there to be at least as high as that in European populations. However, between 1979 and 1983, mortality from coronary artery disease was 36 percent higher among Asian immigrants in England and Wales than it was among the general population for men born in Asia, and 46 percent higher for Asian-born women, in the age group twenty to seventy years. This pattern seems to repeat itself wherever these populations have settled.

Not surprisingly, epidemiologists have had a field day trying to sort out the reason for these striking differences. However, population surveys have shown that risk factors in Asians, including smoking, blood pressure, blood cholesterol levels, or studies of blood-clotting activity, do not explain this greatly increased prevalence of heart disease. In fact, blood cholesterol levels are similar to or lower than the national average in all the main groups of Asian immigrants.

The other observation that sets these Asian immigrant populations apart from the rest of the countries in which they have settled is the very high frequency of non-insulin-dependent diabetes. This disorder is found in about 20 percent of all Asian men and women over the age of forty years in the United Kingdom, compared with about 5 percent in the indigenous population. In effect, the disease is assuming epidemic proportions in this racial group. Similar prevalence rates have been observed in other overseas Asian populations and in some urban populations in India. Although it is known that a high frequency of vascular disease is associated with this type of diabetes, many Asian patients with coronary artery disease are not diabetic, and therefore intolerance to glucose cannot explain all the excess of coronary artery disease in Asian people. Thus it seems likely that the high prevalence of diabetes is merely one manifestation of a syndrome of physiological disturbances related to resistance to insulin. We will return, later in this chapter, to the important question and how the syndrome may predispose to heart disease, diabetes, obesity, and raised blood pressure But, regardless of the mechanism, it seems an inescapable conclusion that these Asian populations have a strong genetic component in their susceptibility to these important "Western" diseases.

For an even more remarkable example of the unusual distribution of disease in a migrant population group, we turn to the constellation of disorders to which anthropologists have given the name "New World syndrome." Although there has been much debate about the origins of the first Americans, it is now generally agreed that they came from Asia across the Bering Strait, which separates Alaska and Siberia. The exact timing is still in dispute.

Some believe that the major migration occurred about 12,000 years ago. Archaeological evidence suggests that the Americas were populated 11,500 years ago, although there may have been population movements into the Americas as long as 30,000 years ago. Whenever they came, these travelers populated both North and South America; by the time Columbus arrived, over a thousand different languages were spoken among the native Indian peoples. The linguist Joseph Greenberg has analyzed the six hundred languages that survive and, astonishingly, has traced them back to just three: Amerind, the most widespread, Na-Dene, and Aleut-Eskimo. Greenberg suggests that these groups represent three separate migrations, and there is evidence that the Eskimos arrived later than some of the other indigenous populations. Over the last few hundred years the Amerindian descendants of the early settlers have been exposed to conditions created by Western industrialization, but their pattern of diseases is quite different from that of the European settlers with whom they are sharing their environment.

The New World syndrome is characterized by a very high prevalence, reaching epidemic levels, of obesity at an early adult age, insulin-resistant diabetes, and gallstones and gall bladder cancer, especially in females. It appears that this epidemic began, or at least spread dramatically, after the Second World War. The anthropologists who describe this extraordinary constellation of diseases in this racial group propose that it reflects a high frequency of genes that cause susceptibility to environmental agents associated with Westernization that arose by virtue of selective advantage during or before the initial peopling of the Americas.

The distribution and frequency of diseases that make up the New World syndrome are so unusual as to leave little doubt that genetic factors are involved. The Pima Indian people have a nineteenfold greater incidence of insulin-resistant diabetes than prevails in Caucasian populations in other parts of the Americas. In fact, this has given the Pima the dubious distinction of having one of the highest-reported incidences of diabetes anywhere in the world. An increased frequency of this type of diabetes has been found in almost all North American aborigines, with the exception of Eskimos. It does not matter from which tribes they have arisen or where they live; this implies that the environment is insufficient to explain the high frequency of the disease. An increased frequency of diabetes is also seen in hybrid Amerindian-admixed Hispanic populations.

Although descriptions by travelers, paintings, and early photographs of Amerindians during the nineteenth and early twentieth centuries suggest a lean and healthy population, there is now a remarkably high frequency of obesity in early-adult Amerindians, a phenomenon that seems to have become particularly common over the last fifty years. In the Pima, and in populations elsewhere, there is a strong correlation between obesity and diabetes. Similarly, many surveys have shown a very high frequency of gallstones in these populations. In the Pima, for example, the prevalence of gallstones reaches 90 percent in women by the age of sixty-five years. Furthermore, the chemical composition of these stones is different from that of the stones in Caucasians in the Americas. There seems to be a strong relation between the high frequency of gallstones and the acquisition of cancer of the gallbladder, a relatively rare disorder in other populations in the Americas.

The relation between the New World syndrome and Westernization. shows another interesting feature. As we mentioned previously, increased rates of obesity, diabetes, and particular kinds of gallstones are all recognized as the consequences of our changed lifestyles. Similarly, our high carbohydrate/low fiber diets and sedentary existences have been implicated in increased rates of coronary artery disease and of certain cancers, including those of the breast, uterus, colon, and prostate. Both in migrants and in Westernized populations the prevalence of these tumors has risen, often dramatically. However, in the Amerindian populations, although the prevalence of obesity, diabetes, and gallstones has increased, "Western" tumors occur at a lower rate than would have been expected. The typical Western pattern of cancer does not seem to be affecting Mexican Americans. The prevalences of these cancers and of heart disease are lagging well behind those of diabetes and obesity, suggesting that the response to the Western environment is different in Amerindian peoples, presumably because of their genetic makeup.

A similar picture is being observed in Australian aborigines. The prevalence of diabetes among Australians of European descent is about 3.4 percent, whereas in the aboriginal population the figure is 10 to 20 percent. The Westernization of these people, who until recently had the lifestyles and diets typical of hunter-gatherers, is also associated with obesity, higher blood pressure, increased levels of insulin and cholesterol in the blood, and, in this case, coronary artery disease. Efforts to maintain groups of aborigines on diets closer to those of their hunter-gatherer forebears have resulted in a reduction of blood sugar and cholesterol levels.

The same pattern of illnesses found in Amerindians and Australian aborigines has been observed in recent years in some of the island populations of Micronesia and Polynesia. Insulin-resistant diabetes is increasingly prevalent in some of these populations, although its distribution is patchy. One of the most remarkable findings is the very high prevalence of diabetes on Nauru Island, which now seems to be affecting over 60 percent of the population. Nauru is a remote atoll with a population of about five thousand Micronesians. A series of colonizations by Britain, Australia, and New Zealand, together with a rapid increase in income from the fruits of phosphate mining, completely changed the lifestyle of Nauruans. Nearly all their food is now imported, and they live on a typically high-energy, Westernized diet. There is a very high frequency of obesity and insulin-resistant diabetes, which was almost nonexistent some years ago and which started to reach epidemic frequencies in the 1950s.

This pattern of a dramatic increase in the frequency of this type of diabetes, being seen in Asian immigrants in many populations, in the New World and the Pacific islands, also seems to be affecting the Chinese of Singapore, Taiwan, and Hong Kong, though not yet those on the Chinese mainland. In fact, it has been estimated that there could well be up to fifty million diabetics in China and India alone by the year 2000.

The American geneticist Kenneth Weiss and his colleagues have pointed out that conditions were ideal for an advantageous gene to have spread through the small population of the Arctic corridor in North America, and for it to have become fixed in the subsequent thousands of years of the hunter-gatherer existence of the early Amerindians. Some years ago, when a dramatic increase in the prevalence of diabetes in some contemporary human populations started to become apparent, one of the founding fathers of human population genetics, James Neel, suggested that it might be due to dietary plenty imposed on what he called a "thrifty" genotype, which had been selected to take advantage of sporadic food availability in primitive societies. This hypothesis has been discussed by many geneticists and anthropologists and extended by Weiss and his colleagues. They propose that the gene or genes related to food storage in the form of fat may have had a particular advantage in allowing women to become fertile or nurse infants in times of Arctic unpredictability. A fast storage capacity, or thrifty metabolic genotype, could have a selective advantage and might explain the New World syndrome and the rapid increase in diseases that result from the overexposure of this genetic constitution to today's excessive diet. Although this attractive notion has met with criticism over the years, quite recently it has been found that a familial form of non-insulin-dependent diabetes of early onset is due to several different defects in an enzyme involved in mediating insulin response to the level of sugar in the blood. This remarkable discovery now opens the way for an analysis of the genes that are involved in generating the diabetes-prone phenotype. No doubt many will be involved.

It is clear, therefore, that non-insulin-dependent diabetes and its complications are assuming epidemic proportions in many parts of the world. If this reflects a particular genetic makeup, it must be extremely widespread. This poses the intriguing question of when it arose and why its distribution is so variable among different populations. We do not know the answer, of course, but we can hazard some guesses that might provide a framework in which to study the problem in the future. But to do so we will have to digress briefly again and consider current thoughts about the origins of human races.

Much of what we know about our ancestors has been gleaned from studies

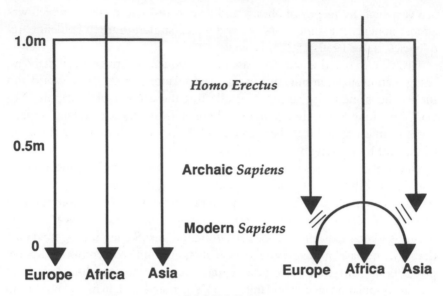

FIGURE 38

The two different models of the evolution of modern man. That on the left is nicknamed the candelabra model, while that on the right is sometimes called the Noah's Ark model. Modified from Lewin (1989).

of fossil remains, although more recently, because our history is written in our DNA, it has been possible to obtain extremely valuable information from comparative studies of the sequence of DNA in different species and racial groups. Two major hypotheses about the origins of modern man are shown in figure 38. One, known as the candelabra model, has it that modern human populations are the direct, geographical descendants of one of our primitive forebears called *Homo erectus,* who moved out of Africa over a million years ago. In this scheme of things modern man developed separately in Asia, Europe, and other parts of the world, directly from *Homo erectus.* This would mean that present-day races all have deep genetic roots, having been separated from one another for a very long time, at least for over a million years. The other model, sometimes called Noah's Ark, suggests that *Homo erectus* moved out of Africa a million years ago and established populations throughout the Old World but that they were replaced by later migrations of peoples who were much closer to modern humans. In the second model, present-day human races would be much more closely related genetically because they would have had much less time to change their genetic constitutions.

Recently it has become possible to compare the DNA sequences of a number of human genes in different races. One of the surprising features of the structure of our DNA is that it varies considerably from person to person, and much of this variability seems to be harmless. Thus by comparing one race or species with another, we can gain some idea of genetic relatedness and estimate the time when species and races separated from one another. Furthermore, we have two types of DNA in our cells. In addition to the DNA that determines our body structure and chemistry and that resides in the nucleus of our cells, every cell in the body contains a small dynamo called a mitochondrion. It seems likely that sometime in our evolutionary past we took on board small parasites and utilized them as our major energy-producing machinery. Our mitochondria have their own DNA, and since mitochondria are not transmitted in the sperm, we receive all of it from our mothers. Remarkably, the rate of change of structure of DNA, or mutation, seems to be faster in mitochondrial DNA than in nuclear DNA. In effect, we have two evolutionary time clocks ticking away in our cells at different speeds. Thus, comparative studies of nuclear and mitochondrial DNA sequences give us a powerful tool for studying our patterns of evolution. Of course, this is all prone to error, not least because we are not absolutely certain at what speed the two clocks are running. But it is possible to make some educated guesses.

One hypothesis based on studies of this type holds that modern man arose from a relatively small stock that emerged from Africa somewhere around 150,000 years ago. Because much of this work is based on mitochondrial DNA sequences that we inherit from our mothers, our African "forebear" is sometimes called the "mitochondrial Eve." It is suggested, therefore, that this Eve evolved in Africa some 150,000 years ago and that all our modern human races evolved from migrations of relatively small populations that started about 100,000 years ago, reaching Australasia 50,000 years ago, western Europe 35,000 years ago, and the New World 12,000 years ago (figure 39).

It is important to point out, however, that these conclusions are the subject of considerable controversy, not least because of uncertainties about the rate of mutation of mitochondrial DNA. Many workers in this field still prefer the notion of parallel local evolution on different continents, beginning with the first migrations of *Homo erectus* from Africa, or a combination of both pathways to the evolution of modern man. But, regardless of how it happened, modern human populations have expanded across the world over the last 100,000 years. By the end of the Paleolithic period, some 10,000 years ago, most areas were peopled, though the numbers of inhabitants were about a thousand times smaller than today. The geneticist Luigi Cavalli-Sforza has noted how, over this time, major genetic differences between human groups would have evolved from a series of local expansions, through selection or the development of new skills.

Given the widespread distribution of the thrifty genotype among peoples of Asian origin, it is possible that it arose very early during human evolution and was carried into many different races because of its great selective value.

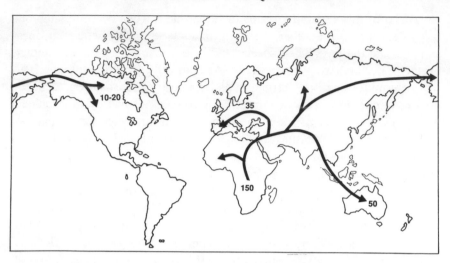

FIGURE 39

*One view of the patterns and times of migrations of modern man. Modified from
Lewin (1989).*

When we consider the extraordinary diversity of environmental conditions
that must have been encountered, it is not surprising that the thrifty genotype
became common and widely distributed. If it attained high frequencies by
selection, we would expect to see considerable variability in its distribution,
depending on particular conditions to which different populations have been
exposed. The Nauruans, for example, reached their island after very long
voyages in canoes, which would have increased the possibility of death by
starvation; the original settlers of the Americas must have had equally ardu-
ous journeys. And its patchy distribution may also reflect random genetic
drift in these small founding populations, that is, variation in gene frequencies as a result of chance fluctuations.

Why, therefore, if this genotype has been so advantageous is it not even
more widespread, and why is it not more common in European populations?
One possibility is that those who settled in Europe did not undergo such
privations. Another, less likely explanation is that it has already been reduced
in frequency by natural selection; in Nauru insulin-resistant diabetes often
affects those of reproductive age. Could it be that what we are seeing in
Nauru and Pima is the telescoping into a few generations of what has been
happening gradually in Europe and elsewhere by the operation of natural
selection for many centuries? Perhaps, but the timescale probably is too short
to account for the great difference in the frequency of diabetes by this mechanism.

Of course, much of this is pure speculation. If the thrifty genotype was of great advantage to us in the past, it is likely that many different genes are involved and many ways in which we could have arrived at an obese and diabetes-prone phenotype. There would be nothing new in this; we have already seen how the genetic blood diseases, which have come under strong selection by resistance to severe forms of malaria over only a few thousand years, have arisen on many different occasions in different populations, and that many different genes and literally hundreds of different mutations are involved. Furthermore, although obesity and diabetes seem to be the diseases common to many of these populations' response to Westernization, the prevalence of heart disease, cancer, and other Western diseases is quite different among them. The complications of the thrifty genotype may well be modified by many other genetic factors. But, whatever the mechanisms involved, there is little doubt that we are seeing an epidemic of a noninfectious disease that reflects a particular type of genetic makeup meeting an inappropriate environment. It seems likely that the high frequency of heart disease and high blood pressure in other populations that have changed their environment recently will turn out to have a similar basis.

These examples of the telescoping of evolution reflect another remarkable experiment of nature. They may be telling us, in effect, that many of the current killers of our rich countries are the result of a genetic makeup inappropriate to the environment that we have created. And they may be warning us that it will be very difficult to find the different genes that constitute the ancient polymorphisms that underlie our inability to adapt to the modern world.

The Genetic Factors Involved in
Our Common Diseases Are Complex

Research into the genetic influences that make us more or less prone to coronary artery disease exemplifies the complexity of studying polygenic disorders, diseases that result from a variation in the function of a number of different genes together with a major environmental influence. To start with, information gleaned from studies of twins suggests that we are not in for an easy time. With a concordance rate of just 20 to 30 percent for coronary artery disease in identical twins, it is clear that our genes play only a modest part in this disease and that our environment is more important. Earlier we saw how, as we get older, the linings of our arteries become silted up due to the deposition of calcified plaques of atheroma, and how, when these plaques break down and blood clots are formed on top of them, our coronary arteries may become blocked, leading to a heart attack. We know very little about the way in which atheromatous plaques are formed, but lessons learned from epidemiological studies suggest that cholesterol and other fatty substances

play an important role. And since the formation of a blood clot on an ulcerated plaque of atheroma seems to be the event that finally triggers a heart
attack, the complex series of chemical interactions that leads to the coagula-
tion of our blood must be another key player in the generation of heart
attacks. Hence many studies of the genetic factors that underlie coronary
artery disease have started by trying to determine the genetic regulation of
levels of cholesterol and blood-clotting factors.

Cholesterol is an important component of the membranes that surround
all cells. As has been pointed out by Joseph Goldstein and Michael Brown,
whose work on cholesterol won them the Nobel Prize in 1985, cholesterol is
a Janus-faced molecule. The very property that makes it useful in the membranes of our cells—that is, its complete insolubility in water—also makes it
lethal when it gets in the wrong place, such as the walls of our arteries. For
this reason it has to be transported in a safe state in our blood. To this end,
it is packaged up in the cores of carrier molecules called lipoproteins. There
are several classes, defined by their behavior in a powerful centrifugal field as
low-density lipoprotein (LDL), high-density lipoprotein (HDL), and so on.
Much of the cholesterol in our blood is found with the low-density lipoproteins.

As we saw earlier, the prevalence of heart attacks rises in proportion to the
plasma cholesterol level or, more specifically, to the plasma level of LDL
cholesterol. It seems likely, therefore, that cholesterol is a major player in
coronary artery disease, although we should bear in mind that associations
of this kind do not prove a direct causal relation. Goldstein and Brown have
shown that blood cholesterol levels are modified by the activity of specific
receptors that bind LDL optimally at cholesterol concentrations of 2.5 mg/dl
(milligrams/deciliter). Receptors are molecules on the surface of cells that
transfer substances like cholesterol to the interior. It has been calculated that
a blood level of 25 mg/dl would be sufficient to nourish our body cells with
cholesterol. But this is roughly one-fifth of the level usually seen in Western
societies. Several lines of evidence suggest that a lower level, 25–60 mg/dl,
might be about right for human beings; species that do not develop atheroma
have much lower cholesterol levels. Furthermore, the level is lower in newborn babies and well within the range that seems to be appropriate for receptor binding. The same goes for humans raised on a diet low in fat.

So what is the mechanism for the high levels of plasma LDL cholesterol
that are so frequent in Western industrialized societies? It is becoming clear
that both diet and heredity are involved. As we have already noted, there is
good evidence for the importance of diet, although when even moderate
amounts of animal fat are introduced into our diets and our plasma cholesterol levels rise, the effect is not the same in everybody. And not all of us with
raised levels of cholesterol suffer from serious atheroma and heart disease. It

is now known that the genes that control the LDL receptor and those responsible for the production of lipoproteins all show remarkable variability, some
of which may make us more (or less) likely to develop vascular disease. For
example, in their pioneering work on the LDL receptor, Goldstein and Brown
discovered that there are many different genetic defects in LDL receptors,
found in about 1 person in every 500, which lead to a markedly elevated
level of blood cholesterol and an enhanced risk of coronary artery disease.
They also realized that diets high in cholesterol and saturated fats can simu-
late these defects in genetically normal persons because these substances, in
effect, signal cells to produce fewer LDL receptors. These complex interactions between nature and nurture are underlined by recent studies in China
that suggest that individuals who have mutations of the LDL receptor gene,
similar to those found in patients in the United States, tend to have lower
cholesterol levels, presumably because their diet is lower in fat.

The work of Goldstein and Brown is seminal to our discussion in several
respects. First, it provides clear evidence at the molecular and cellular level
that human beings are ill adapted to their diets, not just in rich industrialized
countries but even in rural China. In addition, it tells us that it should be
possible to begin to understand some of the genetic factors involved in
determining the variation of cholesterol levels among individuals who are on
a similar diet. In the last few years, stimulated by these discoveries, intensive
studies of the genes for lipoproteins and blood-clotting factors have been
undertaken and have started to hint at the complexity of the genetic regulation of cholesterol levels and blood clotting. And work in this field has provided some clues as to which particular genetic variants of these systems
make us more or less prone to vascular disease. A similar approach has been
taken to try to dissect the different genetic components involved in determining blood pressure, another important factor in the generation of heart disease and stroke. We will return to these important studies in chapter 9.

What about conditions in which the genetic contribution seems to be
stronger than vascular disease—late-onset diabetes, for example? A great deal
is known about the way in which the body regulates sugar and controls
insulin production. It was easy, therefore, to select some good candidates for
genes that might help make us more or less susceptible to developing insulin-
resistant diabetes later in life. For a while progress was slow; all that happened was that many of the likely players were excluded in studies of families
that had several affected members. But recently we seem to have struck lucky.
There is a form of insulin-resistant diabetes that occurs in younger people
and can be traced through families as a single dominant gene. In some of
these families the culprit gene has been found. This important discovery suggests that it should be possible to determine the cause of the other forms of
insulin-resistant diabetes that affect so many of us as we get older.

There has also been some progress in identifying the genes involved in the insulin-responsive type of diabetes. In this case it appears that the major histocompatibility genes play an important role. We saw earlier how the products of these genes are responsible for presenting "foreign" proteins—viruses, for example—to the cells of the immune system. Here, for the first time, is a clue to the way in which environmental agents are involved in the self-destruction of the pancreas, the hallmark of this form of diabetes.

All of us will develop one or another of the many forms of arthritis at some time in our lives. Are they simply the result of many years of wear and tear? If so, why do some of us contract them in early or middle life? Again, studies of our genes may offer us some clues. The serious forms of arthritis that affect young people, such as rheumatoid arthritis, also seem to fall into the group of disorders in which the body mounts an attack on its own tissues, in this case the joints. Some of the genes have already been found, and as we start to understand the functions of their products, and what is different about them in people who are susceptible to arthritis, we should begin to learn more about how these conditions arise. And even in the commoner forms of arthritis that affect us all as we age—osteoarthritis, for example—it is beginning to look as if it might be possible to find some of the genes that are involved, particularly when these diseases strike younger people. A long shot, no doubt, but the only hope we have of learning how these distressing disorders arise.

The genetic approach to cancer research may prove even more rewarding, and complement the environmental studies that we discussed in the preceding chapter. Cancer is not a genetic disease in the sense that it runs true in families; most forms of cancer do not. But there is increasing evidence that most cancers result from "somatic mutations," changes in our genes that we acquire during our lifetime. As we will see later, it appears that many cancers result from the acquisition of several different mutations, which alter the functions of genes important in controlling the division and proliferation of our cells. This new branch of genetics promises to take us to the very heart of what happens when a cell becomes cancerous, and it may help us understand how environmental agents act on our DNA to produce the mutations that underlie common forms of cancer. But there may be other ways in which we can acquire these changes in our genetic machinery, one of which is aging.

THE ROLE OF AGING IN OUR CURRENT ILLS

Since the bulk of the intractable diseases of our richer countries occur in middle life, and later, they are seen against the background of changes in our tissues and organs that are the direct result of the mysterious process of aging.

Is the "pathology" of aging partly responsible for some of these disorders, and, if so, is it, too, modified by our genetic makeup? And to what extent does aging reduce our capacity to cope with the new environment we have created?

As we saw earlier, since the mid nineteenth century there has been a dramatic increase in the life expectancy for those who live in the richer countries of the world. Over this period the expectation of life at birth has almost doubled, from forty to nearly eighty years. This remarkable change has been the result of a rapid decline in neonatal, infant, and maternal mortality. Currently, mortality rates in younger and middle-age groups are extremely low in the United States and Europe, and it has been estimated that the complete elimination of mortality before the age of fifty, which accounts for about 12 percent of all deaths, would increase life expectancy at birth by only three and a half years.

In the last twenty-five years age-adjusted death rates for major cardiovascular disease have fallen by approximately 35 percent in the United States. Interestingly, most of the declines in mortality and gains in life expectancy during this period has been achieved in elderly populations. This phenomenon is completely unexpected and unexplained but has important implications for health care in the future. The average life expectancy at birth in the developed world seems at present to be about ten years less than its potential maximum, largely because of premature deaths from cancer and cardiovascular disease. If we were to eradicate deaths from these disorders, we would add only half a dozen or so more years to our life expectancy at birth. It appears, therefore, that whatever we do, we are programmed to have a maximum life span of about eighty-five years, and as we gradually eliminate the major causes of premature death and morbidity, the aging process itself becomes the major risk factor for seeing us off. But to what extent does aging itself contribute to out current problem diseases?

What is aging? Since different species have different life spans but since within species there is wide variation, it seems likely that aging reflects the action of both nature and nurture. The genetic theory of aging is based on the differential life span among species and on the fact that cell lines—cells grown in culture media outside the body—have a relatively fixed life expectancy. For example, certain types of human fetal cells will, under these conditions, always divide approximately fifty times and then stop. It is not clear what tells cells to stop dividing, and, until recently, it was impossible to define any particular genes that might be involved. However, recent studies in fruit flies and worms suggest that there is a strong genetic component to the determination of life span; a mutation of a gene called *age I* in a certain type of worm can increase its life span by 70 percent. It will be of great interest to determine the product of these genes, although whether it will be possible to extrapolate what goes on in geriatric flies and worms to human beings remains to be seen.

Another way of looking at aging is to assume that it is a random process that results from the deleterious effects of chemicals, which are produced throughout our lives, on our DNA. If this turns out to be the major factor in determining the rate at which we age, sorting out the process of aging, let alone trying to retard it, is going to be an extremely complicated business.

One of the most popular notions about the mechanism of aging has been given the unattractive name "garbage can" hypothesis. As we age, this theory has it, our DNA suffers an increasing amount of damage as a result of the generation and accumulation of toxic by-products of the normal chemical reactions responsible for life. One form of chemical garbage that may be implicated is a family of molecules called oxygen-free radicals. During normal metabolism, oxygen forms molecules called superoxides, highly reactive molecules with an extra electron that allows them to bond with other chemical groups. Free radicals of this kind can damage fats and proteins as well as DNA. But such chemical reactions are highly elusive; their extremely short life span makes it very difficult to measure free radicals directly. However, when they damage cells, they leave visiting cards in the form of the chemicals produced in the damaging reactions. There is quite good evidence that our cells undergo trauma in this way, and it is believed that aging may represent a gradual breakdown of the mechanisms designed to counteract this effect. Studies of cell lines suggest that young cell populations contain a family of housekeeping enzymes called antioxidants, which are capable of disarming free radicals. It could be that aging reflects an increased production of free radicals or, alternatively, a gradual failure of the processes responsible for their inactivation.

These new ideas about the chemical basis of aging link this process to some of our major problem diseases. In the case of cancer, they are starting to help us understand how damage to our DNA, though mediated by external agents, may also reflect changes caused by the aging process that may, in turn, be modified by genetic variability of the various chemical pathways designed to deal with dangerous oxidants. Since recent studies suggest that cholesterol in the form of oxidized LDL is particularly bad for our arteries, and that antioxidants may protect us against heart disease as well as cancer, it is difficult to dissociate the cause of our major killers from the aging process.

Other substances have the potential for damaging tissues and may accumulate as we age. For example, the simple sugar glucose, one of our major sources of energy, has a tendency to attach itself to proteins, in a process called glycosylation. The protein-sugar complex sets in motion a chain of chemical reactions that cross-link adjacent proteins, causing them to stick together in a yellowish brown mess. As we age, these glycosylation end products accumulate in our tissues and make them less elastic, and leathery. And the work of the American scientist Tony Cerami indicates that they can interact with our DNA and cause mutations that interfere with our cells' ability to both repair and replicate their genetic material. In culinary terms, aging may simply mean that we are "overdone."

These ideas about the endogenous production of chemicals that damage DNA have been pursued by the American scientist Bruce Ames, who has identified a number of agents with this potential. The concept that is evolving is that both cancer and the degenerative changes of aging are the results of this type of damage to our DNA. The process may be accelerated, in the case of cancer, by exogenous factors such as tobacco or decreased, at least in rodents, by the restriction of calories. Ames has pointed out that a crucial factor in longevity is the basal metabolic rate, that is, the overall speed of the chemical reactions responsible for life. The metabolic rate is much lower in man than in rodents and could affect the rate and level of production of endogenous mutagens, that is, substances able to damage DNA. This view of aging suggests that the amount of maintenance and repair of our body tissues and DNA is always less than that required for indefinite survival. If this is so, a certain amount of damage to the DNA in our cells, induced by endogenous mutagens, will inevitably accumulate over time.

Many of these disparate observations on the mechanisms of aging are compatible with a general theory proposed by the British biologist Thomas Kirkwood, which he calls the "disposable soma model." Kirkwood suggests that, in evolutionary terms, animals have not evolved maintenance systems that lead to immortality, because such systems would squander energy that is better used to ensure effective reproduction. Since environmental hazards will inevitably see us off, a species should invest in protective mechanisms that guarantee vigor during its reproductive period. In a sense, we have been designed rather like the products of the automobile industry—that is, to function well for a short time but not to last for too long. The idea that we are programmed to reproduce and that what happens after that does not matter too much, while of little comfort to those with grown-up families, is certainly compatible with many of the biochemical changes associated with aging that we have just discussed.

While these new ideas about the mechanisms of aging are complex, and may be a gross oversimplification of what is going on, at least they offer a conceptual framework in which to carry out experiments designed to test the relative roles of nature and nurture in the processes of aging, and the generation of cancer and other diseases as part of the aging mechanism. For example, it has been possible to define many of the genes involved in the repair of DNA and to start to unravel the different mechanisms that control the proliferation and maturation of cells with respect to the aging process. Studies of a disease called Werner's syndrome, a rare condition characterized by premature aging, are also starting to yield information about some of the changes in our proteins, which may be extremely important in the aging process. And because our immune system tends to run down as we age, a great deal of work is being done to find out why.

The field of gerontology has taken an enormous leap forward over the last few years, because it has been possible to start to make use of advances in molecular and cell biology to study the aging process. It is difficult to see where this new field of research may end up, but it is already raising some fascinating questions about the significance of aging. Although it seems self-evident that aging is a natural phenomenon, gerontologists suggest that under some circumstances it may be regarded as a pathological state, with the potential to be reversed. They are also telling us that cancer and vascular disease are, at least in part, by-products of the pathology of aging. Clearly, when we talk of age-related changes in our tisues as being "normal" or "pathological," we are entering a world of semantics for which we are not prepared. But it seems likely that future work in this field has the potential to help us control some of the grosser changes of aging, although whether it will have a genuine effect on how long we live remains to be seen.

Some scientists working on this topic doubtless have it in the back of their minds that it might be possible to interrupt the cascade of genetic and aquired events that leads to aging. In a review in the American journal *Science* a few years ago, one of the leading workers on aging, Samuel Goldstein, was quoted as follows: "It is obvious that an intimate understanding of the events that lead to senescence would tell you which genes are important and how they are regulated, and how you can begin to change the gene regulation to make the cells live longer or restore their normal function." While many of us might think that there are more pressing problems in clinical practice, we cannot ignore what is going on in this field.

WE CAN'T PUT ALL OUR MEDICAL EGGS
IN THE ENVIRONMENTAL BASKET

The emerging view, therefore, is that the major killers in Western industrialized countries reflect the completely new environments and ways of living that we have created for ourselves, set against the background of aging and a genetic makeup that, although it served us well for thousands of years in the wild, may not be adapted to our new lifestyles. Rather, it appears that some of the genes that may have been selected because they were advantageous in our evolutionary past are just those that make us more prone to the hazards of our new surroundings and lifestyles. And since evolution loses interest in us after we have reproduced, the chemical traits that have been selected to this end may not be the ones that help us to grow old gracefully. Since it is likely that the genes that render us more or less susceptible to our new environments reflect ancient and complex adaptations to many different environmental pressures, it may be hard to pinpoint them and to clarify their role in our present-day diseases. But we must try, for the defining of their actions offers us our best hope for understanding and controlling the illnesses of middle life and old age.

Although, as we saw in the preceding chapter, we can do much to improve our lifestyles, we are not likely to revert to those of our hunter-gatherer forebears. Even the acutely health conscious, with their vegetarian health food cultures, which start the day with what could pass as macerated cardboard, and their valiant attempts to obtain a few calories from mounds of foliage, are still a long way from their cave-dwelling ancestors. We can curb the frightening increase of certain cancers and vascular diseases by stopping smoking. In addition, we might further reduce the burden of heart disease with some sensible advice about diet and exercise. Clearly we must do these things; it would be foolish to allow a population the size of North America's to kill itself by the early part of the next century while we wait until we understand why tobacco smoke causes lung cancer or heart disease. But knowledge of the deleterious effects of cigarette smoking reflects a remarkable experiment that we have carried out on ourselves and that has provided an absolutely clear-cut answer for epidemiologists. There is no reason to believe that answers to questions about the environmental causes of other diseases will come so easily.

Recent studies of the relation between cholesterol and coronary artery disease in Shanghai emphasize some of the problems we will have to face up to. Overall, the blood cholesterol levels of the Chinese are low compared with those of Western populations. However, even within this relatively low range, there is a relation between blood cholesterol levels and coronary artery disease deaths. In discussing these results, Richard Peto and his colleagues from Oxford and Shanghai who carried out this large analysis suggest, like Goldstein and Brown, that few people in Western or Oriental populations have biologically "normal," as opposed to population average "normal," cholesterol levels. They point out that even for the Chinese whose blood cholesterol values would be considered low or well within the normal range in Western societies, the levels are too high. They add that they have encountered populations in China with extremely low cholesterol levels and with even lower rates of coronary artery heart disease and suggest that they serve as an "interesting model" for what might eventually be achievable in the West, if methods could be devised for the widespread reduction of cholesterol levels. In other words, most of the world's populations have cholesterol levels that may be too high.

The implications of these findings are far-reaching. At first sight, they suggest that most human populations should drastically change their eating habits or receive cholesterol-lowering agents, preferably ones with no side effects. This news will be of little comfort to anybody except the shareholders of our large pharmaceutical houses. Clearly this will not be feasible. By reevaluating our eating habits and educating populations that seem to be at high risk as they adopt more Westernized cultures, we might make some progress in the reduction of heart disease. But Western society is not going to adopt the lifestyles of Chinese rural communities. It is essential, therefore, that the laboratory sciences move quickly to identify individuals at very high risk for heart disease and determine how the damage to the walls of our blood vessels is mediated. If the epidemiologists are right, their message is frightening. In essence, they are telling us that most, if not all, human beings are unsuited, at least at this stage of their evolution, to life in the modern world.

Another cause for concern is that, as societies become more sophisticated, they are capable of changing their environment or social habits over very short periods. The complexities of a rapidly changing environment for the health of populations was highlighted by the cholera epidemic that hit parts of Latin America a few years ago. Cholera is well suited to spread in populations of low economic status, poor sanitation, and crowding. This epidemic had a devastating effect on the population of Peru, a country that was spared earlier cholera epidemics that struck South America, because most of the population lived in villages scattered high in the Andes. After the Second World War major public health measures sought to eliminate malaria from the Peruvian coast. This well-intentioned activity led to a large wave of migration from the peasant communities of the Andes to newly developed coastal cities. Because of the collapse in the Peruvian economy and the ghastly conditions in which people lived in these cities, the recent cholera epidemic spread very rapidly through this new urban population. Amazingly, the death rate was kept extremely low on the Peruvian coast, at less than 1 percent. The low mortality was a testament to the oral rehydration regimens developed by clinical scientists from the United States working in India some years ago. This form of treatment was used with enormous success in the recent Peruvian epidemic, despite the difficult conditions under which the doctors had to work.

This story highlights some of the problems of modern medicine. The major public health drive to eradicate malaria undoubtedly led to the conditions that were ripe for the development of a cholera epidemic. Had treatment derived from clinical and laboratory science not been available, there would have occurred an enormous loss of life. By altering human environments with the objective of making them healthier, we may create the very conditions that encourage the development and spread of other diseases. We must not ignore the study of disease on the assumption that if we improve our environment, such knowledge will be redundant.

As we saw in chapter 6, the frightening increase in the frequency of asthma, particularly in babies and young children, almost certainly reflects the results of a changing environment working on genetically susceptible individuals. One of the most important environmental agents for precipitating asthma in susceptible children is the domestic mite, which thrives in our ill-ventilated, overheated houses. Why not simply kill off the mites? This is not as easy as it sounds. Simple cleaning and vacuuming of our homes has no effect whatever. It has been difficult to produce bedding materials that are less conducive to colonization by mites or to evolve methods for separating mites from sleepers. The rate of colonization of homes can be modified by reducing the temperature and humidity, but achieving the right conditions is both difficult and costly. Agents are being developed for killing mites, but the problems of maintaining a mite-free household over a long period are formidable. This is just one example of how difficult it may be to deal with what on the face of it appears to be an extremely simple environmental hazard. We will not control asthma without a combined attack on the environment and on the disease process in susceptible persons.

The effects of environmental change may be extremely subtle. Recent epidemiological studies in England suggest that leukemia occurs more commonly in children of higher socioeconomic status. There is increasing evidence that this disease results from at least two distinct genetic events in children's white blood cells, or their precursors. The English scientist Melvyn Greaves speculates that this genetic damage stems from viral infections, the first of which occurs in utero and makes white cells more vulnerable to damage to their DNA sometime in the future. The second event, he suggests, occurs later in childhood, perhaps when an infection stimulates the white blood cells to proliferate and divide again. Greaves proposes that children exposed to viruses later in childhood, which include a high proportion of the offspring of wealthier families who have been sheltered from other children, may be at a higher risk of developing leukemia from the second event, either because their delayed exposure results in a more powerful immunological response or because the virus has a direct effect on their bone marrow. These speculations are further examples of the complexities posed by changing environments and social conditions.

Our intractable killers and causes of chronic ill health reflect, then, an extremely complex mixture of nature and nurture, set against the background of aging, which itself may be modified by both genes and environment. They are likely to have multiple causes, and there may be many different routes to their pathology. Moreover, we have the facility rapidly to control and change our surroundings. Clearly we cannot rely on environmental measures alone to control our problem diseases. As we did at the beginning of this century, we will need a great deal of support from the basic biological sciences. In the chapters that follow we will look at their potential to help us.

PART V

What Does the
Future Hold?

The New Revolution
in the Medical Sciences

It is not easy to convey, unless one has experienced it, the dramatic feeling of sudden enlightenment that floods the mind when the right idea finally clinches into place.

—*Francis Crick (1916–)*

We started our journey through the medical sciences by suggesting that the doctors of today are in much the same position as their predecessors at the beginning of this century. Many of the infectious diseases that were killing their patients could be partly controlled by better housing and hygiene and other environmental improvements. Yet it was not clear whether these measures would eradicate these diseases, and in the meantime there was little they could do for the patients with infections who filled their wards; the seminal advances in the basic biological sciences that promised to solve their problems had been on the move for half a century but still appeared to be of limited practical value. In the event, they had to wait another forty years before the development of antimicrobial drugs and new vaccines altered the whole face of clinical practice by the prevention and treatment of many infectious diseases. So where do we stand today with respect to the scientific developments that might help us with our new set of intractable diseases?

The twentieth century has seen a revolution in the basic sciences that started in physics, spread to chemistry, and, ultimately, transformed biology. The extraordinary developments in physics around the turn of the present century led to an understanding of how atoms are joined together to form molecules and paved the way for the development of a new kind of chemistry, which would start to explain the structure of the molecules that make up living things. The amalgamation of physics and chemistry spawned a new discipline called molecular biology, which was to unravel how genetic information is passed from generation to generation and how individual cells function, both as self-contained units and as parts of the complex communications network that forms the basis of life itself.

In the last twenty years a start has been made in applying the new sciences of molecular and cell biology to the study of human disease. The emphasis in medical research has slowly shifted from the study of disease at the level of patients or their diseased organs to the study of their cells and molecules. Although major scientific achievements do not always have practical benefits for many years, we already know enough about the potential value of molecular biology to predict that the medical sciences are moving into their most exciting and productive period.

Why do the new molecular sciences have so much to offer medical practice in the long term? In the preceding chapter we developed the theme that many of the intractable diseases of our richer countries reflect the effects of aging together with a new environment to which we are completely unsuited after many thousands of years of evolution and adaptation to a much simpler lifestyle. And we also saw how susceptibility to these illnesses may be modified by individual genetic variability in response to the new conditions that we have created for ourselves. Hence an understanding of how our genes work, and how we differ from each other with respect to our reactions to our new environment, offers a novel approach to discovering the causes and mechanisms that underlie our current diseases.

How can human molecular genetics help us achieve this end? If we are to understand disease and the mechanisms of aging at the molecular and cellular level, we need to be able to isolate, purify, and determine the structure and function of the basic molecules of life—that is, the proteins of which we are formed and the enzymes and mediators that drive the chemical reactions responsible for the normal function of our organs. Since the structure of all these molecules is controlled by individual genes, it should be possible, if we can pinpoint them, to determine their structure and that of their protein products, find out how they work, in health and in disease, and hence ferret out the causes of illness at the level of cells and molecules. This approach is applicable both to inherited diseases and to those that result from more complex interactions between our genetic makeup and our environment and lifestyle.

What is molecular biology? As Francis Crick, one of its founders has noted, the term is unfortunate because it has two meanings. In the broad sense, it covers an explanation for any biological phenomenon in terms of atoms and molecules. However, as it is usually used, it encompasses the structure and interactions of the building blocks of living things, particularly proteins and nucleic acids, and studies of gene structure, replication, and expression. Crick claims that he was forced to call himself a molecular biologist because it was more convenient to do so when visiting clergymen asked what he did, rather than explain that he was a mixture of an x-ray crystallographer, biophysicist, biochemist, and geneticist.

All living things are composed of lifeless molecules. Molecular biology and its partner, cell biology, describe the structure of these molecules and how they are made and how they function and interact with one another to produce what we understand as "life." Because these recent developments in the biological sciences have their origins in physics and chemistry earlier this century, and because so many other disciplines also contribute to this new field, it is difficult to know where to start when trying to give a bird's-eye view of how molecular biology evolved and was applied to medical research. Since the molecules of life have to be made correctly and in the right place at the right time, and the information required to carry out this complex business resides in our genes, perhaps the best way through this bewildering maze is first to sketch how genetics, the science of heredity, has developed over the last hundred years.

GENETICS APPLIED TO MEDICINE

Whether we are glancing surreptitiously at the state of preservation of a prospective parent-in-law to predict how our intended might look fifty years hence or engaging in the futile pastime of trying to find some semblance of similarity between the features of a newborn baby and its doting parents, all of us are interested in heredity in one way or another. Hippocrates knew that baldness begets baldness and squinting squinting, but until the middle of the nineteenth century thoughts about the nature of heredity were extremely confused. Through the seventeenth century it was believed that an embryo grows by the sequential production of its parts in the course of time—a process called epigenesis. But this notion failed to explain how the embryo could carry the information to make such a complex transition. During the eighteenth century ideas changed, and it was thought that a complete miniature organism is already in existence from the beginning. The "preformation" theory has been summarized by the historian Peter Bowler as follows:

> The theory of pre-existing germs held that all organisms grow from miniatures or "germs" created by God at the beginning of the universe, stored up one within the other like a series of Russian dolls. The first woman, Eve, literally contained within her ovaries the whole of the rest of the human race, generation after generation of miniatures packed one inside the other, each waiting for an act of fertilization to give it a chance to grow. . . . The male semen merely provided the stimulus that triggered off the expansion of the outermost miniature.

The theory of preexisting germs made complete nonsense of any attempt to trace individual characters from one generation to the next. If the germ defined an individual, the person's characteristics depended on what God had created, not on anything that had been transmitted from his or her parents. Even after the discovery of sperm, the preformation hypothesis was not abandoned, and some even held that an individual was preformed in the sperm, and only nurtured by the mother. This theory never gained much popularity, however, and many eighteenth century physiologists dismissed spermatozoa as by-products of the seminal fluid.

But although the preformation concept continued in one form or another throughout the eighteenth century, the importance of sperm for galvanizing things into action was demonstrated by some elegant experiments carried out by the great physiologist Lazzaro Spallanzani (1729–1799). He chose frogs to investigate the activities of sperm because they fertilize their eggs outside the body. When the female releases her eggs into the surrounding water, the male sprays his semen onto them. It had been observed that eggs that made contact with semen produced tadpoles, whereas those taken directly from inside the female's body were sterile. With extraordinary ingenuity and, one imagines, despite not a little skepticism on the part of his colleagues, Spallanzani constructed pairs of tight-fitting taffeta pants for male frogs to contain their semen. In the event, the frogs that kept their trousers on were unable to fertilise the eggs. He then went on to take semen directly from male frogs and paint it on unfertilized eggs, again stimulating fertilization. Finally, he demonstrated that the liquid portion of semen will not fertilize eggs and that it is the cellular fraction, which contains the sperm, that is effective. These experiments were beautifully executed and showed quite unequivocally that sperm are vital for the normal fertilization of ova. But it was not until the further development of cell biology in the nineteenth century that the true significance of sexual reproduction became apparent.

Three unrelated events in the middle of the nineteenth century, two of them in the same year, together with an increasing understanding of the properties of cells, were to revolutionize our understanding of human biology and heredity in general and the genetics of disease in particular. On November 24, 1859, the first edition of Charles Darwin's (1809–1882) *The Origin of Species* was published, and promptly sold out. On February 8 and March 8, 1865, the Moravian monk Gregor Mendel (1822–1884) presented his studies "Experiments in Plant Hybridization" to the Natural Science Society in Brünn, now Brno, Moravia, and subsequently published them in the society's proceedings. In the same year an eccentric Englishman, Francis Galton (1822–1911), published two short papers entitled "Hereditary Talent and Character." These events laid the ground for an understanding of how species have developed, the genetic mechanisms involved, and the practical applications of the science of heredity for the study of human characteristics and, ultimately, inherited diseases.

Mendel (figure 40) spent most of his working life in the Augustinian monastery in Brünn, combining his clerical duties with a passionate interest in science. Soon after Mendel entered the monastery, in 1844, the abbot realized that he was not cut out for the duties of a parish priest and, appreciating his exceptional intellect, arranged for him to spend two years at Vienna University. Mendel studied both the physical and the biological sciences and was introduced to the expanding fields of animal and plant breeding, as well as to statistics. His early research involved plant hybridization and fertilization, work he continued when he returned to the monastery. Brünn was a thriving center for science in the second half of the nineteenth century, and Mendel was an enthusiastic member of its Natural Science Society. He is sometimes depicted as a lonely monk whose hobby happened to be breeding flowers. This is far from the truth. He was, in fact, part of a lively scientific community in which animal breeding and plant breeding, because of their commercial importance, were of major interest.

Mendel was stimulated to carry out his now famous breeding experiments by observations on ornamental plants, for which he tried to breed new color variants by artificial insemination. In the end he selected peas for his experiments. He crossed varieties with differences in single characters such as color (yellow or green) or the form of seed (round, or angular and wrinkled), and simply counted all the alternative types in the offspring of first-generation crosses, and crosses in later generations. It is estimated that between 1854 and 1863 he studied some 28,000 plants. On the basis of the results of these experiments Mendel formulated the concept of what was later called the gene, a unit of heredity that is passed from generation to generation in a way that obeys two simple mathematical laws. First, genes segregate; members of the same pair of genes, alleles, are never present in the same gamete (eggs or sperm) but always separate and are transmitted in different gametes. Second, genes assort independently; members of different pairs of genes move to gametes independently of each other. To put it in a nutshell, alleles segregate; nonalleles assort.

The consequences of these laws are beautifully simple. Take the first. If a man and a woman marry and have identical genes at a particular locus (position on a chromosome), let us call them A, they can produce only gametes of type A, and consequently they can only have children with an AA genotype (genetic constitution). If, on the other hand, A exists in another form, say a, and the genotype of the father is AA and that of the mother Aa, then although the father can produce only A gametes, half the mother's will be A and half will be a

The possible genotypes of the children can be worked out as follows:

FIGURE 40

Gregor Mendel (1822–1884).

		paternal gametes	
		A	A
maternal	A	AA	AA
gametes	a	Aa	Aa

It follows that half the children will have the genotype AA and half Aa. On the other hand, if both parents have the genotype Aa, then one fourth of the children will each have the genotype AA or aa, and half will have the genotype Aa, as follows:

		paternal gametes	
		A	a
maternal	A	AA	Aa
gametes	a	Aa	aa

Children with the genotypes AA and aa are called homozygotes; those with the genotype Aa, heterozygotes.

Genes are described as "dominant" or "recessive," terms originally invented by Mendel, who defined them as follows: "Those characters which are transmitted entire, or almost unchanged by hybridisation, and therefore in themselves constitute the characters of a hybrid, are termed dominant, and those which become latent in the process, recessive." In other words, a dominant allele is one that manifests its phenotypic (recognizable) effect in heterozygotes (Aa); a recessive allele causes a phenotypic effect only when present in the homozygous state (AA or aa). In the matings that we have just described, if the a allele produced a disease in the heterozygous state (Aa) it would be referred to as dominant; if it produced a recognizable condition only in the homozygous (aa) state, it would be defined as recessive.

Mendel's work had no immediate impact and was forgotten until it was rediscovered independently by several workers at the beginning of the twentieth century. At first it was the subject of great controversy, but thanks to strong protagonists, particularly the English biologist William Bateson (1861–1926), it came to be accepted. The word "gene" was first coined by the Danish botanist Wilhelm Johannsen (1857–1927), in 1911.

Cell theory, which was central to the development of genetics, started to revolutionize biology during the middle of the nineteenth century. In 1888 the German embryologist Wilhelm Roux (1850–1924) suggested that fetal development must be predetermined by the contents of the egg. Furthermore, he postulated that when it divides, a portion of the germinal material is transmitted to the daughter cells. A year earlier Walther Flemming (1843–1905), a German anatomist, had discovered that when cells are dividing, it is possible to see thread-like structures that absorb color from certain dyes.

Flemming called these structures "chromatin," and later they became known as chromosomes. When a cell divides by a process called mitosis, the chromosomes divide and identical copies are passed on to the next generation of cells. The idea that they might carry genetic information was developed by the German zoologist August Weismann (1834–1914). Although his ideas on how this process might occur were vague, he made one critical observation. If the hereditary material from both parents resides in the chromosomes, and is mixed in a fertilized egg, the egg ought to contain twice as much of the hereditary material as the parents' cells. In every succeeding generation, the amount of material would double. Clearly, this could not happen. In 1887 Weismann reasoned that the only way around this conundrum was to assume that during the process of cell division in eggs and sperm there is a reduction in the amount of hereditary material by half. This type of reduction division is now called meiosis.

These ideas, which were further developed by the American biologist Edmund Wilson (1856–1939) and others, led to a clear picture of how chromosomes are inherited. Except during the formation of gametes—that is, ova or sperm—cells divide by mitosis, which is preceded by a doubling of each pair of chromosomes. This ensures that each of the two daughter cells acquires a set of chromosomes identical to the parental cell, a state described as diploid. During gamete formation, however, cell division occurs by meiosis. Here homologous pairs of chromosomes segregate, or separate, to give progeny with half the number of chromosomes, a condition called the haploid state. Fertilization restores cells from the haploid to the diploid state. Because chromosomes segregate during gamete formation, so must genes. Since chromosome pairs segregate independently of each other, so do genes that are not on the same chromosome.

The true significance of sexual reproduction and meiosis became apparent only during the early part of the twentieth century, following a brilliant series of breeding experiments with the fruit fly *Drosophila* by Thomas Hunt Morgan (1866–1945) and his colleagues in the United States. When maternal and paternal chromosomes become closely wound around each other during meiosis, it is possible for genetic material to pass from one to the other by a process called crossing-over, or recombination. Mendel's laws had discussed only the inheritance of a particular gene, but Morgan's group and William Bateson in England pointed out that if two genes are on the same chromosome, and especially if they are close together, they tend to be inherited together; the genes are then said to be linked. When the parental chromosomes become closely opposed at meiosis, crossing-over of genes can occur so that the two characters determined by the genes will part in some of the offspring. The closer together a pair of genes are on the same chromosome, the smaller the chance they will have to cross over. Hence the number of crossovers is a measure of the distance between the genes. These observations provided a way of obtaining maps of genes on chromosomes. Furthermore, from the work of Hermann Muller (1890–1967) it became clear that genes can change their structure—that is, undergo mutation—and that this process may be speeded up under certain conditions, such as exposure of cells to radiation.

These studies in the United States were mirrored by equally important developments in England at about the same time. Here, for the first time, statistical methods were used to study the behavior of genes in large populations. The work of biometricians like Ronald Fisher, J. B. S. Haldane, and Karl Pearson (1857–1936) not only distinguished between simple Mendelian inheritance and more complex systems under the control of many different genes but also paved the way for the amalgamation of Mendelism and Darwinism, which formed the basis for modern evolutionary theory.

These remarkable developments in genetics early in this century laid the foundation for a synthesis of the way in which Darwinian evolution had come about. Darwin (figure 41) based his observations on the changing patterns of species and, influenced by the predictions of Thomas Malthus on the problems for survival that would occur as populations expanded in size, developed his theory of the gradual evolution of species based on the selection of their fitter members—that is, those that, by better adaptation to their environment, can produce more offspring that survive to reproduce themselves. But he had no idea how these changes might have come about. Once it was clear that genes could be shuffled by sexual reproduction and altered by mutations and, as it turned out later, by moving major segments of chromosomes around the genome, it became clear how evolution might have happened.

There had been hints since the time of Hippocrates that inheritance might be an important factor in disease. The medical literature of the eighteenth and nineteenth centuries contains a few observations to this effect. For example, the astronomer and mathematician Pierre Louis Maupertuis (1698–1759) described a family with polydactyly—that is, extra fingers—which ran through four generations. Remarkably, he carried out a calculation showing that the likelihood that chance alone accounted for the concentration of this trait in families is 8,000,000,000,000 to 1. And in 1814 a British physician, Joseph Adams (1756–1818), in a book entitled *A Treatise on the Supposed Hereditary Properties of Diseases,* set out a scheme for what we would now call genetic counseling. This extraordinary work identified different forms of hereditary disease, noted the importance of parental consanguinity, and, long ahead of its time, proposed the establishment of registries of families with inherited diseases.

But the first serious effort to measure heritability in human populations

FIGURE 41

Charles Darwin (1809–1882).

was made by the Englishman Francis Galton at about the time that Mendel was carrying out his experiments in Moravia. Galton was born in 1822, the same year as Mendel, into a wealthy family. He started his career by studying medicine but disliked it so intensely that he changed to mathematics and obtained a degree at Cambridge University. However, in 1844 the death of his father left him with a large inheritance, which freed him from any need to earn a living, and he spent the rest of his life traveling and making important contributions to exploration and to the biological sciences.

Galton was fascinated by how talent appears to run in families—particularly those of lord chancellors. He was interested from the beginning in attempting to improve the human species by selective breeding. His early studies were summarized in 1869 in his book *Hereditary Genius,* a second edition of which was published in 1892. The observation that distinguished people tend to come from distinguished families led him to believe that heredity must determine not only physical features but also talent and character. In an earlier article he advocated a state-sponsored competitive examination for hereditary merit; winners would be wedded to each other in a public ceremony at Westminster Abbey, after which they would be given financial encouragement to spawn numerous distinguished offspring. Much of his thinking about genetics was confused because, unlike Mendel, he was most interested in traits that did not follow simple patterns of inheritance. He was nonetheless the initiator of quantitative human genetics, and those who followed him, notably Karl Pearson, Ronald Fisher, and J. B. S. Haldane, laid the foundations of population genetics.

But it was the clinical and chemical studies of the English physician Archibald Garrod that sowed the seeds for understanding the biochemical basis of human genetic diseases and, in the long term, of how genes function. In June 1908 he delivered a series of lectures to the Royal College of Physicians of London, which were published in *Lancet* in July of the same year. This work was extended and formed the basis for a book entitled *Inborn Errors of Metabolism,* which described several rare diseases that, Garrod realized, were due to inherited defects in the body's chemical pathways. At that time there was still considerable doubt about the importance of Mendel's findings, and hence it is not surprising that Garrod did not immediately appreciate the true significance of some of his observations. However, he noted that there was a high incidence of cousin marriages in some of his families. William Bateson pointed out that this is precisely what would be expected if these conditions were inherited as Mendelian recessive characters.

Garrod's work, like that of Mendel, was largely ignored for many years. Its true value became apparent only when advances in biochemistry started to point to the importance of the genetic regulation of metabolic pathways. In fact, it was not until the early 1940s that the elegant studies by the Americans

George Beadle (1903–1989) and Edward Tatum (1909–1975) on the bread mold *Neurospora* demonstrated that the primary action of a gene is to direct the production of a specific protein, in their case a series of enzymes. In a lecture delivered in Stockholm in 1958 on the occasion of the award of the Nobel Prize to Beadle and Tatum, Beadle described how his earlier work on eye pigments in the fruit fly had led to the experiments with Tatum on *Neurospora,* and how in both these organisms it was possible to produce a series of biochemical mutants, of exactly the same type as those originally envisaged by Garrod to explain the inborn errors of metabolism in man. In his perhaps overgenerous tribute to Garrod he went on to say, "In this long and round-about way, first in *drosophila* and now in *neurospora,* we had rediscovered what Garrod had seen so clearly many years before. . . ."

Like the discovery of insulin and the cure of pernicious anemia, described in chapter 3, Garrod's work is an example of clinical research at its very best. His initial observation was that there were families in which more than one member passed dark urine, first noticed by an unusual staining of their napkins (diapers) shortly after birth. Although this condition was symptomless in early life, these children developed joint disease as they got older. Alkaptonuria, the name given to this unusual disease, had already been described, but Garrod identified the chemical nature of the pigment, observed that the disease tends to occur in more than one family member, and made a reasonable guess at the biochemical defect involved. If he had stopped here, alkaptonuria would have been simply another interesting medical oddity. Garrod's great contribution was to bring together a group of these disorders, to develop a hypothesis concerning human biochemical individuality, and hence to lay the ground for the development of human biochemical genetics, a subject that was to tell us so much about the mechanisms of human disease when it finally came into its own in the 1950s.

Human genetics started to flourish in England during the first half of this century. When Francis Galton died in 1911, University College London was left sufficient money to establish the Galton Eugenics Professorship and the Department of Applied Statistics, which included the Galton and Biometric Laboratories. The first holder of the Galton chair was Karl Pearson. The word "eugenics," which means wellborn, was coined by Galton to describe his concept of the improvement of society by selective breeding. The early years of the development of human genetics were very much influenced by disagreements over the significance of Mendelism and of the application of eugenic principles. Although the eugenics movement thrived, both in England and in the United States, its influence gradually declined, and it was more or less discredited by the end of the Second World War. On the other hand, the work of Fisher, Haldane, and, later, Lionel Penrose and other members of the

Galton Laboratory was to establish the scientific basis for studies of human genetics. Work at the Galton Laboratory, which became the mecca for workers in human genetics from all over the world in the period just before and after the Second World War, put human pedigree analysis on a firm statistical basis, established the first genetic linkages in man, and laid the foundation for the study of genetic disease. However, this early work had very little impact on the medical profession; not until the 1950s did medical genetics start to flourish, in the United States.

If medical genetics was slow in getting off the ground, it could not have chosen a better time to start moving, because in the 1950s the extraordinary developments in physics, chemistry, and the basic biological sciences came together to form the new discipline of molecular biology, which would within a few years describe the structure of genes, how they work, and how they can be isolated. Thus, by an odd quirk of chance and timing, medical genetics and molecular biology developed side by side from the 1950s onwards. But before we continue the story of medical genetics, we must digress briefly to look at the early development of molecular biology.

THE DEVELOPMENT OF MOLECULAR BIOLOGY

The remarkable developments in physics that started in the second half of the nineteenth century led to a new understanding of atoms and subatomic particles. They created the field of quantum chemistry, which opened the way to an understanding of the way in which atoms bond together to form molecules. Ultimately they led to a detailed understanding of the way in which large molecules such as proteins, the basis of all living things, are put together and function. In 1943 the physicist Erwin Schrödinger (1887–1961), in a series of lectures entitled "What Is Life?" and delivered under the auspices of the Dublin Institute for Advanced Studies, presented a novel view of living things that was based on the laws of physics. A new type of biology was evolving, one that was to become the domain of physicists and chemists.

One of the major goals in biology and chemistry in the first half of the twentieth century was to try to comprehend the structure and function of proteins. Proteins are the basic building blocks of all living things. They are extraordinarily diverse in their structure, ranging from the tough collagen that forms the backbone of all our body tissues to the hundreds of enzymes that drive the chemical reactions responsible for life. Proteins consist of one or more long strings, or peptide chains, of amino acids, a family of small molecules, only twenty of which occur in living organisms. It gradually became clear that the enormous diversity of living things must depend on the existence of many different types of proteins and reflect the different sequence of the amino acids in the peptide chains from which they are constructed.

In March 1953 Frederick Sanger and his colleagues in Cambridge published the amino acid sequence of insulin. This work established beyond any doubt that proteins consist of chains of amino acids and that every molecule of a particular protein is made up of chains in which the amino acids are always in the same order. Amino acids themselves have side chains that can undergo complex interactions with each other. Hence their order in a peptide chain determines the way in which it folds to produce the three-dimensional structure of a protein, so essential for its functions. There must thus be an informational system that ensures that when, for example, insulin is made, the order of amino acids is always exactly the same. It follows that the function of the gene that controls the production of insulin is to carry the instructions ensuring that when it is produced in the pancreas, its constituent amino acids are always in the right order. The central question was, how is this information stored and passed from generation to generation, and what kind of complex cellular machinery is required to convert a piece of coded information to a string of amino acids.

The idea that genetic information might be carried by nucleic acids took a long time to evolve. Nucleic acids had been discovered in the midnineteenth century by the Swiss biochemist Friedrich Miescher (1844–1895). By this time it was thought that hereditary transmission from cell to cell might be mediated through proteins, and Miescher set out to try to identify them by means of the most easily available source, which turned out to be the white blood cells, which, incidentally, are the main constituents of pus; because so little could be done for infections at the time, he had plenty of material to work with. Although he was not successful in identifying the proteins he was looking for, he discovered in these cells a new substance that appeared to come from their nuclei, hence the name nucleic acid. During the first half of the twentieth century work on nucleic acids moved in two directions. It was found that one of the major forms of nucleic acid, deoxyribonucleic acid, or DNA, consists of a sugar, deoxyribose, combined with four so-called bases—adenine (A), guanine (G), cytosine (C), and thymine (T). By 1950 it was clear from the work of the Austrian-born American biochemist Erwin Chargaff that the total amount of G + A is always equal to the amount of C + T, that the amount of A is the same as that of T, and that the amount of G is the same as that of C.

Meanwhile, evidence was accumulating that DNA is the informational molecule that everybody was looking for. The first clue came from the work of an English bacteriologist, Fred Griffith, working for the Ministry of Health in London. Griffith was trying to understand how different strains of pneumococci, the organisms responsible for severe forms of pneumonia, differ in their virulence when injected into mice. The results of his experiments suggested that the factor that causes virulence could be transferred between virulent and nonvirulent strains. The transforming factor was identified as DNA in a series of beautifully executed experiments by the Americans Oswald Avery (1877–1955), Colin MacLeod (1909–1972), and Maclyn McCarty in 1944.

The story of the frenetic years that followed, during which the structure of DNA was described, the nature of the genetic code determined, and a general model for protein synthesis established, has been told on many occasions. In 1953 James Watson and Francis Crick (figure 42), in a short letter to the scientific journal *Nature,* described a model in which DNA is a double helix consisting of two chains of the bases adenine, guanine, cytosine, and thymine, wrapped around each other. It turned out that the building blocks of each chain are complex molecules called deoxyribonucleotides, each of which consists of a base, the sugar deoxyribose, and phosphate, joined by relatively weak chemical bonds (figure 43). The critical property of this molecule as an informational replicator lies in the constraints of the pairing between bases; A always has to pair with T, and C with G. Thus when the two strands come apart during cell division, each acts as a template for the synthesis of a new DNA molecule, which, because of these rules, must be identical to its parent molecule.

Over the next few years the genetic code was deciphered by groups in the United States and Great Britain. It turned out that genetic information is encoded by the order of bases in a DNA strand and that it is a triplet, nonoverlapping code in which three bases (codons) determine a particular amino acid. A gene is therefore a length of DNA that has a particular order of bases that constitutes the triplet code for the order of amino acids in its protein product.

But a major problem remained. Most of our cells consist of an outer covering, or membrane, and an inner body, or nucleus; in between the two lies a region called the cytoplasm. It was known by then that proteins are made in the cytoplasm, while DNA is confined to the nucleus. Furthermore, since amino acids cannot interact directly with nucleic acids, a number of intermediary steps would be required before the message encoded by DNA could be translated into the sequence of amino acids in a protein. Clearly some kind of "messenger" molecule was required. It turned out that a number of key steps and players are involved.

When a gene is activated to make a protein product, one of the strands of DNA is copied to produce a mirror image molecule of very similar structure, called messenger RNA. The messenger moves from the nucleus to the cytoplasm, where it acts as a template for protein synthesis. The constituent

FIGURE 42

Watson and Crick in front of their model of DNA.

amino acids are brought to the template attached to a type of molecule called transfer RNA. There are several different transfer RNAs for each individual amino acid. A transfer RNA molecule binds to its particular three-base codon on messenger RNA because it has a group of three nucleotides, called an anticodon, which is able to find its complementary partner sequences on messenger RNA. Successive transfer RNAs carry their amino acids to the messenger RNA template, the amino acids are joined to each other by the formation of simple chemical bonds, and hence a chain of amino acids is gradually built up, with each of its constituent amino acids in the correct position.

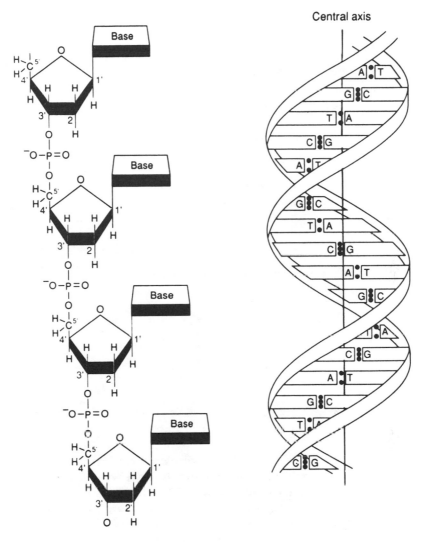

FIGURE 43

The structure of DNA. The double helix with the bases adenine (A), thymine (T), guanine (G), and cytosine (C) are shown on the right. On the left the chemical structure of the sugar phosphate backbone of the molecule is shown in more detail.

Thus the flow of information is from DNA to messenger RNA, and thence to protein. These complicated steps are outlined in figure 44.

Once the general mechanisms of gene action were worked out, in the 1960s, the field of molecular biology developed in many different directions. Much was learned about the structure of DNA and about the way genes are

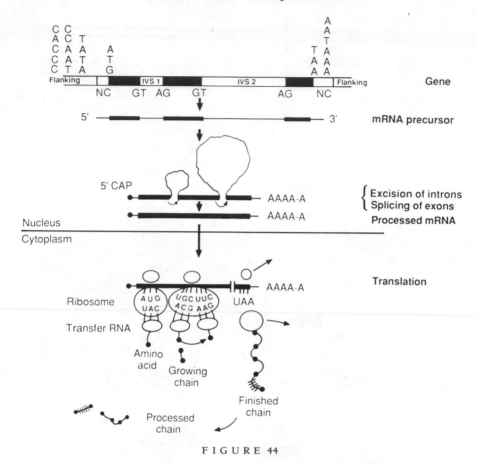

FIGURE 44

The way in which a gene is transcribed to provide a template for building a chain of amino acids. The gene is shown at the top of the figure with some of the critical regulatory regions marked CACCC, TATA, etc. It is split up into coding regions (exons) and noncoding regions, or introns (IVS 1 and IVS 2). When the gene is to be transcribed, one strand is copied into a mirror image, called the messenger RNA (mRNA) precursor. While still in the nucleus of the cell, the exon sequences are removed and the intron sequences are spliced together to deform the definitive messenger, or template. This moves into the cytoplasm of the cell, where it acts as a workbench for the synthesis of a chain of amino acids. The latter are brought to the template attached to molecules called transfer RNAs, which find the appropriate codons on the template, thus ensuring that the order of amino acids is always the same.

regulated and switched on and off at appropriate times in particular tissues. It was found that, in most mammalian cells, messenger RNA has to be processed; long stretches of apparently useless bases, called introns, have to be removed before the molecule is transferred to the cell cytoplasm to act as a template for protein production (figure 44).

By the 1970s there thus already existed a fairly clear picture of the structure and function of genes and how their genetic information is used to determine the order of amino acids in proteins. It was a period of extraordinary achievement in the biological sciences, which gave birth to a completely new technology that promises to revolutionize human biology and that has enormous implications for medical practice.

THE NEW TECHNOLOGY

The discovery of the mechanism of gene action provided a great incentive for the development of methods for isolating genes, determining their structure, and trying to understand how they are regulated. We do not need to consider any of these complex new techniques in detail, but if we are to explore the possibilities of molecular biology for the study of disease, we must grasp, at least in outline, what is possible.

One of the earliest practical advances in this field came from the discovery that some bacteria produce enzymes that cut DNA at very specific sites. This finding was the result of curiosity-driven research that asked why the growth of certain viruses is restricted in particular strains of bacteria. The isolation and purification of these enzymes, which became known as restriction enzymes, made it possible to fractionate DNA and marked a seminal advance in the practical applications of molecular biology.

Another major step forward resulted from the observation that when DNA is heated or treated with alkali, its two strands come apart. They will come together again, or re-anneal, only if the two strands are more or less identical with respect to the sequences of their constituent bases. Hence it is possible to produce DNA / DNA or DNA / messenger RNA hybrids. These findings led to the technique of molecular hybridization and to the construction of what are called gene probes. The idea is to make a short length of DNA or RNA that is complementary to the gene being investigated. If the short gene probe is radioactively labeled, it can be used as a marker to "find," or "hybridize," to its complementary sequences. This approach has been invaluable for finding genes and central to the development of the most important technique of molecular genetics, the cloning of genes.

The recombinant DNA era started in the mid-1970s, when it was found that it is possible to insert pieces of animal or human DNA into small circles of DNA called plasmids, which normally live and replicate in bacterial cells. Plasmids are responsible for transferring information—antibiotic resistance, for example—between bacteria. Bacteria grow and divide very rapidly. Hence, if we wish to isolate large quantities of a particular gene or fragment of DNA, we insert them into a bacterial plasmid to produce a recombinant molecule. The plasmid containing the foreign DNA is then persuaded to enter bacterial cells, where it replicates, a process called cloning (figure 45). By cutting up DNA into appropriate sizes, one can construct gene libraries, representing almost the entire DNA of an individual, growing in plasmids in bacterial cells on culture plates. By means of a gene probe the particular colony of bacteria that contains a gene of interest can be isolated and grown, and relatively large amounts of its DNA can be obtained. Gene cloning has since been modified by a variety of ingenious tricks of genetic engineering. In recent years it has been possible to clone much larger pieces of DNA, inserted into bacterial viruses or even into small artificial chromosomes derived from yeast, in bacterial cells. These advances have been backed up with equally spectacular progress in the development of methods for the rapid sequencing of DNA, some of which have now been automated (figure 46).

It is now possible to transfer genes into cells in tissue culture or even into living animals and to persuade them to function in their new environments. These techniques have enabled us to learn a great deal about the way in which genes are controlled. Another major advance has been the discovery that if particular genes are inserted into fertilized eggs, they may be expressed in the offspring. Such transgenic animals, as they are called, have provided another valuable approach to understanding the regulation of genes, particularly during different phases of development. And, of course, genes that are functioning in a foreign environment, whether in bacteria, in tissue culture, or in living animals, are a potential source of almost any of their products that we wish to make in large amounts.

EARLY MEDICAL APPLICATIONS OF MOLECULAR BIOLOGY

In the 1990s, when there seem to be few areas of medical research that are not being influenced by molecular and cell biology, it is difficult to believe that this field all started less then fifty years ago and that it is only fifteen years since human genes were cloned for the first time.

While molecular biology was developing in the 1950s, medical genetics was beginning to find its feet. The quantitative basis of human genetics had been developed in Great Britain and the United States in the period before the Second World War. But this new mathematical approach to the study of the behavior of genes, both in families and in populations, had little impact

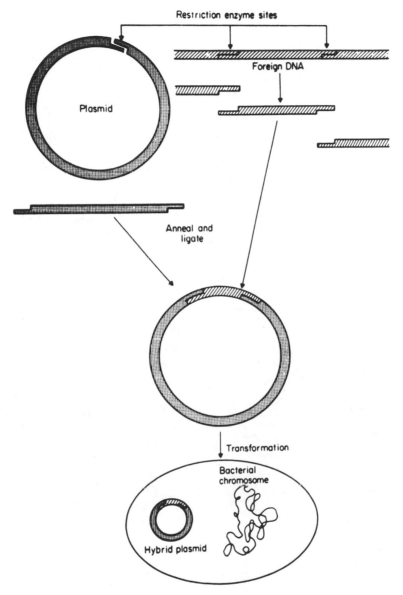

FIGURE 45

Cloning human DNA. The foreign DNA—that is, human DNA—is chopped up into small fragments, which are mixed with the DNA of a bacterial plasmid, which has also been cut with the same enzyme. The mixture is treated under conditions that encourage re-annealing. If we are lucky, the plasmid will incorporate pieces of human DNA. The now recombinant molecule is then persuaded to move into a bacterium, where it replicates both itself and its human DNA separate from the bacterial genetic machinery. In this way it is possible to generate large amounts of any piece of human DNA that we require.

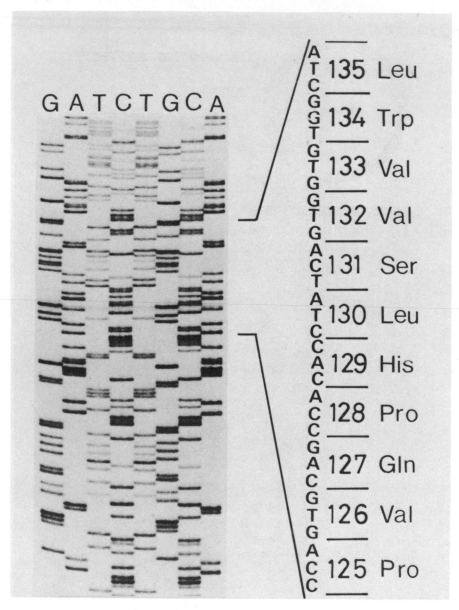

FIGURE 46

Rapid sequencing of both strands of human DNA. After appropriate chemical treatment the bases are separated in an electric field, and their sequence can be read directly from the gel. On the right is shown a short length of sequence with the appropriate bases (codons) for the particular amino acids of the protein that is encoded by the gene being sequenced (Leu = leucine, Val = valine, and so on).

on clinicians. It was not until the 1950s that clinical genetics energed as a new specialty.

The development of clinical genetics in the United States just after the Second World War can be attributed, at least in part, to concerns about the potential ill effects of the atomic bombs that were dropped on Hiroshima and Nagasaki. The U.S. Atomic Energy Commission sent a team to establish a long-term study to assess the rates of appearance of mutations in the Japanese populations exposed to atomic radiation. Among those who became involved in this work was a young physician who had also been trained in genetics, James Neel. On his return to the United States, Neel and a group of talented young clinical research workers set out to investigate the genetic transmission of several common inherited blood diseases. By the early 1950s it was clear that genetics was likely to have an important role in medicine, and several other physicians, including Victor McKusick in Baltimore and Arno Motulsky in Seattle, established new departments of medical genetics.

Soon similar departments sprang up all over the United States and elsewhere, and genetics was applied to many different aspects of medical practice. A few of this new breed of clinicians soon became aware of what was going on in molecular biology and understood enough of both the clinical and the basic-science worlds to see the promise that the molecular field held for a better understanding of human disease.

Of the numerous advances in our understanding of the cause of disease that have already resulted from the application of the techniques of molecular biology, the most spectacular have been in medical genetics and cancer. A brief description of what has happened over the last quarter century in these two areas of medical research should help the reader to appreciate both the excitement and the the enormous potential of this new branch of medical research.

Genetic Disease at the Molecular Level

In 1945 a conversation took place on an overnight train journey between Denver and Chicago that was to have far-reaching consequences for human genetics. The participants were Linus Pauling (1901–1994), a distinguished protein chemist, and the hematologist William Castle. Castle told Pauling that he and his colleagues had noticed that the red blood cells of his patients with sickle-cell anemia had an unusual appearance when viewed under polarized light. Sickle-cell anemia had been described some twenty years previously by the American physician James Herrick, who was one of the first to describe coronary thrombosis. Herrick had noticed that the red cells of some black patients who were under his care had a curious sickled appearance, particularly under conditions of reduced oxygen (figure 47a). It was already evident that this might be an inherited disease, although, at the time this conversation took place, nothing was known about its cause.

As a protein chemist, Pauling was intrigued by Castle's observations because he realized that the changes Castle had noted might mean that the defect in sickle-cell anemia is within the hemoglobin molecule. On returning to his laboratory, he suggested to one of his young colleagues, Harvey Itano, that this might make an interesting research project. How right he was! In 1949 Pauling, Itano, and their colleagues announced that the hemoglobin of patients with sickle-cell anemia has a different rate of movement in an electric field from that of normal persons (figure 47b). This, they reasoned, must mean that the charges of the two molecules are different and hence that there must be a difference in their amino acid constitution. Pauling and his colleagues thus coined the term "molecular disease" to describe sickle-cell anemia.

These predictions were found to be correct when, in 1956, Vernon Ingram, then a young protein chemist working in Cambridge, observed that sickle-cell hemoglobin differs from normal hemoglobin by a single amino acid substitution, valine for glutamic acid. At about the same time it was learned that normal human hemoglobin is made up of two different pairs of chains of amino acids, which were called alpha and beta chains. Ingram subsequently found that the amino acid substitution in sickle hemoglobin involves the beta chain. This was an absolutely seminal discovery because it provided direct evidence for the idea that a gene, in this case the beta globin gene, directs the synthesis of a single string of amino acids, or peptide chains, thus providing confirmation of Beadle and Tatum's classic studies on *Neurospora*.

In the late 1950s these important findings led to the widespread use of a technique called electrophoresis, the study of the movement of proteins in an electrical field, as a tool for examining human proteins, both in health and in disease. It was soon seen that many of them show variability, some of which is harmless but which in other cases is associated with disease. Over the next few years many hundreds of hemoglobin variants were discovered, and methods were developed for identifying the way in which their structures differ from that of normal hemoglobin, usually by only a single amino acid substitution. Remarkably, in each case the particular substitution was consistent with only a single nucleotide change in the genetic code, the nature of which was first derived from work in microorganisms. It appeared, therefore, that both man and bacteria use the same genetic code, a clear verification of the proposed mechanism for Darwinian evolution.

As these different amino acid substitutions were traced to particular parts of the hemoglobin molecule, it gradually became evident why they cause different diseases. In some cases they alter the stability of the molecule and lead to its premature destruction, and hence to anemia. In others they cause

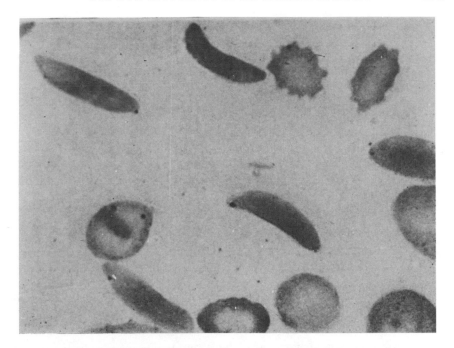

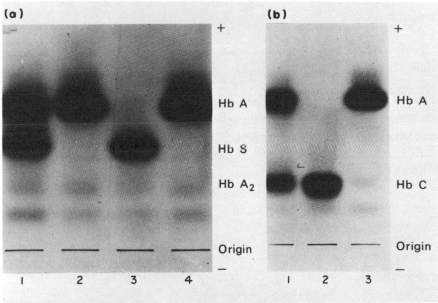

FIGURE 47

(A) Sickle cells in the blood of a patient with sickle-cell anemia. (B) The separation of different human hemoglobins by electrophoresis. (a) The patterns in patients with (1) sickle-cell trait, (2) blood from a normal person, (3) sickle-cell disease, and (4) sickle-cell trait. (b) The patterns from individuals with (1) hemoglobin C trait, (2) hemoglobin C disease, and (3) normal hemoglobin.

the hemoglobin to hang on to its oxygen more avidly than is normal. Patients with these hemoglobins are, in effect, constantly starved of oxygen, as though they were living at a high altitude. The usual adaptive response to living on the tops of mountains is to increase the number of red blood cells, and this is precisely what happens to these patients at ground level. On the other hand, the substitution that causes sickle-cell anemia makes hemoglobin molecules pile up on top of each other when blood is exposed to low oxygen tensions. This, in turn, results in the red blood cells' becoming sickle shaped and in their premature destruction. Linus Pauling's prediction had been correct; it was now possible to relate events at the molecular level to the clinical disorders seen in patients who inherited these hemoglobin variants.

But the abnormal-hemoglobin field was to yield information of even greater importance for medical practice. In chapter 7 we saw how, because symptomless carriers are protected against severe malaria, thalassemia has become the commonest genetic disease in humans. Studies carried out in the 1950s suggested that thalassemia is not a single disorder but a group of diseases inherited similarly to the way the other abnormal hemoglobin variants are inherited. Subsequently it was found that they result from the defective production of either the alpha or beta globin chains, the protein subunits that together constitute hemoglobin, and hence they were called alpha or beta thalassemias.

It turned out that, unlike the hemoglobin variants, the alpha or beta globin chains in thalassemia usually have no structural abnormality, and hence these diseases are caused by a different mechanism, one that leads to a reduced rate of production of a globin chain. It was clear, therefore, that if we could determine the reason for defective production of the alpha and beta globin chains, we might start to understand the mechanisms involved in the defective production of any protein or enzyme involved in a genetic disease. Work carried out over the last fifteen years has shown that this prediction was correct.

Once it became possible to isolate human messenger RNA and to clone and sequence the globin genes, it was relatively easy to work out the defects that cause thalassemia. As was predicted on the basis of their genetic transmission, this group of diseases results from a variety of different mutations that involve the genes responsible for the production of the alpha or beta globin chains. There is no one cause of thalassemia; there are literally hundreds. In some cases all or part of the genes are missing, or deleted. In others there are single nucleotide base substitutions in the DNA of the genes that inactivate them. For example, all genes must have a code word that reads "stop," so that the synthesis of a protein on its messenger RNA template ceases at the appropriate place. Some base substitutions generate a stop word in a gene in the wrong place, and hence a truncated and useless product is

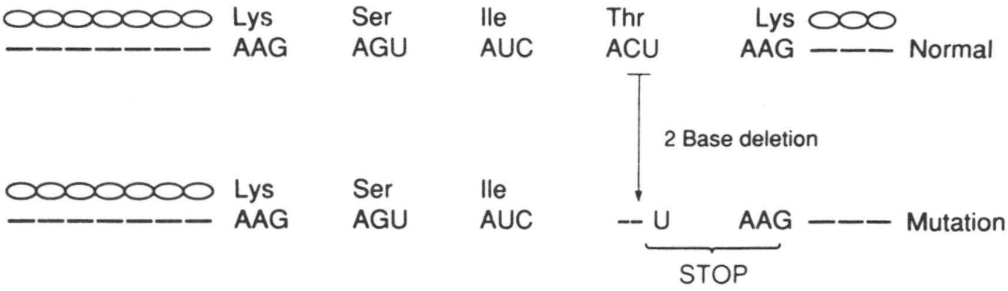

○○○○○○○ Lys   Ser   Ile   Thr   Lys ○○○
———————— AAG   AGU   AUC   ACU   AAG ———— Normal

2 Base deletion

○○○○○○○ Lys   Ser   Ile
———————— AAG   AGU   AUC   -- U   AAG ———— Mutation
                                        STOP

FIGURE 48

A common mutation found in human genetic disease. The normal gene sequence with the corresponding amino acids in the protein chain is shown at the top of the figure. At the bottom is shown the mutation that involves the loss, or deletion, of two nucleotide bases, A and C. This causes a shift in the reading frame of the genetic code, and the next three-letter word becomes UAA, instead of ACU. This is a stop codon—that is, protein synthesis ceases on the messenger RNA at this point. If this occurs in the middle of a gene, it causes premature cessation of the production of the protein, and the synthesis of a useless truncated fragment.

made by its messenger RNA. Other substitutions scramble the genetic code. As we saw earlier, the code consists of triplet of bases; that is, three bases determine the position of each amino acid in a peptide chain. Hence if one, two, or four bases are lost or inserted, the reading frame is thrown out of sequence. Such frameshift mutations occur quite commonly as the cause of different forms of thalassemia (figure 48).

It was not all that surprising to find these different mutations in human beings, because they had all been observed some years earlier in bacteria. Quite unexpected, though, was the discovery that many forms of thalassemia result from mutations that interfere with the processing of messenger RNA. Nearly all human genes are broken up into coding regions, called exons, and regions of apparent junk, called introns. When a gene is transcribed, the messenger RNA product contains both exon and intron sequences, and the introns have to be removed and the different exons spliced together to form the template for protein synthesis (see figure 44). It turned out that many forms of thalassemia result from mutations that either partially or completely inactivate this splicing mechanism and hence lead to a reduction of normal messenger RNA for globin. Many other novel mutations that cause the thalassemia have been found, and the clarification of the molecular pathology of this disease has given us a fairly clear idea of the repertoire of molecular changes that can cause genetic diseases.

The same approaches were used over the next few years to study other diseases. For example, hemophilia, the disorder that leads to abnormal blood clotting and bleeding, is due to the defective function of a gene carried on the X chromosome, one of the sex chromosomes we will discuss in more detail in the next chapter. When it was isolated and sequenced, it was found that it is much larger than the globin genes and that it contains many exons and introns. Its product is called factor VIII; blood clotting involves the interactions of many different proteins, some of which are designated by Roman numerals. When the factor VIII genes of patients with hemophilia were examined, similar defects to those in the thalassemias were discovered. Many other disorders that are transmitted by a single defective gene have since been studied in this way and the same types of abnormality found.

These early successes in human molecular genetics were restricted to diseases for which there was already some knowledge of both the underlying cause and the genes involved. For example, the hemoglobin disorders and thalassemias were known to be due to defects in the structure or synthesis of the globin chains of hemoglobin, and a great deal had been learned about the genes that control hemoglobin production, some years before the advent of molecular technology. Likewise, hemophilia was known to result from the defective production of factor VIII, and the gene involved was already assigned to the X chromosome. However, there were many important genetic diseases for which the underlying cause and the chromosomal location of the gene involved were not known. It became apparent in the early 1980s that the new techniques of molecular genetics might offer a way of tackling this problem. The idea behind this new approach to gene hunting was not new, but the methods of molecular genetics enabled it to be explored in a novel and, as it turned out, extremely productive way.

As we have seen already, our 50,000 to 100,000 genes are distributed among our twenty-three pairs of chromosomes. In 1927 J. B. S. Haldane reasoned that if it were possible to map fifty or more inherited characters—that is, to place them in the appropriate place on their chromosomes—they could be used as markers for predicting whether children would inherit genes for important diseases. The idea is beautifully simple. Suppose we want to follow the progress of a particular genetic trait through a family but have no way of identifying it. The thing to do is to find a gene whose product we can easily identify and which is linked to the gene for the trait that we are looking for. If the two are so close that they always pass together through successive generations, we now have a "handle" on the gene we cannot identify; if the marker gene is inherited, so must be the gene that is closely linked to it. And, of course, if we know the chromosomal location of our marker gene, the gene we cannot identify must be close to it on the same region of the chromosome.

The first gene to be assigned to a human chromosome was the one for color-blindness, which was found to be on the X chromosome by workers at Columbia University in 1911. Fifty-seven years were to pass before the first gene was assigned to anything but the X chromosome; a particular blood group was assigned to chromosome 1 by a team at Johns Hopkins University in 1968. But for the next twenty years or so, progress in assigning genes to chromosomes was very slow, largely because there were so few markers for linkage studies. By 1976 at least one gene had been assigned to each of the twenty-three chromosomes, and by 1987 at least 1,215 genes had been assigned, 365 of which are known to be the site of mutations that cause disease. These remarkable successes stemmed from the use of new mapping techniques, first involving the fusion of human and animal cells and later the advances in the recombinant DNA field.

The ability to fuse human cells with various rodent tumor cells opened a completely new field of genetics. Cells with more than one nucleus had been observed as early as the middle of the nineteenth century by microscopists who were examining tumor or inflammatory tissue. Later on in the century it became evident that viruses might be involved in the phenomenon of cell fusion; cells containing several nuclei were observed at the periphery of smallpox pustules, for example. This observation did not cause much interest in the scientific world, although John Enders and his colleagues, in their seminal studies on the growth of viruses in tissue culture, noticed that the measles virus induced cells to fuse together into multinucleated groups. In the early 1960s it became clear that fusion between different but related mouse tumor lines occurs when these cells are grown in culture. A few years later methods were developed for isolating these fused cells, but the techniques for encouraging fusion were extremely inefficient. The problem was finally solved in 1965 by Henry Harris and his colleagues at Oxford; they found that it was possible greatly to enhance the rate of fusion of cells by treatment with an inactivated virus.

It turned out that if human cells are mixed with rodent tumor cells, and grown in culture together with an organism called sendi virus, they tend to fuse together. After fusion the chromosomes of each of the cells become mixed together, and subsequently many of them are lost from the now hybrid cell; human chromosomes are shed preferentially. Furthermore, they are lost in a random fashion. It is therefore possible to propagate hybrid cells containing specific human chromosomes. The particular genes on these chromosomes can be identified in a variety of ways. For example, it may be possible to detect a specific product in a hybrid cell line and hence relate its gene to the chromosome that is unique to that line. In recent years many other ingenious ways of identifying individual genes on human chromosomes in hybrid cell lines have been developed.

The other approach to mapping human genes followed a completely different route. Once it became possible to chop DNA with bacterial restriction enzymes, it soon grew apparent that all of us vary considerably in the structure of our DNA. Much of this variation is harmless and simply reflects the presence of a different nucleotide base, which either produces a new cutting site for an enzyme or removes a previously existing one. Thus the hallmark of this type of genetic individuality is variation in the length of DNA fragments produced by a particular enzyme. Such restriction fragment length polymorphisms (RFLPs) have become valuable genetic markers, particularly if their position on a chromosome is known. And further exploration of human DNA showed the existence of even more useful markers. For example, certain regions of DNA vary considerably in length in different individuals, and there are even more subtle structural variations of this type. These findings have revolutionized human genetics, for they have made it possible to find genes involved in causing genetic diseases, the nature of which used to be unknown.

In linkage studies of this kind, families are examined both for the presence of a particular marker—an RFLP, for example—and for the disease or other trait that is being studied. If the disease and marker are on different chromosomes, they will be found together as often as they are apart in different generations. If, on the other hand, the two are close together, independent assortment of this kind will not occur, and they will stay together unless separated by a crossover when the maternal and paternal chromosomes pair at meiosis. On the average, there are about fifty-two crossovers at each meiosis—that is, about one to six per chromosome, depending on its length. By using some relatively simple mathematics, one can derive an approximate idea of how close the gene and marker are to each other.

After establishing a linkage of this type, we still have some work to do; we may still be several million bases away from the gene that we are looking for. However, by some ingenious genetic engineering, rather picturesquely called chromosome walking or hopping, it is possible to move from the marker toward the gene of interest and, finally, if we are lucky, to find it. Once having identified the gene, we can sequence it. This information then yields some idea of what the protein product might be like, and what it might do.

This unlikely activity, which was originally called reverse genetics but which has since been rechristened positional cloning, represents one of the most important developments in molecular medicine. It has led to the identification of the genes for important diseases like muscular dystrophy and cystic fibrosis, among many others. Muscular dystrophy is a relatively common genetic disease that causes progressive muscular weakness after the first few years of life. The gene involved is on the X chromosome. Positional cloning was used to identify it and to determine the nature of its product. It turned out to be a large protein, called dystrophin, which seems to be absent in children with severe muscular dystrophy. Many different types of mutation are involved in this disease, just as was found for the thalassemias. At the time of this writing, the precise function of dystrophin is still unknown.

Another success was the discovery of the gene involved in cystic fibrosis. This is the commonest genetic disease in Western societies and is characterized by the production of thick, viscid sputum and chronic respiratory infection, as well as defective production of the digestive juices of the pancreas, which may lead to serious gastrointestinal dysfunction in infancy. The gene for this condition was isolated by linkage and positional cloning, and many different mutations have been found, although one particular type predominates in northern European populations. Its product has all the earmarks of a transport protein involved in moving various critical minerals in and out of cells.

The discoveries of the gene for muscular dystrophy by Louis Kunkel and his colleagues in Boston, and of that for cystic fibrosis, by Lap-Chee Tsui and Francis Collins and their colleagues in Toronto and Ann Arbor, were the fruit of several years of intensive research by these groups and many other teams around the world. This work was particularly important for human molecular genetics because it showed that the linkage approach, followed by some deft genetic engineering, is capable of isolating genes for conditions for which there is no inkling of the underlying cause. Over the last few years the finding of novel genes has become so commonplace that rarely a week goes by without the announcement of the isolation of another gene for an important genetic disease. But it is not always as easy as it sounds. The gene for Huntington's disease, a serious disorder of the nervous system, was pinpointed to chromosome 7 in 1983; it was not isolated until ten years later.

Although the function of the gene that is changed in Huntington's disease is not yet understood, the elucidation of the novel type of mutation that causes it adds strength to several recent discoveries of this kind that are throwing some light on a group of mysterious inherited diseases of the brain and nervous system. It turns out that the genes that are involved, including those for Huntington's disease and the fragile X syndrome, a common cause of mental retardation in boys, have, either within them or close by, long tracks of what are called trinucleotide repeats. These are simply lengths of DNA that consist of multiple repeats of the same three bases. They appear to be unstable and may increase in length in successive generations. When they reach a critical size, they interfere with the function of the gene, and the disease is manifest. This quite unexpected finding is starting to make sense of the curious patterns of inheritance of these distressing disorders.

It is now clear that almost any interesting gene can be found and that we will soon know a great deal about the molecular basis of most important genetic diseases. Analyzing them at the DNA level enables us to identify carriers—that is, those who have only one copy of a defective gene and, though not themselves sick, can pass the defect on to their children—and also to detect genetic disease in fetal life in those cases in which the parents do not

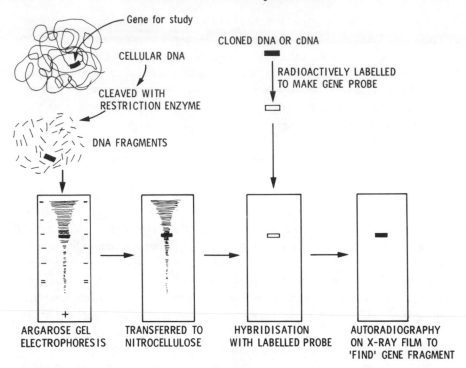

FIGURE 49

*Searching for human genes by Southern blotting (so called after its inventor, the
Oxford scientist Edwin Southern). The gene that we are looking for is marked as
a black box, hidden away in the whole of the genome. The latter is chopped up in
small pieces with enzymes called restriction endonucleases and separated by size
in an electric field. The pieces of DNA are then blotted from the gel onto
nitrocellulose paper. To find the position of the gene on the paper, a radioactive
counterpart, or probe, is constructed and sent to search for its partner; the two
will anneal together only if they have an identical structure. Everything else is
washed off the paper, which is then placed on an x-ray plate. The radioactivity
produces a band on the x-ray film that points to the position of the gene, or gene
fragment.*

wish to have a seriously disabled child. Thoughts are even turning to correcting genetic diseases by placing "good" genes into patients' cells, a theme to
which we will return later.

This new field has already yielded important practical applications for the
control of common genetic diseases. The first applications of DNA technology
for screening and prenatal detection of genetic disease came from the thalassemia field. It became possible to detect thalassemia in small samples of fetal
blood obtained at about eighteen weeks' gestation, and later, by means of
DNA analysis (figure 49), defective globin genes could be identified as early

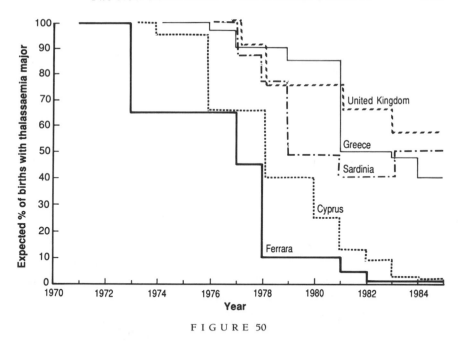

FIGURE 50

The results of prenatal diagnosis programs for thalassemia in different populations. There has been a steady decline in the expected number of births of new cases of beta thalassemia in these countries since the instigation of prenatal diagnosis programs (WHO data).

as eight to nine weeks' gestation in minute amounts of tissue removed from around the fetus. By applying these new methods, we have been able to offer parents the option of terminating pregnancies that carry babies with lethal forms of thalassemia, and hence allowing them to have normal children. This approach has been used extensively in countries in which the thalassemias are common. In the populations of Sardinia, Cyprus, and Greece, and in some of the immigrant populations of Great Britain and the United States, there has been a major reduction in the frequency of these diseases. In particular, some highly effective education and public health measures, together with the use of these new techniques, have almost eradicated serious forms of thalassemia on Sardinia and in parts of Italy (figure 50).

Cancer

With the exception of a few familial forms, cancer is not a genetic disease in the usual sense of the word. Most cancers cannot be traced through families in a way that obeys Mendel's laws. However, it has become apparent over the last few years that cancer results from a series of events in the life of a cell that so alter its genetic makeup that it loses its ability to divide and mature in a normal manner. In this sense cancer is a genetic disease, because it is due to acquired defects in the genetic machinery of cells, defects that they pass on to their progeny.

The last few years have brought some remarkable discoveries concerning the molecular basis for the disordered proliferation, growth, and maturation of cells that characterizes all the common cancers. These advances stem from an amalgamation of information obtained from the study of viruses that cause tumors, alterations in the chromosomal makeup of cancers, and pedigree analyses of families with rare familial cancers. In addition, it should not be forgotten that these discoveries in the basic laboratory sciences have been backed up by equally spectacular successes in studying the epidemiology of cancer, that is, the factors to which individuals may be exposed that tend to make them more likely to develop a tumor.

In 1911 Francis Peyton Rous (1879–1970), an American pathologist working at the Rockefeller Institute, discovered that certain cancers of chickens are caused by a transmissible agent with the properties of a virus. It took a while for the scientific world (and even longer for the Nobel Prize committee) to grasp the importance of this discovery; fortunately Rous lived long enough to receive the Nobel Prize, fifty-five years after he had first made this critical observation. Rous's agent turned out to be an RNA virus, or retrovirus, which, when it invades cells, has its RNA genome transcribed into DNA, which then becomes inserted into the host's chromosome. Many retroviruses have been identified since then, as well as several types of tumor virus made of DNA.

Tumor viruses can cause cancer in several ways. For example, the insertion of the viral DNA into the host's chromosomes may cause an alteration in the behavior of their genes, simply by the presence of the foreign DNA close to them. Or, and more interestingly, many tumor viruses carry one or more genes responsible for their ability to produce cancer. These nasty genes are called viral oncogenes.

In 1989 Michael Bishop and Harold Varmus of San Francisco won the Nobel Prize in medicine for their remarkable discovery that all living organisms, including human beings, have genes that are very similar to viral oncogenes. These homologs of viral oncogenes are called cellular oncogenes, an unfortunate term as it turns out, because they are part of our cells' normal genetic machinery, responsible for the control of their proliferation, differentiation (specialization), and development. It appears, therefore, that at some time in their evolutionary history, some retroviruses "acquired" normal cellular genes, presumably by recombination, and these, in their new home, are capable of causing cancer under appropriate conditions. It must be emphasized that this complicated saga is not telling us that human tumors are caused by viruses. Rather, it was the discovery that viral oncogenes have their counterparts in the cells of living things that led to the discovery of a family of genes that are responsible for a wide variety of normal cellular functions but that, if altered by mutation or in other ways, can cause a cell to change its behavior and become cancerous.

Thus, by a roundabout route, the notion evolved that many tumors may be caused by mutations involving cellular oncogenes. The story of the discovery of how changes in human oncogenes can cause cancerous transformation of cells, worked out in several laboratories in the early 1980s, is one of the most interesting in modern biology.

The introduction of DNA isolated from normal or cancer cells into certain cell lines that can be grown in culture has turned out to be a valuable method for studying the action of oncogenes. One line that has been used extensively for this purpose is derived from mouse cells. DNA sequences from human tumor cells were found to have the capacity to induce what cell biologists call "transformation," or disordered growth, of these cultured mouse cells. It turned out that, after several rounds of growth, the mouse cells contained only the minimal amount of the human DNA required for transformation. The techniques of molecular biology made it possible to focus on this small amount of residual human DNA and to show that it had sequences very similar to those of viral oncogenes. Even more remarkably, when one of the human cellular oncogenes was isolated, it was found to differ from normal by only a single nucleotide base. Thus it was reasonable to suppose that this mutation was responsible for the transforming properties of the human DNA. Since these initial observations many human oncogenes have been discovered and found to contain mutations in different forms of cancer. Particular oncogenes seem to be involved with cancers of different tissues, suggesting a causal relation between oncogene mutations and malignant change.

Another thread in this story appeared after the long-standing observation that some human cancers are associated with specific abnormalities of chromosomes. For example, in some forms of human leukemia the same chromosome defect is always observed. These changes often involve the breakage and translocation of pieces of chromosome to the wrong place. It turned out that the regions of the abnormal chromosomes involved in these breakages and the sites of abnormal association of different pieces of chromosome may involve particular oncogenes. Thus it appears there may be an abnormal activation or defective activity of oncogenes as a result of their being involved in the chromosome breakages or translocations found in these cancers.

The final twist to this tale was the discovery that some tumors may result from the recessive inheritance of single genes. It will be recalled that by "recessive" we mean that we may carry a defective gene, but because we have a good one on the other chromosome, the bad gene shows no effect. The cancer-producing properties of certain recessive genes may become unmasked by a variety of mechanisms, all of which have in common the inactivation of the normal partner, which somehow is able to repress the action of its abnormal allele. Although at first this phenomenon seemed to be restricted to certain childhood cancers, it soon became apparent that it may occur in many common adult tumors.

Studies of rare childhood tumors, particularly of the eye and kidney, have led to the concept of anti-oncogenes. It turns out that if one of a pair of genes of this type carries a mutation, which may be either inherited, as in the case of some of the rare childhood tumors, or acquired later in life, there may be no effect on the cell while its normal partner, or anti-oncogene, remains active. If, however, by some genetic accident, the normal partner is lost or otherwise inactivated, the oncogene that carries the mutation may cause the cell to become cancerous. Loss of anti-oncogene activity by deletion has now been observed in many common cancers, including those of the lung, breast, bowel, and bone. In the last few years it has been possible to isolate the products of these genes and to start to understand how tumor suppression is mediated by anti-oncogenes.

Does this mean that we now know the cause of cancer? Not yet. But these remarkable discoveries have made cancer a much less mysterious condition than it was ten years ago, and a primitive picture of what goes wrong in a cancer cell is emerging.

It seems likely that the first event on the road to a cell's becoming cancerous is a mutation of an oncogene; usually this mutation is acquired, although occasionally it may be inherited and we not know anything about it for many years. Sometimes oncogene activity may be altered by a major chromosomal rearrangement. Many substances such as tobacco smoke and chemicals that we are exposed to in our lifetime can increase the rate of mutation. In other words we have now established a link between what the epidemiologists have been telling us for the last twenty years and what happens in our cells. But it appears that most cancers require mutations of more than one oncogene. Recent studies by Bert Vogelstein of Johns Hopkins University indicate that some common cancers of the bowel may require multiple mutations involving many different oncogenes before the tumor starts to grow and spread. His elegant work suggests that a minimum of six to eight mutations involving different oncogenes, and in a particular order, may be needed. Many of the genes involved and their products and functions have now been identified. These remarkable discoveries will soon have their application in the clinic, for both the diagnosis and the treatment of our common cancers.

HOW DID HUMAN MOLECULAR GENETICS COME OF AGE?

As we have surveyed the development of modern scientific medicine, we have tried to explain why discoveries were made, or why completely new fields evolved, at a particular time. Major advances usually reflected a critical phase of progress in a number of different disciplines. Human molecular genetics offers a particularly good example.

Among the first diseases to become amenable to study at the molecular level were the common genetic disorders of hemoglobin—sickle-cell anemia and thalassemia. In trying to understand why, we need to examine how several different areas of research came together at just the right time, in the mid-1970s.

After William Harvey's work on the circulation of the blood Robert Boyle and Richard Lower helped to show that the function of its passage through the lungs is to aerate venous blood. In the nineteenth century the mechanism of the binding of oxygen to blood became a subject of intense interest, and in 1862 the German biochemist Felix Hoppe-Seyler (1825–1895) first used the term "hemoglobin" to describe the oxygen-carrying pigment of red blood cells. Early in the twentieth century physiologists came to understand the way in which oxygen binds to hemoglobin and some of the remarkable adaptations needed to meet the oxygen requirements of the tissues. It became clear that hemoglobin is a complex molecule that consists of an oxygen-binding complex, or heme, attached to a protein, or globin.

We have already touched on the discovery of sickle-cell anemia and thalassemia, how their genetic transmission was worked out, and how it was recognized that they are the commonest single-gene disorders in humans. By the late 1950s it was apparent from work based on simple family studies that all these conditions are genetic disorders of the globin part of the hemoglobin molecule. At about this time hemoglobin was turning out to be a fascinating protein for curiosity-driven research by protein chemists. Apart from the importance of its function there was a pragmatic reason for their interest. Many proteins are difficult to study simply because they make up only a tiny part of a particular tissue. Before they can be analyzed, long and complex purification processes have to be carried out. This is not so with hemoglobin; it is only necessary to burst open a red cell to obtain a hemoglobin solution that is over 98 percent pure. Thus after Frederick Sanger had led the way by determining the structure of insulin, several protein chemists tackled hemoglobin and soon established that it comprises two different pairs of peptide chains, alpha and beta. At the same time Max Perutz and his colleagues at Cambridge, after years of painstaking work, gradually arrived at a solution for its three-dimensional structure by x-ray analysis and, with others, started to describe how it works as an oxygen carrier.

During the 1960s the research by clinicians who were studying families with hemoglobin disorders combined in a very fruitful manner with the discoveries of the protein chemists. It became apparent that different genes are involved in the production of the alpha and beta globin chains and that they are the site of the mutations that result in hemoglobin variants like sickle-cell hemoglobin or the many forms of thalassemia. Furthermore, by studying the transmission of these disorders in large families, scientists could make a reasonable guess about the precise arrangements of the different hemoglobin genes on their respective chromosomes. At the same time they were becoming curious about how proteins are synthesized and assembled. Radioactive labels soon made it possible to reproduce this process in the test tube. Again, and in this case because it is easy to obtain young red cells that are busy making it, hemoglobin was the protein studied in the greatest detail. When this approach was applied to the thalassemias, it was found that they result from defective alpha or beta chain production.

Thus by the early 1970s the human hemoglobin field was custom-built for the new developments in molecular biology. A great deal was known about the structure and function of hemoglobin, its genetic control and how it is made had been worked out in some detail, and a large family of diseases, known to be caused by mutations of the globin genes, was sitting and waiting for methods to characterize them.

As soon as gene cloning and sequencing arrived on the scene, the deep freezes of several laboratories in the United States and England were raided and DNA extracted from thalassemic blood samples. Between 1975 and 1990 the molecular basis of the thalassemias was worked out in great detail, and a very clear picture emerged of the relation between what goes wrong at the genetic level and what we see in our patients. And just a few years after the discovery of the first thalassemia mutation the new technology was being used in the clinic for the control of the disease.

Although some "professional" molecular biologists moved in and out of the human hemoglobin field at the time when the molecular basis for these common diseases was being worked out, many of the key discoveries were made by groups of medical scientists in university clinical departments who had educated themselves in the new technology of molecular biology. The level of sophistication of this work marked a new phase of development for clinical research. Although it was a natural extension of the more physiological measurements that had been the basis for the success of clinical science for many years, it undoubtedly reflected the beginning of an era in which the emphasis in the study of disease changed from patients and their organs to events at the molecular level. Within a few years many other diseases were being studied in the same way.

The story of the evolution of this one small part of molecular medicine over the last thirty years illuminates how medical science works. Its success depended on fundamental, curiosity-driven science in many fields and on the development of new disciplines—first protein chemistry and then molecular biology. The careful collection and collation of information about the hemoglobin disorders by clinicians all over the world, from the 1920s into the 1960s, led to their being among the best characterized of genetic diseases. It required only a few people to communicate between the clinical and basic-science worlds, and the two fields, which were ripe for each other, came together in a very productive way. From the vantage point of a research worker in the field during this exciting period, the whole thing seemed to happen quite naturally, but it could not have been planned in this manner by an administrator or a government body that supported scientific research. It is yet another example of the value of encouraging curiosity-driven basic science as well as good clinical research; no one can guess the outcome, or when by chance several fields of knowledge will amalgamate like this.

THE HUMAN GENOME PROJECT

Our bird's-eye view of the way in which genetics and molecular biology have developed has shown how they have been applied spectacularly to two important areas of clinical medicine—inherited disease and cancer. The choice of these two topics was arbitrary because the application of this new technology has started to make major advances right across the field of human pathology. Indeed, it has been so successful during the last fifteen years that thoughts have turned to what might be possible if we had a complete map of the human genome and knew the location of all our genes on our chromosomes.

The notion of the Human Genome Project, as this ambitious scheme came to be known, originated in the United States in the mid-1980s, mainly as the result of two meetings. In May 1985 Robert Sinsheimer, chancellor of the Santa Cruz campus of the University of California, invited several American and European molecular biologists to discuss the technical feasibility of the project. A similar workshop was convened at Los Alamos early in 1986 by Charles De Lisi, director of the Office of Health and Environment at the Department of Energy (DOE). The DOE has roots going back to the Manhattan Project and the development of the atomic bomb, but it has long sponsored research in the biology of radiation and human genetics as well. It was the unanimous opinion of the scientists who attended these meetings that the project was feasible. The same conclusion was reached in June 1986 at a meeting organized by James Watson at Cold Spring Harbor. By now there was great enthusiasm for the idea in many of the richer, industrialized countries; the scene was set.

The Human Genome Project has gained momentum over the last few years, although it is not without its problems. It has become the subject of a heated debate over expense, priority compared with more pressing problems of medical research, and, above all, the best way to proceed. Much of this controversy has been clouded by institutional and national pride, commercial concerns, and the usual parochial attitudes that tend to color international collaborative activities in science, or in anything else, for that matter. However, genuine progress is being made, and it seems very likely that in the foreseeable future we will have some kind of map of our genes.

Earlier in this chapter we saw how genetic linkage studies can help us find our way around chromosomes. One type of map of our genome would consist of linkage markers of this kind. In effect, it would be like having a road atlas with the towns, in this case our genes, dotted along the chromosomes, or roads. It would tell us nothing about the state of the roads in between the towns, but should lead us very quickly to our destination, which, in a medical context, would be genes involved in disease. Of course, we might want to go a step farther and describe the state of the roads, that is, to obtain a precise sequence of all the nucleotides that make up the human genome. This tour de force, called a physical map, might contain a lot of uninteresting DNA; we simply don't know. On the basis of these arguments and uncertainties, major efforts are being made to develop a linkage map and establish the technology for producing a complete physical map.

Work on the Human Genome Project has already resulted in a linkage map of most of our chromosomes, although at present some of the markers are still too far apart from one another. A great deal more work is required before we have the kind of map that can lead us quickly to genes in which we are interested. The sequencing part of the project is riding on the back of technological advances. Methods for DNA sequencing are improving all the time, and the computational sciences are turning up valuable leads as to how the enormous amount of data that will be generated can be stored and disseminated. This part of the program is becoming automated and will depend heavily on robotics.

But how will we know what the thousands of genes that we sequence actually do? A major effort is afoot to find methods for determining the functions of proteins from a knowledge of their amino acid sequence, as surmised from the nucleotide sequence of their parent genes. In this important step we are currently bedeviled by difficulties in understanding the factors that determine how long chains of amino acids fold into the shapes required for proteins to function as enzymes or other biological mediators. But though progress is slow, we will surely get there in the end.

The Human Genome Project has been called the Holy Grail of modern biology. It is likely to cost more than three billion dollars and is arguably the most ambitious human endeavor since the Apollo program landed a man on the moon. Its protagonists believe that it will transform medical practice in the twenty-first century; others believe that the knowledge we will gain from the enterprise will raise such serious ethical issues for our great-grandchildren that we should not be embarking on it at all.

But I suspect that, having gone this far, we cannot turn back and that we will have a dictionary of the position of all our genes and much of the sequence of the human genome completed by the second or third decade of the twenty-first century. It will not be light bedtime reading; if published as a book, it will require six or seven thousand volumes of the size of this one. Is it all worth it? Will it, as its champions suggest, completely change the face of medicine? And if so, when? And what ethical problems will it pose for us? These are some of the questions that we will address in the remaining chapters.

What Might We Expect from Basic Medical Research in the Future?

Basic research is not the same as development. A crash programme for the latter may be successful; but for the former it is like trying to make nine women pregnant at once in the hope of getting a baby in a month's time.

—*Richard Doll*

We have seen how the sophisticated new tools of molecular and cell biology have revolutionized research on genetic diseases and cancer and how they are being used to pinpoint the position of all our genes on our chromosomes. But it is already clear that this new field has much more to offer. With any luck it should start to tell us something about what causes the common disorders of Western society and how to tackle some of the stubborn problems of the developing world. It may help us understand the mysteries of human development and the ways in which things can go wrong and lead to congenital malformation and mental retardation. It also has the potential to revolutionize the pharmaceutical industry and to offer us novel approaches to the prevention, diagnosis, and treatment of our diseases. In short, this remarkable new technology should lead to major advances across the whole spectrum of clinical practice. And it will make an impact on broader issues of human biology, such as evolution, aging, and human behavior.

But before we look into the future, we must remind ourselves that the molecular sciences, exciting though they are, may take many years to achieve their full potential. It is important not to become wholly seduced by their attractions, and thereby neglect the less newsworthy areas of scientific medicine. Epidemiology and medical statistics, the other fields in the medical sciences that have made great advances in the second half of this century, will continue to unravel the relation between our environment and disease, lay a firmer and more critical basis for preventive medicine and the delivery of health care, and, incidentally, point the way in which the molecular sciences must go in sorting out the causes of our common killers. Almost every branch of clinical practice could be improved by the application of properly designed clinical trials and scientifically based audit. And, while we may have been critical of our current high-technology practice, directed at prolonging the lives of those suffering from the intractable diseases of middle and old age, we must continue to pursue research at the bedside to try to improve it; it would be foolish to down tools in the expectation that preventive medicine and molecular biology will solve our problems overnight.

In this chapter we will hazard a guess at how the new basic medical sciences will develop over the next few years. We will start by taking up the story of genetic disease from where we left it in the preceding chapter, for here there seem to be more certainties than in many branches of current medical research.

DISEASES ENTIRELY GENETIC IN ORIGIN

In his remarkable compendium of genetic diseases, *Mendelian Inheritance in Man,* the Baltimore medical geneticist Victor McKusick lists no fewer than four thousand diseases inherited in a way that suggests they are due to single defective genes. If the gene is dominant, only a single dose inherited from an affected parent is required to cause the disease; if it is recessive, defective genes must be inherited from both parents, each of whom, having only a single dose, is unaffected. In addition to our twenty-two pairs of what are called autosomal chromosomes, we all have a pair of sex chromosomes; females have two X chromosomes; males have an X and a Y. Many inherited diseases are carried on the X chromosome. Because females have two X chromosomes, they are usually unaffected by X-linked recessive diseases; the defective gene is "masked" by the function of its normal partner. However, a mother may hand down an X chromosome with a defective gene to her son, who, because this is his only X chromosome, will then express the disease.

All of us carry about eight to ten recessive genes for serious genetic diseases. This does us no harm unless we happen to have our children with a partner who is carrying a gene for the same disease. In this case, Mendel's rules of inheritance predict, each time we have a child, there is a one-in-four chance that it will receive two doses of the defective gene. Of course, if we marry a close relative there is a greater chance that he or she will carry the same recessive genes that we ourselves carry and hence a greater likelihood of our having affected children. It has been realized for centuries that consanguinity is associated with a higher prevalence of inherited diseases, and this is one of the main reasons why marriage to close relatives is forbidden in many societies.

Fortunately, most genetic diseases are extremely rare. However, because there are so many of them, they cause a considerable amount of ill health and mortality; it has been estimated that about 1 percent of all children carry some kind of genetic disability. But the load of genetic disorders is greater than this because of the related family of diseases that result from major abnormalities of chromosomes. Many different kinds of mistakes can occur in germ cells that may lead to defects of this kind. The frequency of chromosome abnormalities at birth is approximately 5.6 per 1,000. The abnormalities are of two kinds. First, there is a variation in their number, which accounts for sixty or more different diseases. The commonest is Down's syndrome, previously called mongolism, which in many cases is due to an extra chromosome 21. In addition, there are structural abnormalities, many of which are due to translocations, or exchanges of pieces of various lengths between different chromosomes. Or we may be born missing particular regions. Because these abnormalities involve pieces of our chromosomes that carry hundreds or thousands of genes, it is not surprising that they cause severe disability.

These diseases will always be with us because new mutations occur all the time. Their frequency depends on the rates of mutation and, in the case of recessive diseases, on whether carriers are (or were in the past) at any advantage. We have already seen that the genes for sickle-cell anemia and thalassemia have come under intense selection by malaria, and that cystic fibrosis may be common because one particular mutation that causes it may also have afforded protection to carriers, perhaps against one of the epidemics of infectious disease that devastated European populations in the last thousand years.

How can we control genetic diseases? Very few can be cured. Some congenital malformations can be corrected by surgery, but for the majority the only form of therapy is to try to control the symptoms. There exist many varieties of symptomatic treatment, which differ in expense and unpleasantness: replacement therapy for missing clotting factors in hemophiliacs; long-term blood transfusion for thalassemia; growth hormone for some inherited forms of dwarfism; and so on. It may be possible to bypass inherited biochemical defects by designing diets that prevent the accumulation of toxic substances. For example, the common condition called phenylketonuria, if untreated, leads to severe mental retardation. Affected children can, by rigid adherence to a special diet, grow and develop normally. Many genetic diseases are due to defective enzymes, and efforts have been made to replace them, although not with much success. A few can be cured by organ transplantation—of bone marrow or liver, for example.

How successful is the treatment of genetic disease? In 1985 a joint research program was set up between McGill and Johns Hopkins universities. Altogether, 351 diseases caused by single defective genes were studied and the effects of treatment assessed by means of three variables: life span, reproductive capability, and social adaptation. It was found that symptomatic treatment resulted in a normal life span in 15 percent of the diseases, allowed reproduction in 11 percent, and led to what was judged to be adequate social adaptation in only 6 percent. When 65 of the diseases for which the cause was known were examined in more detail, it was concluded that treatment was completely successful in only eight, or 12 percent of those studied, moderately effective in another 26, and completely useless in the remainder, or nearly 50 percent.

So what are the options for this group of distressing diseases? The recombinant DNA field promises to lead us in two directions. First, it will help us develop programs for their avoidance. In the longer term, it may enable us to devise ways to treat them by a completely new approach called gene therapy.

We talk about "avoiding" genetic disease, rather than "preventing" it, for a good reason. The only way to prevent it would be to stop mutations from occurring altogether. Yet it is mutations that with natural selection, have made us what we are; single-gene disorders are, in effect, the battle scars of that painful process. We can try to ensure that the mutation rate does not increase, by avoiding exposure to radiation or other mutagens, but simply by the nature of things we will always have to cope with deleterious mutations.

There is a variety of ways in which we might try to avoid genetic disease. Particularly now that we can isolate defective genes, we can develop rapid diagnostic methods for identifying carriers. This means that it is possible to screen those with a family history of a particular disease, or to offer population screening—in the antenatal clinic, for example. It is only practical to screen populations for diseases that are unusually common, such as cystic fibrosis and thalassemia and sickle-cell anemia in those of appropriate racial backgrounds.

If a couple are found to be at risk for having children with a genetic disease, they have several options. They can decide to forgo having any, to adopt, or to have a family and accept that some of them will have a serious genetic disease. However, an alternative is becoming increasingly available as more is learned about the molecular basis for genetic diseases. It involves prenatal detection and termination of the pregnancy if the fetus is found to be affected.

Prenatal diagnosis is not new. For many years it has been possible to identify mongol babies by examining chromosomes, or by other tests, after removing a sample of amniotic fluid. Several other genetic diseases can be identified in this way. In the mid-1970s methods were developed for obtaining blood samples from the fetus in order to diagnose inherited blood diseases like sickle-cell anemia and thalassemia. But they could be carried out only in the middle of pregnancy. This meant that if parents wished to terminate a pregnancy because the fetus had a serious genetic disease, it had to be done in about the twentieth week of pregnancy, which was extremely distressing for many women. However, now that we can identify many genetic diseases by means of a small sample of DNA, prenatal diagnosis can be done much earlier in pregnancy. Thanks to a method called chorion villus sampling, which entails removing a tiny piece of tissue from the membrane that surrounds the fetus early in pregnancy, many genetic diseases can be diagnosed between the ninth and fourteenth week (figure 51). And by amplifying DNA, we can even identify mutations in a few cells taken from eggs after in vitro fertilization.

There is no doubt, therefore, that the application of DNA technology will make it possible for many families at risk of having a child with a serious genetic disease to undergo screening and prenatal detection of disease early in pregnancy. They will then have the option of terminating affected pregnancies and, of course, of having normal children through pregnancies in which their babies do not inherit the defective genes.

But however successful we become at avoiding inherited diseases, children will continue to be born with them. So what are the prospects for improving our unsatisfactory methods of treatment? Because we can now isolate human genes and are starting to understand which parts of them are required for their control, thoughts are turning to correcting genetic disease. There are two ways in which this might be done. The first is called somatic gene therapy. This means that a "good" gene would be placed into a particular tissue so that it replaced the function of a defective gene during the life of the individual. It would not be passed on to future generations and hence, in principle, would be little different from organ transplantion, which is now an accepted part of day-to-day medical practice. The other way is called germline gene therapy. This technique would place our "good" gene into a fertilized egg, where it would become dispersed throughout many tissues and organs, including the cells that would form eggs or sperm during later development. In this case the "foreign" gene could be passed on to future generations. In most countries germline gene therapy is banned, largely because so little is known about its safety and because it is felt, at least at the present time, that it is not right to interfere with the genetic makeup of future generations.

Somatic gene therapy is certainly in the cards, although how long it will take to perfect is anybody's guess. It is relatively easy to isolate human genes, and we are beginning to understand how much DNA is required for a particular gene to be properly controlled, that is, for it to function in the right tissues and to make its product in appropriate amounts. The chief technical problems that remain are how to transfer genes into cells and how to obtain sufficient cells that the treatment will be successful.

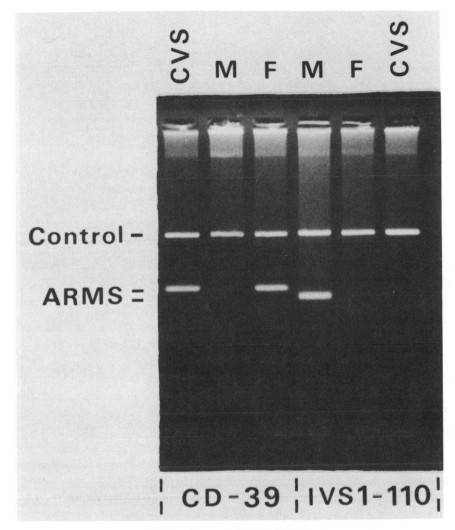

FIGURE 51

The rapid prenatal diagnosis of genetic disease by DNA analysis. The ARMS method allows the detection of individual mutations. This baby was at risk for two different thalassemia mutations, one from each parent, marked here as CD-39 and IVS1-110. This technique allows both mutations to be searched for in the mother (M), the father (F), and the DNA from the fetus obtained by chorion villus sampling (CVS). The bands are generated by both the normal (control) chromosomes and those containing the mutation (ARMS). It will be seen that the father and mother are carriers for two different mutations, which produce different bands in the ARMS test. If the baby had been affected with both mutations, and therefore been at risk for serious disease, it would have had both of these bands. In the event, it inherited the father's abnormal chromosome but not the mother's and therefore was a carrier. The whole investigation was completed within an afternoon.

Currently many research teams are trying to insert genes into the cells of the bone marrow. The marrow is the source of the cells that make up our blood: red cells, which carry oxygen; white cells, which help us combat infection; and tiny cells called platelets, which heal gaps in blood vessels and prevent bleeding. The marrow is an extremely busy factory. Every twenty-four hours we destroy approximately 1 percent of our red blood cells, which have a life of about 120 days. Our white blood cells vary in their life span, from a week or so to months or even years in the case of cells called lymphocytes, which are vital components of the body's immune system.

Many genetic diseases involve red or white blood cells. However, since they have a limited life span, it is no use putting genes into them, because we would have to repeat the exercise every few weeks. All our blood cells are the products of a small, self-renewing population of stem cells in the marrow. The problem is that they constitute only a tiny percentage of bone marrow cells, and currently we have no reliable way of identifying or isolating them. Despite this, attempts are being made to insert genes into marrow cells in the hope of correcting genetic diseases, and into other types of cells, such as liver, skin, and vascular endothelium (the lining cells of blood vesels).

How do we transfer genes into cells? It is impractical to inject them one cell at a time. We need to insert our genes into many thousands of cells if we are to have any hope of correcting a genetic disease. Currently the most promising approach is to use nature's way—that is, to attach the genes to retroviruses, which have been designed specifically to insert their genes into cells and to incorporate them into the genetic machinery of the recipient cell. Of course, retroviruses are dangerous creatures that cause disease, but it is possible, by some extremely ingenious genetic engineering, to remove the genes that make them what they are and to replace them with human genes that we wish to insert. This is turning out to be a fairly efficient way of moving genes around, although there are still some doubts about its safety and about the life span and effectiveness of the genes that are inserted.

Because many scientists believe that it will never be possible to be absolutely sure about the safety of retroviruses, other methods for gene therapy are being explored. The use of different types of viruses, with a particular liking for the tissue in which we wish to insert a gene, or joining genes to molecules that attach themselves to cell-surface receptors as a way of gaining entry, seems to be particularly promising. Many different viruses are being explored as candidates for acting as carrier vectors. Adenovirus, for example, specifically invades the cells that line our bronchial passages. This vector is being used to deliver the gene that is defective in cystic fibrosis, in an attempt to correct the pathology in the lungs. Experiments in humans have run into some unexpected problems of infection, but expression of the normal counterpart of the protein that is defective in cystic fibrosis has been observed in the bronchial linings of rats for up to six weeks after viruses containing the gene were installed. Promising results have also been obtained by inhalation of the genes packaged up in small, fatty envelopes, to avoid the use of a viral vector.

In the preceding chapter we saw how genes are exchanged when chromosomes line up with each other in our germ cells. Attempts are being made to reproduce this effect in the test tube, with the objective of exchanging normal for defective regions of genes. This has the attraction of precision; genes inserted into cells attached to retroviruses may end up anywhere in our DNA, and there is always the outside chance that they will interfere with the activity of an adjacent gene. And even more novel avenues are being explored, in which artificial minichromosomes are constructed with DNA from both yeast and human sources, or genes are transferred in tiny, fatty envelopes. In an attempt to treat muscle disease DNA is even being injected directly into the tissue.

Some progress has already been made. For example, it has been possible to insert a gene into the white blood cells of children that have an inherited defect in combating infection. These cells, now carrying a "good" copy of the gene that is defective, appear to restore their ability to fight infection, at least during their limited lifetime. And as we will see later, gene therapy has also been used for the treatment of advanced cancer. These are early days, and I have cited here only a few of the many different routes being explored to try to perfect this novel form of treatment.

THE INTRACTABLE DISEASES OF RICH COUNTRIES

At first sight it is not clear how molecular genetics could help us understand how the common intractable diseases of our richer countries are caused, let alone how to treat them. After all, if coronary artery disease is the result of the complex interplay between our lifestyles and our genetic makeup, and we cannot do much about the latter, why should we be interested in the genes that are involved? Even if we could identify those at particular risk from the effects of their way of life, would we really want to know that a newborn infant was very likely to have a heart attack in middle age, unless, that is, we could do something to keep it from happening? Surely it would be better to concentrate our efforts on trying to control the environmental factors we already know are important, and forget about our genes.

But, as we argued earlier, although we may reduce the frequency of some of these diseases, it is unlikely that we will be able to eradicate many of our leading killers in this way. And in some cases, such as those of psychiatric and rheumatic disorders, there is no evidence that our environment plays a major role. We must therefore follow any leads that the molecular sciences have to offer into understanding the causes and mechanisms of these conditions so that we can try to develop more logical ways of controlling them.

Another reason for trying to understand the genetics of common disease is to determine which of us is particularly prone to the environmental factors that cause them. This would certainly be a useful bonus so that we could concentrate our efforts in preventive medicine on special subsets of people. But considering the great complexity of these conditions, and their possible ties with the mysterious mechanisms of aging, I suspect that it will be a long time before we reach this level of diagnostic sophistication.

Ultimately, just as we got to grips with our infectious killers only by understanding the disease mechanisms involved, we will have to explain our present problem diseases in terms of defective cellular function. Suppose it turns out that we can identify a handful of genes that make us more or less likely to have a heart attack or to develop diabetes. It follows that their products must be closely involved in the disordered chemistry of the cells and organs that are affected. Hence these genes have the potential to point the way toward a better understanding of why these diseases occur; they may not, of course, but we have no other clues about where to start looking. In other words, molecular and cell biology have given us new tools to study how cells function and to purify the molecules involved. Their scope for research into illnesses of this type thus goes much further than investigating diseases that are purely genetic in origin.

Susceptibility or resistance to these common diseases is not usually determined by single genes. We can seldom track heart attacks or strokes through families in a way that obeys Mendel's laws and that follows the simple pattern of inheritance that we see in diseases such as cystic fibrosis or hemophilia. Nevertheless, there is an inherited component. Hence these diseases are called polygenic, meaning that they are thought to result from the interplay of several different genes with the environment. Until the advent of DNA technology this was an extremely difficult area of human genetics to investigate.

In the preceding chapter we saw how geneticists are progressing toward obtaining a linkage map of the human genome. The idea is to have a series of easily identifiable genetic markers at regular intervals along our chromosomes, so well characterized that we would know at exactly which point on a particular chromosome the marker could be found. With this information, how might we go about looking for the genes involved in a disease like diabetes?

It is easy to find families in which brothers or sisters are afflicted with common diseases, and even some with several affected generations. With a reasonable number we might progress in two ways. First, we can make an educated guess at which genes might be involved in a particular disease. This "candidate gene" approach is particularly relevant to diseases like diabetes for which we already know many of the players involved. Hence we can screen for identifiable variants of our candidates, say the genes for insulin or other hormones that regulate the levels of sugar in the blood, and see whether they stay together, or segregate, with diabetes in our families.

The other way of finding our elusive genes is called "blind" linkage. Here we simply take all linkage markers that we can find and study families to see whether any particular marker segregates in different generations with the disease we are studying. If it does, the gene involved in the disease must be close to the marker; if we know where the marker is, the culprit gene must be close to it. There are several clever variations on this theme. One entails looking to see if there is a disturbance in the expected patterns of segregation of markers in affected siblings. This is turning out to be of particular value in diseases of middle life for which it is difficult to obtain multigeneration families.

There is another, less obvious way to study the genetics of common diseases. Because it is so easy to breed mice, we know a great deal about their genes and chromosomes and already have extensive maps of many of them. Mice and rats share many diseases with humans, such as diabetes and high blood pressure. Our knowledge of the genetic makeup of these animals enables us to carry out breeding and linkage studies to determine which genes are involved. It turns out that many mouse chromosomes have human homologs. Having determined that a particular gene causes a disease in mice, we can then search for its equivalent in humans.

Linkage studies for human diseases require sophisticated DNA technology to define disease markers, and daunting mathematics to work out linkages. However, with the help of ingenious computer programs this field has moved very quickly over the last few years, and there are already hints that it will yield some successes. It must be admitted that it has also thrown up a few false starts, which is perhaps not surprising when we consider the complexity of the analytic methods required to identify important genes in these "noisy" polygenic systems.

Diabetes

Diabetes provides an excellent example of how these new methods are being used. It will be recalled that there are two main types: type 1, which often starts in early life and is characterized by a deficiency of insulin due to self-destruction of the islet cells of the pancreas, and type 2, which occurs later in life and is associated with obesity and resistance to insulin. Twin studies have suggested that type 1 diabetes has a relatively small inherited component but that genetic factors play a major role in type 2.

Type 1 diabetes, like multiple sclerosis, rheumatoid arthritis, and many other common diseases, almost certainly has an autoimmune basis—that is, it results from the production of antibodies that destroy the pancreas. It has been known for some time that it has a strong association with certain genetic subtypes of the HLA-DR gene family. This gene complex seems to have evolved as part of the body's defense mechanisms against infection. Many autoimmune diseases are related to particular genetic subtypes of this system, both in mouse and in man. Once it was possible to examine it at the molecular level, it was found that genetic susceptibility or resistance to diabetes resides in only a few amino acid differences in molecules of the HLA-DR system that are involved in presenting foreign proteins, or antigens, to lymphocytes, vital cells of the body's defense system. And, even more remarkably, the same amino acid differences were found in the corresponding molecules in diabetic mice.

There are several classes of lymphocytes—some called B cells, which produce antibodies, and others called T cells, which directly kill virus-infected cells or activate other cells of the immune system. The HLA-DR system presents antigens to particular classes of T cells. There is increasing evidence that genetic variability of structure of the receptors on the surfaces of these cells may also help determine susceptibility to autoimmune diseases.

In case this short outline of the genetic approach to insulin-dependent diabetes leaves the reader with the idea that the answers to the question why people contract it will be straightforward, let me hasten to say that they will not. Recent studies in mice suffering from a similar type of diabetes have shown that at least five, and possibly twice that many, genes may be involved. At least one major player is the mouse equivalent of the human HLA-DR genes, but others that are being identified seem to have nothing to do with immunity. And recent work in humans suggests that at least one other gene is of great importance for the development of type 1 diabetes and is close to the insulin gene.

But work in this field has already taken us to the hinterland between the environment and the particular cells that may mediate immune damage of the pancreas. We still do not understand why subtle inherited changes in the structure of the molecules that are involved in our defense mechanisms can, given the appropriate environmental signals, lead to self-destruction of an organ. Nor is it clear why not everyone with this genetic makeup becomes diabetic. Extensive epidemiological studies have provided no clue as to the external agent that may be the culprit. It could be a ubiquitous virus or even something in our diet, or many different ones may be involved. Until we identify the environmental triggers, we cannot predict whether we will be able to prevent diabetes and other autoimmune diseases. But, given further progress and a better understanding of the immune mechanisms involved, it should be possible to interfere with the complex processes that lead young diabetics to destroy their pancreas, or patients with multiple sclerosis irreparably to damage the supporting tissue of the nervous system.

Type 2 diabetes presents a completely different problem. Immune factors do not play a role in this type, which appears to be due to an inherited peculiarity in how we handle glucose that becomes manifest only later in life, possibly in response to our diet and lifestyle. Which genes might be involved? Apart from insulin, many other hormones help us convert glucose into energy. Chemical messengers of this type are recognized by the cells on which they act by specific receptors on their surface, the production of which is also controlled by genes. In addition there is a large family of complex chemical pathways involved in glucose storage and utilization that are driven by many different enzymes. Thus there are plenty of candidate genes to explore. Recently strong linkages between type 2 diabetes and genes for several enzymes and proteins involved in glucose utilization have been found.

An enzyme called glucokinase appears to be the culprit in at least some families with type 2 diabetes of an unusually early onset. Evidence that the glucokinase gene, and not one closely linked to it, is responsible for diabetes has come from studies of its structure in diabetic family members. It turns out that they carry one copy of a defective glucokinase gene. At least twenty distinct abnormalities have been found in this gene in different families. Glucokinase has a major role in regulating insulin production by the pancreas in response to different levels of sugar in the blood. But it is already clear that other genes are involved. A few families have been found in which diabetes of this kind results from mutations of the gene that controls the receptor for insulin on the surfaces of cells. In another family the gene for glycogen synthase, another important enzyme that helps control the way we handle sugar, seems to be involved. These recent findings are extremely encouraging because, for the first time, we are starting to gain some real insight into the cause of at least some types of insulin-resistant diabetes.

But what of their practical applications to a disease that affects approximately 4 percent of the populations of north European origin as they get older and that is reaching epidemic proportions in many populations of the world? The high frequency of this disease perhaps reflects a number of ancient genetic polymorphisms maintained at a high level by earlier adaptations to periods of starvation and climatic extremes. If this notion is correct, many different genes are likely to be involved. As we discover them, it should be possible to identify individuals at very high risk before the illness is established, and to use this information for their management by appropriate diet and other modifications of their lifestyle. As we gain a deeper understanding of the chemical changes that result from variability in the function of the many genes involved, we should be able to devise more definitive forms of prevention and treatment. But because so many different mechanisms may be involved, it it may be necessary to tailormake therapy to particular genetic subsets of the disease.

It is difficult to predict which areas of diabetes research will lead to major improvements in prevention or treatment. Currently pancreatic islet cell transplants have aroused much interest, and although many difficulties remain, some limited successes have already been achieved. Thoughts are also turning to the development of glucose-sensing systems that would entail some fancy genetic engineering using the insulin gene. But genuine progress toward prevention or radical cure seems unlikely until more is understood about the mechanisms that underlie the different forms of the disease.

Diseases of the Heart and Blood Vessels

The main problem with the genetic approach to diseases of the heart and blood vessels is their sheer complexity. The changes in the vessel walls themselves are complicated enough. While it is clear that cholesterol levels in the blood are of great importance, we know very little about why cholesterol seeps into the linings of our blood vessels, and even less about the ensuing events that lead to the formation of an atheromatous plaque. A heart attack results from the ulceration of these plaques and the formation of a blood clot. Clotting involves the interaction of blood platelets, many different proteins at work in the formation of the clot, and others that make up the fail-safe system for dissolving the unwanted clot. And what of the heart itself? When its blood supply is cut off, as when a coronary artery is blocked, many factors must play a role in determining the amount of injury to the heart muscle.

The problem of stroke is no less daunting. We know that degenerative changes in the blood vessels to the brain are involved, but a raised blood pressure is also an important risk factor; we know very little about why our blood pressure rises as we get older. And we know even less about the factors that may be brought into play to try to limit the damage to our brain after its blood supply has been cut off.

Thus, although it is clear that our genes are important when it comes to deciding whether we are going to have a heart attack or a stroke, the very complexity of these diseases should be telling us that many different genes will be involved, together with numerous environmental agents. It is not surprising, therefore, that twin studies in coronary artery disease have shown that the concordance rate between identical twins is only 19 percent and that between nonidentical twins 8 percent; this compares with a rate for identical twins in insulin-dependent diabetes of 50 percent and a rate in non-insulin-dependent diabetes of nearly 100 percent. The concordance rate for high blood pressure among identical twins is only 30 percent. In the light of these unpromising figures, not to mention the vagaries of twin studies, we should not expect quick answers from research that tries to understand the genetic factors in heart disease or high blood pressure. But since we have made singularly little progress in understanding their causes and mechanisms by any other approach, we should not be put off by these difficulties.

We have already touched on the centrality of cholesterol in current thinking about the genetics of heart disease. We have seen how this fatty substance, an essential component of all the cells of our body, may be lethal if it gets into the wrong place, such as the linings of blood vessels, and how, to prevent this, it is packaged up with different protein carriers that deliver it to specific receptors on our cells, mainly in the form of low-density lipoprotein (LDL). The seminal work of Joseph Goldstein and Michael Brown told us that there are many different genetic defects of the LDL receptors, found in about 1 person in every 500, which lead to a greatly elevated level of cholesterol in the blood and an enhanced risk of coronary artery disease. Encouraged by these discoveries, research workers throughout the world are trying to find some of the other genes that may affect the level of cholesterol in our blood.

A large family of proteins are involved in transporting cholesterol and other lipids. One that has been particularly well characterized is called apolipoprotein B-100. As many as 1 in 500 subjects in Europe and North America are carriers for a particular mutation of the gene for this protein, which causes a marked elevation in the level of blood cholesterol and a propensity for premature coronary artery disease. Other variants of this protein seem to cause unusually low cholesterol levels. The gene for another carrier protein, apolipoprotein E, has also been found to be the site of several different mutations. The various alleles are estimated to account for about 16 percent of the variability in blood cholesterol levels in different populations.

Another interesting player in the cholesterol story is a transporter called lipoprotein (a). Its concentration varies nearly a thousandfold among different individuals but remains remarkably constant through life. There appears to be a good correlation between high levels of lipoprotein (a) in the blood and the development of heart attacks in young men. Lipoprotein (a) has a structure similar to LDL's. It consists of cholesterol, one molecule of apolipoprotein B-100, and another large protein, called apolipoprotein (a). The structure of the last is of special interest because it links lipids to blood clotting in causing heart attacks.

The end result of the chain of events leading to the formation of a blood clot is the precipitation of a protein called fibrin, the main component of the clot. The body has a complex fail-safe mechanism to remove an unwanted clot, one of the major components of which is a protein called plasminogen, which is able to chop up fibrin. Plasminogen normally circulates in an inactive form but, when a clot is formed, is activated at the site by a family of enzymes called plasminogen activators. It follows that the presence or absence of plasminogen at the site of a blood clot may be important in determining the size of a clot and hence whether or not a blood vessel is blocked. Remarkably, the structure of apolipoprotein (a) and that of plasminogen are similar, particularly with respect to parts of the apolipoprotein (a) molecules called kringles, so named because they faintly resemble in shape a Danish pastry of that name. Their similarity makes it possible that high levels of apolipoprotein (a) compete with plasminogen for access to fibrin and hence interfere with its role in removing a blood clot.

There are other hints that genetic variability of the efficiency of our blood-clotting systems may play a role in determining whether we have a heart attack. The fibrin of blood clots is the product of a soluble precursor called fibrinogen. Studies of large populations suggest that the level of fibrinogen in the blood is of major importance in determining the likelihood of a heart attack. Some progress has been made in pinpointing the changes in the fibrinogen gene that causes its product to be made more or less effectively. Similarly, inherited variability in the proteins that activate fibrinogen may also affect the likelihood of a heart attack. And it appears that genetic differences in the structure of the proteins that remove clots render us more or less prone to form them in our veins.

Given the importance of cholesterol and blood clots in blocking a coronary artery, these discoveries are not entirely unexpected. But what is surprising is that inherited variability of other proteins, which do not appear to be involved directly in the events that damage our arteries, may be equally important.

Very recently it has been found that variability of the structure of a gene for a protein called angiotensin converting enzyme (ACE) may be yet another major factor in determining whether we have a heart attack. ACE is involved in converting a protein called angiotensin I to angiotensin II, a potent constrictor of blood vessels. And it perhaps also plays a role in degrading several other agents involved in regulating their size. Work carried out in France suggests that people with a particular type of ACE gene are more likely to develop heart attacks. Even more interestingly, the French workers believe that this genetic variant may account for up to 30 percent of all heart attacks in those who have no other risk factors—that is, do not smoke, have low blood cholesterol levels, and show other features making them less likely to have coronary artery disease. Remarkably, this news broke only a few months after a clinical trial had shown that drugs that inhibit ACE have a beneficial effect in preventing recurrent heart attacks. If these unexpected findings are confirmed, they will open up a completely new chapter in coronary artery disease research.

Many of these recent discoveries raise more questions than they answer. Certain rare mutations of genes that control cholesterol clearly make us prone to coronary artery disease. And then a broad spectrum of genetic variability of proteins involved in handling cholesterol, blood clotting, and other systems may make it more or less likely for us to have coronary artery disease. What we do not know is how all these different inherited variables interact with one another and with our diets and lifestyles. Patients with a mutation of their LDL receptor, which normally would make them highly prone to coronary artery disease, may be protected if they have a particular mutation of another gene that tends to lower the cholesterol level. But these are rare occurrences, and for the most part we still do not understand the complex interplay of nature and nurture that sets the level of our cholesterol or clotting factors and that determines whether we will have a heart attack.

Despite all these uncertainties, genetic information of this kind is already starting to make its mark in the clinic. We can now identify people at particularly high risk for heart disease because they have a mutation of a gene for their LDL receptor or one of their apolipoproteins. We can test their families and institute intensive cholesterol-lowering treatment early in life. Ultimately work of this sort may help us to understand the mechanisms of heart disease and to assess an individual's risk much more effectively.

High blood pressure also has a genetic basis, although twin studies suggest that, as in heart disease, the environment may be more important. The distribution of the levels of blood pressure in any population is continuous, and hence the definition of raised blood pressure is, like height or weight, arbitrary. In this sense hypertension is not a disease; it merely reflects the upper 15 to 25 percent of a biological variable, a level at which there is undoubtedly an increased risk of stroke and heart and kidney disease. Because blood pressure control involves several hormones and is closely tied up with how the body handles salt and water, there is no shortage of candidate genes to study. Yet progress has been slow.

Many different systems are involved in the control of the salt and water balance. For instance, a substance called renin is produced in the kidney and secreted into the bloodstream, where it generates a small protein called angiotensin I from a precursor. Angiotensin I is then split by an enzyme, angiotensin converting enzyme (ACE), to produce a protein called angiotensin II, which can directly contract blood vessels and hence raise blood pressure. Angiotensin II also controls the way the kidney deals with salt and water, and it has a wide range of other actions. Studies of genetic variability of this complicated system have given inconsistent results in humans. However, it has been found that, in a breed of rats with high blood pressure, a gene close to that which controls ACE, and which could be the ACE gene itself, is responsible for at least 20 percent of the blood pressure variation in response to a high sodium diet. Homologies between rat and human chromosomes made it possible to find out whether this same gene plays a role in causing high blood pressure in humans. In the event, the ACE gene is not a major player in human hypertension, although it may be very important in coronary artery disease. This little story illustrates how extraordinarily complex this field is and how little we know, despite our powerful technology.

But there are many other candidate genes to explore. For example, the behavior of muscle cells, which maintain the tone of the walls of our small blood vessels, is of great importance in the regulation of blood pressure. The function of these cells is modified by their concentration of calcium and sodium, which is in turn controlled by a variety of complex transport systems. Their genes are being isolated, and it should soon be possible to see whether variability in their structure or function is related to the likelihood of developing high blood pressure. There is already some evidence from large-scale family studies that this is so. Thus, although we are still far from any understanding of the causes of high blood pressure, at least we have a few hints as to where future work should be directed.

Research on the genes that may be involved in common forms of vascular disease continues to turn up findings that would have been difficult to predict. As we age, our blood vessels tend to develop bulges in their walls, called aneurysms. This occurs most often in the main vessel that leads from the heart to the abdomen and lower limbs, the abdominal aorta. Abdominal aortic aneurysms tend to enlarge progressively and, if not repaired, may burst, with catastrophic results. It has long been suspected that aortic aneurysms tend to run in families. Quite recently it has been learned that in a few cases they may be associated with mutations of one of the genes for collagen, the tough protein found in bone, connective tissue, the walls of blood vessels, and throughout the body. There are many different forms, each with its own genes. Mutations of some of them may have disastrous effects and lead to intrauterine death or the birth of children with dreadful malformations. But others, involving genes for different types of collagen, may cause several common conditions—notably, osteoarthritis and, in the present context, weakening of the walls of arteries and the formation of aneurysms. And another set of genes involved in the musculoskeletal system, in this case controlling one of the key proteins of the heart muscle, has been implicated in a common genetic disorder in which there is an enlargement of the heart, with a tendency to early death due to abnormal rhythms or heart failure.

Finally, we should remember that pinpointing the genes that may be involved is by no means the only way in which the basic sciences are helping to study vascular disease and high blood pressure. The vascular endothelium, the tissue that lines our blood vessels, has become the meeting place for disciplines ranging from pharmacology and more traditional biochemistry to molecular and cell biology.

Some remarkable discoveries are being made about the way in which the tone of our blood vessels is regulated. For example, an unexpected finding, made as recently as 1987, is that the simple substance nitric oxide, which is made in blood vessels and other tissues from the amino acid L-arginine, is involved in their relaxation. It now seems likely that the loss of nitric oxide–mediated dilatation of vessels plays an important role in the genesis of high blood pressure. And 1988 brought the discovery of a small peptide consisting of twenty-one amino acids, called endothelin, which turns out to be a potent constrictor of small blood vessels in many organs, including the kidney. Endothelin is only one member of a family of small peptides of this type. Their genes have been isolated, and a great deal is being learned about their properties and regulation. In short, the whole face of cardiovascular research is being changed by our newfound ability to isolate regulatory molecules and to study their action on heart muscle and blood vessels.

These few examples of the applications of the new techniques of molecular and cell biology for the study of cardiovascular disease will, I trust, have given some inkling of the excitement and complexity of the field. There is a feeling abroad that we may, at last, be getting close to the solution of problems that have puzzled us for so long.

Psychiatric Disease

Psychiatric disorders cause an enormous drain on health resources, and virtually nothing is known about their cause. As we saw earlier, the psychoses fall into two major groups: the manic-depressive diseases, which are characterized by bouts of irrational behavior or by extreme depression and suicidal tendencies, and schizophrenia, which takes many forms, all of them marked by bizarre thought processes, hallucinations, and other distressing features. Twin and adoption studies suggest that the familial aggregation of these disorders is caused predominantly by genetic factors. They would thus appear to be excellent candidates for the new techniques of molecular genetics. However, the outcome of the first applications of our powerful analytical methods to the study of schizophrenia shows the importance of asking the right questions.

A major problem soon emerged in defining what we mean by schizophrenia. When some of the earlier twin research was repeated, with much tighter criteria for the diagnosis, it was found that, particularly in adoption studies, the genetic component of the disease might not be as strong as previously thought. And a few years ago at least two groups of scientists claimed that, using molecular linkage methods, they had been able to localize to a particular chromosome a gene that might be involved in susceptibility to schizophrenia. Unfortunately, when this work was repeated with more stringent diagnostic criteria, these results could not be confirmed. Similar problems have arisen in studies of manic-depressive disorders. These early setbacks have caused some gloom and despondency in the psychiatric community. Until it is possible better to define the phenotypes—that is, the characteristic manifestations of the disease—it will be difficult to apply modern methods of molecular genetics to the major psychoses.

At the same time other clues have appeared about the possible cause of schizophrenia. For example, some patients have structural abnormalities of the brain that can be identified by sophisticated scanning techniques or at an autopsy. These observations, important though they are, are also difficult to interpret, because they cannot be made in all cases of schizophrenia. Overall, though, they may be telling us that schizophrenia is a neurodevelopmental disease with a strong genetic component.

What kind of genetic factors should we be looking for? The more obvious would encompass the many genes involved in the regulation of the chemistry of the brain. But some recent epidemiological evidence suggests that the answer could lie in a completely different direction. At least three independent studies have related schizophrenia to epidemics of influenza at the time that the mothers of the patients were pregnant. Might the genetic component of the disease reflect an unusual immune response to infection that resulted in intrauterine damage to the fetal brain? This intriguing possibility has sent scientists scurrying back to review studies carried out some years ago that suggested that schizophrenia is associated with certain variants of the HLA-DR genes, a gene family that keeps cropping up in diseases that feature self-destruction of tissues. Might schizophrenia be yet another autoimmune disease?

Some readers may be surprised to hear of all these efforts to find genes linked to psychiatric diseases. The searches are based on the belief that many of the diseases result from underlying biochemical derangements in the function of the brain. Hence, if there is an inherited component, we may, by identifying the products of the genes involved start to understand the chemical basis for at least the more serious forms of psychiatric disturbance. Armed with this knowledge, we should be able to develop more logical treatments for them. These ideas may be naive, and the genetic component of mental illness may be much more subtle. It seems unlikely that there will be one "schizophrenia gene"; and environment doubtless plays a major role in generating psychiatric disease. This is certainly true in the less serious forms of depression and anxiety from which all of us suffer to some degree.

Many scientists worry about the implications of this type of research and the dangers of stigmatization that it will pose if we are able to identify one or more genes that make us more likely to develop these distressing illnesses. These concerns are highlighted by a recent study, carried out by Dean Hamer and his colleagues at the National Institutes of Health, on the genetics of homosexual behavior. Hamer's team investigated the pedigrees of 114 families of homosexual men. It found increased rates of what it calls "same-sex orientation" in the maternal uncles and male cousins of these subjects, but not in their fathers or paternal relatives. This raised the possibility that it is transmitted by a gene, or genes, on the X chromosome. The team then investigated 40 families and carried out molecular genetic linkage studies. It concluded that there may be a gene, or family of genes, on a particular region of the long arm of the X chromosome that predisposes toward homosexual behavior.

This is the first report of the apparent genetic transmission of a behavioral trait, and it has caused consternation in many quarters. But what does this work actually tell us? First and foremost, if it is confirmed, it is saying that there is *not* a single gene for homosexuality. Linkage to the X chromosome was not found in all the families; even in those in which it was, though statistically significant, it was not all that strong. Indeed, these findings should not surprise us. Almost every condition we have discussed on our journey through human variability and disease reveals a complex mixture of nature and nurture. Hamer and his colleagues have not, as the media suggest, given us the opportunity to abort potential homosexuals. What they are saying is that there could be a gene, or group of genes, on a particular chromosome, which may have some influence in determining the pattern of sexual behavior. But it is clear that many other genes will be involved, not to mention the crucial influences of our upbringing and environment.

But we should not denigrate this interesting study, particularly if further work substantiates its findings. Provided we can interpret information of this kind in its true scientific and social perspective, it offers us a valuable new approach to trying to understand the complex interactions of our genetic makeup with our environment that help determine our patterns of behavior.

Dementia

Dementia—that is, progressive loss of memory, lack of judgment, and deterioration of intellect, temperament, and behavior—is a distressing disease and causes a large drain on our health resources. Although it results from many different pathological changes in the brain, one form, Alzheimer's disease, accounts for about 70 percent of cases. It is characterized by a loss of nerve cells, together with the formation of curious plaques and bodies called neurofibrillary tangles in particular parts of the brain. It is frighteningly common.

As our population ages, it will become one of the most frequent causes of death, affecting at least 20 percent of those over sixty-five and as many as a quarter of those over eighty. Because patients with dementia often live for many years, it poses grave problems for society as the number of old people increases.

The relative roles of our environment and our genes in the cause of Alzheimer's disease is a subject of intense debate. Although several toxic agents have been implicated—current interest is focusing on the role of aluminum—there is increasing evidence that genetic factors also play a role. Whereas about one in a thousand of the general population under the age of sixty-five has Alzheimer's disease, about one in three patients has a similarly affected first-degree relative. This could reflect the action of an environmental agent. However, twin studies have suggested that identical twins are much more likely to contract the disease than nonidentical twins are, emphasizing the important role of genes.

One of the most tantalizing observations in patients with Alzheimer's disease is that the changes in their brains are indistinguishable from those of middle-aged mongols. Because mongolism is associated with an extra chromosome 21, it seemed possible that a gene or genes on this chromosome might be linked to Alzheimer's disease. A main pathological feature of the brain in this condition is the presence of plaques made of a waxy protein called amyloid. One of the many types of amyloid turns out to be the product of a gene on chromosome 21. Could it also be involved in Alzheimer's disease? After several false starts, recent work has shown that members of at least some families with early-onset Alzheimer's disease have mutations of this gene that are probably responsible for the abnormal accumulation of amyloid in the brain.

These genetic studies have encouraged several groups of scientists to search for genes involved in the more common forms of Alzheimer's disease of older age groups. Very recently teams at Duke University and elsewhere seem to have struck gold. They have found that a gene called ApoE-4 may play an important role in susceptibility to the disease. And, quite unexpectedly, the age of onset seems to be related to the number of copies of the variant gene; one dose reduces it from eighty-five to seventy-five years and two doses to sixty-eight years. But not everyone with this genetic makeup develops Alzheimer's disease, and some who are affected do not have it. Although it is known that ApoE-4 is involved in lipid chemistry in the nervous system, it will take a lot more work to find out what this discovery means and how the product of this particular gene contributes to the characteristic changes in the structure of the brain.

Meanwhile, a great deal of effort is being directed at the symptomatic relief and more definitive prevention or treatment of Alzheimer's disease. For example, the particular pattern of damage to the brain that causes intellectual decline and loss of memory leads to a deficiency of certain chemical messengers. This is reminiscent of the kind of pathology that is thought to underlie Parkinson's disease and that has led to its partial control by the replacement therapy. Using a variety of ingenious approaches, the pharmaceutical industry has developed several promising agents of this type, which are currently being assessed by clinical trials. The main hope lies in a neurotransmitter called acetylcholine, which is essential for memory and which is deficient in Alzheimer's disease.

However, in the long term it will be important to try to get to grips with the basic pathology of the disease. If, as is now believed, this reflects the deleterious effect of amyloid protein on the brain, it may be possible to modify its course. Much has been learned about the amyloid gene and how some of its abnormally processed products cause serious damage to nerve cells. Thoughts are turning toward finding ways of interfering with these abnormal patterns of gene function. These are early days, but at least some progress is being made in this extremely difficult branch of neurobiology.

Mental Handicap

The term "mental handicap" is used to describe a mixed group of conditions that have in common developmental delay in childhood, delay or failure in acquiring the basic skills required for employment and social adaptation, and impaired memory and intellect. In some cases the mental impairment is associated with physical disability or behavioral disturbances. Disorders of this type affect approximately 1.6 million people in Great Britain. They often cause great distress to affected individuals and their families and pose a considerable burden on the community.

Since it appears that nearly half of all cases of mental handicap are due to specific genetic or chromosomal disorders, it is clear that the molecular approach has much to offer for the further elucidation of the cause and treatment of these distressing diseases. We have already alluded to the work that led up to the discovery of the gene involved in the fragile X syndrome, a common cause of mental retardation, particularly in males. A long list of other single-gene disorders may give rise to mental retardation. Recent studies in my own laboratory have shown how it is possible, by means of the tools of molecular biology, to define hitherto unrecognized small deletions at the tips of chromosomes as the likely cause of mental retardation. Similarly, we have found that, by concentrating on one aspect of the abnormal physical phenotypes found in these children, we can trace back to the gene that is involved.

We now possess some of the tools to attack this serious problem. It will not be enough just to define the genes involved. We must determine how they function and how this function breaks down in early development. Only then will we start to have some clues as to how these conditons might be prevented or managed. This work will require a multidisciplinary approach in which teams of workers from fields as widely separated as molecular biology, neurobiology, neuropathology, cognitive psychology, and behavioral genetics combine forces and bring their differing skills to this difficult and extremely important field of medical practice.

A New Concept of Degenerative Diseases of the Brain

While the rest of us were hoarding stamps or works of art, neurologists were collecting eponymous titles for chronic diseases of the brain and nervous system, a pastime that has helped make their speciality obscure to the outside world. Many of these disorders occur at particular times of life. For example, multiple sclerosis usually starts in the second or third decade, whereas the Gerstmann-Sträussler syndrome (GSS), a condition characterized by widespread degeneration of the nervous system and dementia, occurs in slightly older people. Other important degenerative disorders, such as motor neuron disease, Parkinson's disease, Creutzfeldt-Jakob disease (CJD), another condition associated with widespread degenerative changes of the nervous system, and Alzheimer's disease, occur generally between the ages of forty-five and eighty-five.

Over the last few years it has been discovered that certain animal and human degenerative diseases of the nervous system are caused by infections that may take many years to produce clinical manifestations. The concept of "slow infections" was first introduced by a pathologist working in Iceland on neurological diseases in animals. There is a growing consensus that some of these human diseases represent "slow infections" caused by a completely new class of infectious agents. Furthermore, they seem to produce similar pathological changes in the brain, marked by an appearance that pathologists call spongiform degeneration, together with deposition of amyloid, and gliosis, or scarring of the nervous system.

This exciting new field grew from the study of a disease of the nervous system of cattle called scrapie. It was found to be transmitted by a virus-like agent for which the term "prion" was coined. A prion has been defined as a small proteinaceous infectious particle that is resistant to inactivation by the usual procedures that modify nucleic acids. Indeed, there is increasing evidence that prions consist entirely of protein. In addition to three diseases of cattle, including bovine spongiform encephalopathy (known in Britain as mad cow disease), there are at least three human diseases in which prions probably play a major role—kuru, CJD, and GSS. These conditions may be variants of the same disorder. Very recently a fourth, a fatal form of inherited insomnia, has been added to the list. They can all be transmitted to animals by inoculation. Kuru, a degenerative disease of the nervous system found in Papua New Guinea, is thought to have been spread exclusively by ritual cannibalism. A few other examples of the possible exogenous transmission of prions have been described, including CJD contracted from contaminated human growth hormone used to treat dwarfism, but, in general, it is not known how these agents gain access to the nervous system. However, evidence is growing that they accumulate there and lead to spongiform degeneration and gliosis.

It turns out that the prion protein is the product of a highly conserved gene found in organisms as diverse as fruit flies and humans. Its function is still unknown. The abnormal form of prion protein that is found in brain extracts from patients with spongiform encephalopathies, and which is the main constituent of the amyloid plaques found in the brain in these conditions, has in a few cases been traced to mutations of the prion gene. Point mutations of this gene leading to amino acid substitutions have been found in eighteen affected individuals in families with GSS or CJD.

Thus we have the extraordinary situation of a family of diseases associated with the production of abnormal proteins that is caused by mutations at a gene locus that, being so highly conserved through evolution, probably has an important function in the nervous system. At the same time there is good evidence that these disorders are transmissible; yet the altered prion proteins do not appear to contain any nucleic acid. If they are indeed infectious, they represent a completely new class of agents of this type. Clearly this will be an exciting field over the next few years and one in which the techniques of molecular biology will contribute directly to the elucidation of important disorders of the brain.

The Diagnosis and Treatment of Cancer

We have already reviewed recent progress in trying to understand what goes wrong in a cancer cell. The emerging picture indicates a gradual acquisition of different mutations involving oncogenes, genes that play a central role in determining how cells divide, mature, and interact with other cells. Some of these mutations may be inherited and remain silent until there is a mutation of the anti-oncogenes on the opposite pair of the chromosomes on which they reside. Presumably, the acquisition of all these different mutations reflects both the effect of agents in our environment and the action of chemicals we produce as by-products of our natural metabolic processes, particularly as we age. A few human tumors, including those involving the liver or lymph glands, may be due to infections with viruses, although the way in which the tumor develops is extremely complex and many steps away from the original virus infection.

There is thus a great deal more work to be done before we really understand the biology of cancer. Although we have some inkling of the different genes involved, and what they do, in most cases we still do not know why it is necessary for a cell to accumulate more than one mutation before it becomes cancerous. We do not yet understand the genetic changes responsible for a cell's breaking away from its normal surroundings and successfully housing itself somewhere else, an important feature of many malignant tumors. And we need to learn more about how we damage our DNA with chemicals produced in our bodies, particularly as we age, and whether we can prevent its happening. But considering what has been achieved in the last few years and what tools we now possess, it should not be long before many of these questions are answered.

But where will all this new knowledge leave us with respect to the diagnosis and treatment of cancer? For those cancers that we cannot prevent, the next-best thing is early diagnosis while treatment is still possible. Currently screening programs are being carried out for cancers of the cervix, breast, and, in a few populations, bowel and prostate. In Japan, where cancer of the stomach is still very common, attempts have been made to institute regular screening, but with only limited success. Despite some early problems, screening for cancers of the cervix and breast has been shown to be of genuine value. Because cancers of the bowel are particularly common, screening programs that test stools for blood have been set up in several countries, but the overall success for early diagnosis has been disappointing. It is too soon to say whether screening for prostate cancer will make major inroads into the management of this increasingly common disease.

Are the new tools of molecular biology likely to improve cancer screening? Again it is too early to say, despite some promising leads. Recently, for example, analyses of the stools of patients with tumors of the bowel have shown that it is possible to amplify tiny amounts of DNA from cells carrying altered oncogenes. The first results were disappointing; this new technique turned out to be less sensitive than simple examination of the motions for blood. But the fact that it works at all is extremely encouraging. The ability to amplify tiny quantities of DNA offers a new way of screening for tumors of the urinary tract, cervix, bowel, oral cavity, and even lung. And the specificity of some of the abnormal products of oncogenes may also lead to the development of methods for imaging tissues to find small tumors.

The cancer research field is buzzing with new ideas about ways to treat tumors. Cancer treatment now follows three routes. If a growth is amenable to surgery, it is cut out. If, on the other hand, it cannot be removed completely but is likely to be sensitive to radiation, it is treated with x-ray therapy.

And, finally, a battery of drugs can be used to slow the rate of growth of the tumor or even destroy it completely. The problem with x-ray treatment and chemotherapy is that they damage normal cells, and it is very difficult to achieve a balance between the dose required to destroy the cancer and spare the patient. This is why cancer therapy has such distressing side effects. Hence the major goal is to discover properties that are unique to cancer cells. This has been the dream of the cancer field for many years, but it is only recently, with the discovery of oncogenes and their abnormal products, that there seems some genuine hope of turning it into a reality.

Nearly all this work is at an early experimental stage, and very little of it has reached the clinic. Several ways to "turn off" oncogenes that are producing abnormal products are being explored. Some of them are already effective in cancer cells growing in the test tube, although whether they can be scaled up to treat patients remains to be seen. For example, when DNA is transcribed, only one of its strands, the "sense" strand, is used as a template to make messenger RNA. The mirror image of the sequence of this strand, the antisense strand, can be synthesized and introduced into cells to bind to and inactivate the messenger RNA of the altered oncogene while it is still in the nucleus, or to prevent the transcription of the gene.

Other novel approaches to cancer treatment are directed at the protein products of cancer cells. Cells divide and proliferate in response to external messages but also to proteins they produce themselves and release locally. Autocrine signaling, as this is called, offers the opportunity to try to interfere with the proliferation of cancer cells. For example, some lung tumors produce small peptides that have now been identified as autocrine growth factors. Various antagonists are being synthesized, some of which are able to block the biological effects of the growth factors produced by the tumor cells. And, by a different approach, attempts are being made to produce antibodies to some of the abnormal oncogene products.

Gene therapy, which entails changing the genetic makeup of cells, is well along the way to clinical trial. Some of the first applications are in the cancer field. For example, Steven Rosenberg and his colleagues at the U.S. National Institutes of Health have found that a certain class of white cells, which they call tumor-infiltrating lymphocytes (TIL cells), can be extracted from patients with cancer, grown outside the body, and returned to the patient's body, where they sometimes cause regression of tumors. The first step in the genetic manipulation of these cells was to insert a "marker" gene to find out whether they really do return to the site of the tumor. Having convinced themselves that this is what happens, Rosenberg and his team designed a new set of experiments, in which they inserted into these cells a gene that codes for a protein called tumor necrosis factor, which kills cancer cells. The danger is that a few of these white cells might get in the wrong place and damage normal tissue. Much will depend on the specificity of TIL cells for particular tumors.

Gene therapy is being adapted for several other ingenious approaches to the management of cancer. For example, some of the most intractable malignant tumors affect the brain. Attempts are being made to insert genes for enzymes that will activate drugs able to destroy the tumors. This notion is particularly promising because of the potential specificity of the retroviruses used to insert the foreign genes for the rapidly dividing cells of tumors. It has the further advantage that the drugs employed are relatively harmless to normal cells, unless acted on by the enzymes being inserted into the tumor cells. This novel method has already shown promise in animal experiments and is currently being used to treat patients with brain tumors.

Attempts are being made to persuade tumor cells to express molecules of the histocompatibility genes on their surfaces, which might make them appear to be "foreign," and then to persuade the body to mount an immunological attack on them. In essence, the idea is to try to copy the natural way of rejecting incompatible tissue grafts and convert a tumor into the equivalent of an organ from an incompatible donor.

A variety of other experimental methods are being developed for dispatching therapeutic agents directly to cancer cells. It is difficult to predict where all this highly sophisticated science will take us. But considering that only a few years ago we hadn't the faintest idea what goes wrong in a cancer cell, it is hard to imagine that it will not, ultimately, lead to the better control of some of our intractable malignant tumors.

Diseases of Joints and Bones

As we age, all of us suffer to some extent from diseases of our bones and joints. Recent research suggests that the more serious forms of arthritis that affect younger people, particularly rheumatoid arthritis and a condition called ankylosing spondylitis, which causes painful rigidity of the spine, are, like insulin-dependent diabetes, due to a self-destruction of tissues, in this case the linings of our joints. The time is approaching when we will have the tools to find out why this happens and to isolate and control the many different inflammatory proteins involved in damaging our joints.

But what of the commoner forms of arthritis and bone disease that cause so much misery in our elderly populations? Twin studies have shown us that osteoarthritis has an important genetic component. At the moment we know very little about the identity of the particular genes involved. Some clues are emerging, however. For example, it has been found that in families in which severe osteoarthritis strikes early in life the affected members have mutations of one specific class of collagen genes. Collagen, of which there are many different types, forms the major scaffolding for all our tissues and is therefore a very good candidate for variability that might make us more or less prone to diseases of our joints and connective tissues.

Severe thinning of the bones, or osteoporosis, is another major public health problem, particularly common among postmenopausal women. Although this may be due partly to a deficiency or hormones, it has never been clear why severe osteoporosis affects only a proportion of postmenopausal women. Recently research workers in Australia have shed some light on this important question. They have found that the density of our bones and the likelihood of our developing osteoporosis as we age is related to an inherited variability in the way in which we handle vitamin D.

Like so many chemicals that are vital for the normal functioning of our cells, vitamin D is transported into cells by attaching itself to specific molecules on the surface of a cell called vitamin D receptors. The Australian scientists believe that much of the variation of the density of our bones as we age depends on genetic variability of the function of these receptors. This unexpected discovery, if substantiated, has very important implications for the medical care of our aging populations. It offers us a way of identifying people who are at high risk of developing osteoporosis and, at the same time, may help us develop more effective strategies for preventing this distressing condition.

Other Common Diseases

These examples must suffice to show some of the directions in which research in human molecular genetics and biology will move over the next few years. Work of this kind is being applied to many other diseases, including allergy and asthma, individual variability in response to drugs, alcoholism and other behavioral disorders, and all the common autoimmune diseases. In each case the idea is the same. Find the major genes that are involved, determine what they do, and hence try to understand the pathology of the disease. It is complicated work and progress will be slow, but we know enough already to suggest that it offers a promising approach to a genuine understanding of the cause of at least some of the common diseases of Western society. It is far too early to predict how far all this will help us in preventing and curing them. But we can be sure that it will be a very productive area of medical research over the next fifty years.

BIOTECHNOLOGY AND DEVELOPMENTS IN DIAGNOSIS AND TREATMENT

Until the mid-1970s the pharmaceutical industry was wedded to the traditional approaches of medicinal chemistry and pharmacology. New drugs were discovered by screening large numbers of compounds for a desired biological activity and, after promising candidates were found, by ingeneous modifications of their chemical structure. The bulk of them turned out to be small organic molecules, many of which could be synthesized by chemists and then packaged in a form that could be administered to patients. The arrival of recombinant DNA technology, together with the development of new techniques for studying the molecular structures of proteins, has raised the expectation that we are about to see a major revolution in the pharmaceutical and biotechnology industries.

This new kind of pharmacology will evolve in several different directions. We have already seen that it is possible to insert human genes into microorganisms where they express and produce their products. Thus we have the facility to mass-produce any small protein, the structure of which is controlled by a single gene. Many biological functions require the interaction of proteins with receptors on cell surfaces. Large numbers of genes for receptors and their partner proteins have now been cloned, as have those for enzymes and regulatory proteins. And diverse new techniques for studying the molecular structure of proteins, sophisticated computer graphic modeling for example, are being developed. In this way molecules can be designed that have the custom-built potential to modify a wide variety of biological functions. We are, in effect, approaching the time when we can improve on nature and make precise blueprints for drugs for any function that we wish to alter.

And thoughts are even turning to modifying the ways in which genes function. Some of them, such as antisense technology, are relatively crude, but others, such as the design of proteins or nucleic acids that mimic nature's way of switching genes on and off, are extremely sophisticated.

To back up all this technology, methods are being devised for rapidly screening chemicals for their desired pharmaceutical effects. This new field encompasses disciplines ranging from semiconductor technology to chemical synthesis and biology. One company, for example, has developed a process based on microchip technology that can synthesize and screen 65,000 compounds in forty-eight hours on a one-centimeter square chip. By means of special light-hypersensitive labels it is possible to discover which drug candidates are bound most strongly to these sequences. The same approach could be used for genetic screening; it has been claimed that the two hundred or more mutations that cause cystic fibrosis could be incorporated into a single chip, which would tell us immediately whether a particular sample of DNA contained any of them.

As might be expected from an industry that is moving so quickly, there have been some spectacular successes and some equally impressive disasters. And we can be sure that it will want to recoup some of the enormous costs of research and development. Our new pharmacopoeia will not be cheap.

Diagnostics

DNA technology has opened up a whole new world of diagnostic possibilities. For example, bacteria, viruses, and parasites have their own particular genetic makeup: By rapidly amplifying their DNA and using gene probes that are specific for regions of it, we can now identify a wide variety of infections at a level of sensitivity hitherto unattainable. Similarly, as we have just seen, cancer cells often have telltale changes in their genetic machinery. With similar methods it is possible to identify very small numbers of these cells and hence to diagnose cancer and to monitor the course of treatment much more effectively.

DNA technology is already being used widely for the diagnosis of single-gene disorders. However, it may have a much broader application in clinical practice if it turns out that it is possible to identify persons susceptible to common diseases—heart attacks, diabetes, psychiatric disorders, and so on. This is, in fact, one of the principal reasons why many scientists support the Human Genome Project. The idea is that we would all have our DNA profiles analyzed early in life, after which our various disease susceptibilities would be assessed and we would be given appropriate advice about how to avoid certain environmental triggers for particular diseases. Readers who have struggled through the discussions about the complexity of the genetics of common disease in earlier chapters will not need convincing that it may be a while before we achieve this goal.

All of us are unique in the constitution of our DNA, a reality being used in a variety of clever ways to help catch criminals. Certain stretches of DNA vary in length from person to person. By means of gene probes that identify these regions, it is possible to build up what is called a DNA fingerprint—a map of this variable DNA that is unique to each of us. This technique was pioneered by Alec Jeffreys in Leicester, England. Recently he has devised methods for converting this information to a digital printout form, which can be stored or easily transferred between different parts of the country, or the globe.

DNA fingerprinting is now being used widely in police work. Even more discouraging for the criminal community, DNA can be extracted from many different sources. In addition to blood, semen, and any other body tissue, it can be obtained from saliva and even from hair roots. Furthermore, DNA is remarkably tough and can be extracted a long time after it has been deposited, even from bones from ancient burial grounds.

Therapeutics

Recombinant DNA technology has enormous potential for the development of therapeutic agents. The most obvious application is for the generation of human proteins in microorganisms. Insulin was one of the first drugs to be made in this way. This product, developed by the fermentation of micro-organisms containing human genes, is now being used to treat insulin-dependent diabetes. Similarly, human growth hormone produced in microorganisms is being used to correct dwarfism due to growth hormone deficiency. This treatment came into its own when suspicions were aroused that patients who had received growth hormone of animal origin might have developed a degenerative disease of the nervous system, possibly as the result of infection with an agent derived from the raw material.

Because of the danger of transmitting AIDS from infected blood, there has been great interest in producing blood products by recombinant DNA technology. Human clotting factor VIII, a deficiency of which causes hemo-philia, has already been made in this fashion. Several proteins that interfere with blood clotting have been manufactured in microorganisms, notably an agent called tissue plasminogen activator, which is used for the treatment of heart attacks and other conditions in which clots have to be lysed. Quite recently an artificial hemoglobin has been engineered that may be of value for the treatment of acute anemia and shock.

But perhaps the most spectacular success so far is the production of recom-binant erythropoietin. This hormone, normally made in the kidney, is neces-sary for red cell production. If the amount of oxygen being carried to the tissues by the blood is reduced—as a result of anemia, for example—the kidney increases its production of erythropoietin, which then stimulates the bone marrow to make more red cells. Erythropoietin must be produced con-tinuously to maintain a sufficient output of red cells to balance their destruc-tion at the end of their life span. Its deficiency in patients with kidney diseases causes severe anemia; those who receive recombinant erythropoietin show a rapid improvement. In fact, this new therapy has transformed the lives of patients who depend on kidney machines and who, before it was available, survived only on regular blood transfusions. Genetically engineered growth hormone and erythropoietin have broken the £300 million per annum barrier for world sales, and several other products are rapidly approaching this level of commercial respectability.

Molecular technology is turning up a bewildering list of biologically active products, whose therapeutic value is only just starting to be explored. The cells of the immune system, blood cells of the bone marrow, and, indeed, all the cells that make up our tissues are under the control of a complex network of chemical regulators and, in many cases, produce their own mediators, which act at a distance from them. Recombinant DNA technology allows us to isolate and manufacture these molecules, both for research and for clinical use. A few examples should serve to emphasize the potential and complexity of this new field of therapeutics.

In 1956 there was discovered a family of agents that are produced by many different cells in response to infections by viruses and that have the property of protecting other cells from attack by viruses. Hence they were called interferons. The early years of interferon research were bedeviled by difficulties with production and purification, a problem finally overcome a few years ago by the use of recombinant DNA technology. The interferons show a wide range of activity; they confer resistance to infection by many viruses, change the body's immune response, and, perhaps most interestingly, inhibit the growth of certain tumors. Their full potential is only starting to be appreciated, and at the moment very little is known about how they work. But one of them, interferon alpha, has already been found to be of value in the treatment of chronic viral hepatitis and to control the growth of certain cancers, notably those of the blood and lymphatic glands.

Many years ago Donald Metcalf and his colleagues in Australia discovered a substance in urine and blood that is able to stimulate the precursors of white blood cells to form small colonies on particular growth media. Hence it was called colony stimulating factor (CSF). A family of these regulatory molecules has now been identified, each member of which has a powerful stimulatory effect on the bone marrow, encouraging it to produce the kinds of white blood cells that are essential to combat infection. It turns out that the production of these proteins is the body's normal way of regulating white blood cell production. Their output is increased in response to infection. At first sight it might seem that by injecting proteins of this kind we would be unlikely to do better than our own natural defense mechanisms in combating infection. But this is not the case. Several of them are now used in the clinic, and it is already clear they are able to speed the recovery of bone marrow after it has been depressed by drugs used to treat cancer. The discovery of these new agents is opening up a completely new field of treatment for patients whose bone marrow is failing.

Another of these biological mediators produced by recombinant DNA technology seems to have both good and bad properties. The story of the different routes that led to its isolation and purification is remarkable enough in itself. It had been observed for over a hundred years that cancers occasionally shrink in patients with severe bacterial infections. Around the turn of the century a New York surgeon, William Coley, tried to treat patients with inoperable cancers with cultures of bacteria. Later he went on to see whether toxins produced by bacteria might be effective. Using this is very unpleasant treatment, he found that tumors would regress, though not often. Many years later it was determined that the active ingredient of these toxins is a substance called endotoxin, a major constituent of bacterial cell walls.

In 1975 the American scientists Elizabeth Carswell and Lloyd Old discovered that it is a substance produced by cells exposed to endotoxin that causes tumors to shrink, not endotoxin itself. They named this substance tumor necrosis factor (TNF). In 1984 the gene for TNF was cloned in bacteria, and hence sufficient quantities could be made to study its properties. While all this was going on, a scientist then at Rockefeller University, Anthony Cerami, was trying to find out why cattle with sleeping sickness gradually waste away. He isolated the protein responsible and called it cachectin, since "cachexia" is the term for pathological wasting. When he determined the structure of cachectin, it turned out to be identical to tumor necrosis factor.

But there were to be a few more twists to the TNF story. It happened that its effects on tumors were extremely disappointing. On the other hand, it was found that too much TNF can produce many of the changes of severe toxic shock that are seen in patients with bacterial infection. Overproduction of this agent was soon implicated in a wide variety of diseases, ranging from septicemia to severe malaria. And so after years of pursuing this elusive protein, researchers did a great deal more work in an attempt to develop antibodies to inactivate it in order to treat septic shock and severe malaria, particularly the form that affects the brain. Several trials are now in progress to see whether antibodies against TNF can limit these life-threatening diseases. So what started out as a hunt for the panacea for cancer treatment has turned into a chase for something to inactivate it in patients with completely different diseases.

The proteins we have been describing are just a few of the new biological mediators that have been isolated in the last few years. They are known collectively as cytokines. At a recent count there were at least fifty; more will surely be discovered. Some of them are already in clinical use. But there is no doubt that, in administering these agents, doctors are trying to run before they can walk and that their premature use is being encouraged by pharmaceutical companies that have put large amounts of money into their development. Further work on their biological actions is needed, as well as a much cooler look at their therapeutic possibilities. As the English molecular biologist Sydney Brenner has pointed out, the world of pharmacology has changed over recent years; we are rapidly approaching the time when we will have a battery of drugs looking for diseases.

The development of antibiotics, together with vaccination for diseases like smallpox and poliomyelitis, tended to lull us into the belief that infectious disease no longer poses a serious problem. However, worldwide, infectious disease remains the major killer; the emergence of new epidemics like AIDS is a reminder that novel infectious agents are always around the corner. Recently there has been considerable interest in the development of new types of vaccines, stemming mainly from advances in recombinant DNA technology and protein engineering.

These techniques offer many novel approaches for vaccine production. Of course, the basic principles have not changed. The idea is to immunize an individual with an organism, or part of it, and evoke an immune response but not to produce a disease in the process. The immunizing agent is called an antigen. After settling for a promising antigen, one can clone the appropriate genes and then express them in animal or insect cells, or even in yeast.

One of the earliest successes was the production of a vaccine against hepatitis B, a serious liver infection. This was a remarkable tour de force because initially scientists could not grow this small DNA virus at all; they solved the problem by cloning the gene for the appropriate antigen into yeast. So far the vaccine appears to be safe and effective. The engineered yeast cell contains information for only one of the virus genes, and hence there is no way for a complete virus or any other infectious agent to arise during the production of the vaccine. This is only one of a variety of ingenious approaches to vaccine development that are being explored at the moment.

At the same time that it is helping us invent new ways of making vaccines, recombinant DNA technology is also revealing the enormous difficulties that will face us as we try to control some of the major killers of the Third World. Malaria is a prime example. As soon as it was possible to clone the genes for different malarial antigens, it became apparent that the malarial parasite, and many other important tropical parasites, have developed myriad techniques for changing their immunological makeup. It will take even greater ingenuity to overcome some of these problems if we are to develop vaccines to control some of the main killers in the developing world.

In 1984 the Nobel Prize in medicine was awarded to the Cambridge scientists Georges Köhler and Cesar Milstein for a discovery that was to have major implications for the diagnosis and treatment of disease. Immunization results in the production of molecules called antibodies in response to "foreign" proteins, or antigens. Antibodies are produced by white blood cells called B lymphocytes. Köhler and Milstein found that lymphocytes, each of which is able to make one particular antibody, will grow in culture when fused with certain tumor cells obtained from rats. From such hybrid cells, or hybridomas, specific clones can be isolated, each of which contains the descendants of one lymphocyte and therefore produces only one particular antibody. The isolated clones can then be cultured, and in this way it is possible to produce large amounts of individual, or monoclonal, antibodies, which have many clinical applications.

Recent research is tackling the problem of toxic shock by trying to inactivate mediators such as tumor necrosis factor and bacterial endotoxin; it has developed monoclonal antibodies capable of binding to both of these agents. And there is considerable interest in tailor-making tumor-specific antibodies, with the objective of either killing tumors directly or attaching diagnostic markers or tumor-killing drugs to them, so that they can be targeted directly at cancer cells.

One of the chief problems in using antibodies produced in rodent cells as "magic bullets" to kill or image tumors, or for other purposes, is that the antibody may be recognized as "foreign" by the patient, and the resulting immune response may interfere with therapy or cause side effects. Ideally, therefore, it would be better to generate human monoclonal antibodies, but this has proved almost impossible, for a variety of technical reasons. However, using a combination of protein engineering and DNA technology, scientists have been able to process these rodent monoclonal antibodies, in essence to humanize them so that they retain their activity but lose all the important immunological features of their rodent origins. And very recently it has become possible to persuade genetically engineered mice to produce human antibodies.

Finally, recombinant DNA technology offers a novel approach to a problem that has dogged the pharmaceutical industry for many years. Quite often, when a new drug is developed, there is no small-animal model of a human disease on which to test it. If, however, a human gene is injected into the fertilized egg of a mouse, and provided it includes the appropriate control regions, it may be expressed in the offspring of the mice. Such transgenic animals, as they are called, have been bred to express many different human genes. In this way, or by a related approach in which certain mouse genes are inactivated, lines of mice with cancer, sickle-cell anemia, and, most recently, cystic fibrosis have been developed. These animals may be of considerable help in the testing of new drugs to treat these conditions.

BROADER ISSUES OF HUMAN BIOLOGY AND MEDICINE

In the previous sections we touched on a few examples of how molecular and cell biology will be applied to medical research over the next few years. But the remarkable technological advances in these fields have much wider applications to the study of human disease.

For example, an area of biology with particularly important implications in the future for medicine is animal and human development. How does a fertilized egg with its full complement of genes from its parents develop into a human being? How are genes switched on and off at particular periods during development, and why are they expressed only in the appropriate tissue? What is the nature of the biological time clock that regulates these extraordinarily complex interactions?

One of the many new areas of research in developmental biology may lead to a better understanding of congenital malformation. As with so much of the "new biology," the idea is not new, but the availability of the powerful analytical tools of recombinant DNA technology has allowed it to be explored in a novel fashion. In 1894 William Bateson suggested that the study of chance deviations in normal developmental patterns might provide clues about the rules that govern the regulation of development. This idea opened up a new field, the science of positioning and body patterns. In recent years it has been found that a class of genes discovered in the fruit fly *Drosophila* regulates the development of whole body segments. These genes, called homeotic genes, have sequences in common with many other species, including worms, frogs, birds, mice, and humans. Defects of the homeotic genes in insects result in major developmental abnormalities, which include substitutions of one or more segments normally found elsewhere along the body axis. All vertebrate embryos go through a stage of development at which the body is composed of a series of segmental units from which the skeleton, nervous system, and other systems are eventually developed. It has been found that humans have an equivalent of the homeotic genes, probably related in the distant evolutionary past to those of insects. There is evidence that their products play an important role in the expression of other genes during development.

And we are gaining some insight into the ways in which cells find their way around in the early embryo and how things start to get organized. Many of these early events depend on concentration gradients of chemicals called morphogens. These molecules, which are produced at one site and gradually diluted as they migrate, seem to form a chemical environment, or gradient, that is critical for the correct development of tissues—limb buds, for example. In addition it appears that there are batteries of different regulatory proteins able to activate or inactivate DNA at different stages of development. Allowing for the extremely complex genetic regulation of these developmental networks, we should be able to start to understand some of the things that go wrong in early development and lead to malformed babies.

Another field of basic research in biology, closely related to understanding how living things develop, also promises to have implications for clinical practice in the future. Although most of us are aware that when a tadpole turns into a frog it loses its tail, I doubt if many readers will have stopped to think what happens to the tail, let alone how it knows when to drop off. This process, like much of the remodeling of our tissues that goes on during early development, occurs because, in effect, our cells are programmed to die. It is becoming clear that all cells have inbuilt instructions to self-destruct, and that they do not do so all the time because they receive chemical signals from other cells that tell them to survive and proliferate. The mechanism of programmed cell death seems to depend on the interaction of genetic information generated within cells with messages from other cells in the environment. As further research on this remarkable phenomenon starts to tell us how it is regulated, and allows us to isolate some of the molecules involved in the decisions about whether cells die or survive, we will learn a great deal about the causes of important medical problems such as cancer and aging.

Our evolutionary history is also written in our DNA. We can now make detailed comparisons of gene structure between different groups of organisms and study the changes that have occurred during evolution. We can also determine how different species are related by comparing the structure of their DNA. As we saw in chapter 7, work of this type has already given us some valuable clues about the origins and dissemination of some of the early groups of humans. At first it is difficult to see how evolutionary studies at the molecular level could ever have any value for medical practice. However, as we start to understand how our genetic makeup has evolved and made us more or less resistant to bacterial, viral, or parasitic infections, or to other environmental hazards during our evolutionary history, we can perhaps devise new strategies to control these diseases. Why did some individuals survive major epidemics? What made them different? Evolution has left many tantalizing clues for doctors; it is up to them to follow them up, now that the tools are at hand.

These examples must suffice to provide some inkling of how the molecular sciences are being used to tackle some of the broader problems of human biology. We could have explored many related questions, including aging, population genetics, and the behavioral sciences. At the moment all these areas of research seem a long way from the clinic, but as we learn more about human biology at the cellular and molecular levels, much of this knowledge will doubtless yield a better understanding of the causes and mechanisms of disease.

It is clear, therefore, that molecular biology and genetics will dominate human biology and medicine in the next millennium. But it is also apparent, even from this brief outline of their future potential, that many of their benefits for medical practice will not reach the clinic for a long time. Before these fields fulfill their potential, we may have to take a more holistic approach to human biology and try to understand how all the molecules we have discovered are orchestrated in an intact human being, or patient.

In the next chapter we will try to anticipate what we may hope to achieve over the next few years, as we gradually incorporate these new developments into our current patterns of medical research and practice.

T E N

Hopes and Realities

A wonderful fact to reflect upon, that every human creature is constituted to be that profound secret and mystery to every other.
—*Charles Dickens (1812–1870)*

Observers of the medical scene in affluent Western societies might conclude,
not without justification, that all is not well. Whether the system is based on
the marketplace economy of the United States or the government-funded
health services of some European countries, and regardless of the percentage
of gross national product (GNP) that is spent on health, nobody seems to be
able to get it right. The consumption of money by high-technology medicine
appears to be limitless.

It is estimated that in the United States more than 37 million people do
not have ready access to health care. Yet between 1960 and 1993 the proportion of the GNP spent on health increased from 4.4 percent to about 14
percent; at the present rate of progress this figure will increase to about 20
percent by the year 2000, representing some $1.5 trillion in expenditure each
year. And although these inflationary changes have led to many attempts at
cost containment, the overall benefit seems to have been minimal. Richard
Wenzel, summarizing the scene, writes, "Sentiment is that, with respect to
runaway costs for health care, nothing so far has worked."

The situation in the rest of the developed world is not much better. In
Britain, a country that spends considerably less of its GNP on health than do
the United States and many other European countries, the government-
funded National Health Service has undergone one reform after another, all
based, apparently, on the notion that many of the difficulties of providing
health care could be overcome by greater efficiency. The latest reform, which
attempts to make a clear division between purchasers of health, that is, the

government-supported health authorities, and providers, the doctors, nurses, and other medical staff, has again put the accent fairly and squarely on improved managerial efficiency. Nurses and doctors are disappearing to business colleges, and management has been included in the curriculum of medical schools; interviews for medical appointments appear to be directed more at assessing candidates' administrative skills than at evaluating their clinical competence. But, just as seems to have happened in the United States, new tiers of management bureaucracy have emerged and increasing amounts of clinicians' time is spent on committees, in an attempt to raise efficiency and competitiveness in the marketplace of hospital and community practice.

The dilemma for doctors in this increasingly depressing scene was summed up succinctly by Philip Caper in an editorial in the *New England Journal of Medicine* in 1988: "It is unreasonable to expect physicians—or patients—to trade immediate benefits at the bedside for longer-term benefits to society, without clear signals about what limits have been or will be set." In other words, the first responsibility of doctors is to an individual patient. If they know they can do something for them, even if it involves extremely expensive high technology for what may be of marginal benefit, they feel they must try. Can they deny a period of expensive treatment on a kidney machine to an eighty-year-old because they know that the costs incurred might pay for cancer screening for fifty young women?

But who is to set the limits? Following the recent health reforms in Britain it is purchasers of health—the local, government-funded health authorities—who will try to determine the priorities for their communities, on the basis of their perceived needs and the money that is available. But it is far from clear how they will achieve this difficult balancing act. Because neither governments nor health economists who have tried to put a value on the quality of life have found a solution, this kind of impasse is resulting in the runaway costs of medical care. It is becoming evident that these problems have no easy solutions that will be acceptable to a caring society.

In the example just cited, the only way forward is for the medical sciences to try to discover why the eighty-year-old patient's kidneys failed in the first place, and how it might have been prevented. The efficient and humane delivery of health care depends on a genuine understanding of the causes, prevention, and management of disease, both in the hospital and in the community. And as the events that led to the control of infectious disease in the early part of this century have told us, this can be achieved only by research based on solid science.

The problems of the affluent countries are magnified in the developing world. The 1991 report of the WHO Commission on Health and Environment found a strong relation between GNP and infant mortality and life expectancy. There are a few exceptions, however. For example, in Sri Lanka and China, two relatively poor countries as measured by per capita income, the level of health was well above the norm for countries with comparative income levels. It appears that their governments have used their limited resources more effectively in developing health care programs for their populations. The report suggests, therefore, that the main priority for the developing world is to establish more efficient primary medical care, backed up by sanitation, nutrition, and education. But, like a more recent assessment of the health of the nations by the World Bank, it stresses that the control of disease in the developing countries, particularly as they go through the transition toward Western culture, will also depend on a scientifically based attack on the infectious and parasitic diseases that are their major killers, on population control, and on the medical consequences of Westernization.

FUTURE GOALS FOR MEDICAL RESEARCH AND PATIENT CARE

Recent demographic research, we noted earlier, indicates that if we were to eradicate the bulk of premature deaths from cardiovascular disease and preventable cancer, we would add only a dozen or so years to our life expectancies at birth. It seems that we are programmed for a life span of about eighty-five years, and, as the demographers put it, the aging process itself becomes the major risk factor for death. These recent studies on the demography and mechanisms of aging, while of small comfort to those of us who are becoming a geriatric statistic, offer a useful background for our thinking about the future goals of medical research and care. In effect, they are telling us that, unless we make unexpected progress in slowing down the process of aging, we cannot hope to add a great deal to our life expectancies, even if we control vascular disease, cancer, and our other common killers. At the same time they are warning us that the richer countries are about to see a new epidemic—that is, an enormous expansion of their aged population, with all the medical problems that plague older people.

Of course, mortality statistics on vast populations tell nothing of the misery and suffering caused by the intractable disorders that attack what, in numerical terms, appears to be the small proportion of our population that dies prematurely. Our major goals therefore must be to reduce further the number of premature deaths and, at the same time, to deal more effectively with the chronic disorders that, though not apparent in these statistics, mar the quality of life, particularly for our aged populations.

If we take a more global view, and base our life expectancy figures on the richer countries, over one-third of the world's people, about 1.6 billion of those living in Africa, Asia, and Latin America, die prematurely. In the case of the developing world, therefore, our major goal must be to apply such preventive measures as we can, together with basic scientific research, toward improving these disgraceful statistics.

THE HEALTH OF THE RICHER COUNTRIES

Preventing Premature Death

In Britain a hundred years ago about half the population died before middle age and three-quarters before the age of seventy, whereas now only 3 percent are expected to die before the age of forty, although about 30 percent still die before the age of seventy. It appears, therefore, that death before middle age has been largely avoided but that death in middle age is still relatively common. It follows that, while we may be able to reduce death rates in early life a little, and must try, our first priority is to do all we can to tackle the problem of serious illness in middle life and, at the same time, to improve the quality of life for the elderly.

As we saw in chapter 5, infant mortality ranges from just under 5 per 1,000 in Japan to just over 11 per 1,000 in Portugal; Britain and the United States, with rates of 7 to 9 per 1,000 are not doing quite as well as they might. Furthermore, within each country there is a significant difference in infant mortality between different regions and social classes. At least some infant deaths thus seem avoidable by the more effective use of medical services.

In Britain about one-third of the deaths in the first month of life are due to congenital anomalies. Although some are preventable—by controlling maternal infection and vitamin deficiencies, for example—the majority are not, and this is an area in which the basic sciences offer the best hope for some genuine progress. Another 30 percent of the deaths are due to birth injuries and reduced oxygen supply to the baby, complications that could be reduced by better obstetric services and care. About 10 percent are due to low birth weight, which might be controlled, to some extent, by better maternal care—discouraging mothers from smoking, for instance. By concentrating on the health of pregnant mothers, on education, and on improved obstetric services, we should therefore be able to reduce further the neonatal death rate in the developed world. On the other hand, a great deal more epidemiological and basic laboratory research is required if we are to make major inroads into the problem of congenital malformation and prematurity.

If we turn to the causes of deaths in infants between the age of twenty-eight days and one year, we see a different pattern. In this group nearly half the deaths in Britain are due to a strange disorder best known as the "sudden infant death syndrome." Despite a vast amount of research its cause is not known, but there is recent evidence, based on at least twenty controlled studies, that the risk is reduced significantly if it is ensured that infants sleep on their backs. Factors such as maternal smoking and overheating also increase the risk. But a great deal more research into these tragedies is required. Of the other causes of death in this age group, genetic disease and congenital abnormalities are the most important.

In older children the pattern of illness changes again. In the five to fourteen years, over 40 percent of all deaths are due to accidents. In 1991 some 4,500 children were killed on the roads in the United Kingdom. The bulk of the other deaths in this age group are caused by congenital and genetic abnormalities and childhood cancers. Again there are differences in death rates between social classes, suggesting that at least some of the deaths are avoidable. Many of the deaths could be prevented by a major education program directed at drivers and parents. The remainder will require research in the basic sciences before they can be controlled.

It is when we come to the causes of premature death in middle age that the possibilities for preventitive medicine appear most challenging. Here we are dealing mainly with vascular disease, particularly heart attacks and stroke, and cancer. We have already discussed the relative roles of genes, lifestyles, and environment in the genesis of these conditions. As we saw, modern epidemiology has provided a few certainties and many hints for preventing at least a portion of the deaths due to vascular disease and cancer. Cigarette smoking is the most important preventable cause of cancer and vascular disease. It is killing about three million people a year and, if present smoking habits continue, will be causing about ten million deaths a year a few decades from now.

Other sensible approaches to reducing the prevalence of our middle-age killers entail screening for their early detection. In Britain it is estimated that it should be possible to reduce death rates from breast cancer by at least 25 percent over the next ten years, and a similar figure could be achieved for invasive cancer of the cervix. Although some aspects of cancer screening remain controversial, these are reasonable goals. Similarly, it is sensible to encourage the middle-aged to have regular checks of their weight and blood pressure; evidence that a reduction in blood pressure reduces the prevalence of strokes, and that obesity is a risk factor for cardiovascular disease, is good enough for us to act now. We know less about the dangers of high blood cholesterol levels; on the basis of current information it seems sensible to try to reduce the levels of those with particularly high values and to screen individuals with a strong family history of heart disease. But we need better evidence from properly conducted, large-scale clinical trials before we make a wider effort in this direction.

What of other risk factors for vascular disease and cancer? As we saw, there are few established facts. It is reasonable for health educators to continue to press for healthier lifestyles, including more physical exercise at all ages and the establishment of modest targets for a reduction in the intakes of total fat and saturated fatty acids and of sugar. This may be difficult because, in Great Britain at least, there has been little change in recent years in the proportion of total energy derived from fats. Our health education programs should also include clear advice about sensible limits for alcohol consumption. But a great deal more research needs to be carried out before unequivocal advice about the level of consumption of sugars, complex carbohydrates, and salt can be given. We must not offer advice that is based on inadequate data.

If we also aim to reduce sexually transmitted diseases and the use of addictive drugs, we will be setting ourselves objectives that are both reasonable and attainable. As for the fight against the other common killers and chronic diseases—psychiatric illnesses, diabetes, rheumatism, dementia, and the rest, and a good deal of vascular disease and cancer—we must rely on progress in the basic and clinical sciences.

Improving Our Current Practice

Since it may be a while before we reap the benefits of the new developments in preventive medicine and in the basic medical sciences, we must try to improve our current approaches to the avoidance and management of disease. For although there are huge gaps in our knowledge of almost every aspect of pathology and therapeutics, we could undoubtedly help our patients by applying what little we know more effectively.

We have gone at least some way toward learning how to assess the quality and effectiveness of current medical practice. We have seen how clinical trials, either alone or combined by meta-analyses, can be used to improve many aspects of the delivery of patient care. Controlled clinical trials still produce spurious results, but as the number of patients involved increases, and our requirements for the statistical significance of the results become more stringent, they will be of even greater value. And we need to extend these scientific principles into almost every branch of medicine. Even the manner in which we examine patients has, by and large, been passed down from generation to generation. Most of what we do at the bedside has never been subjected to critical analysis. We assume it is the right way to do things, and fail our students if they do not follow suite. We have some idea of what a routine clinical examination should entail, and which investigations should be carried out on an individual patient, but for much of our day-to-day clinical practice we lack clear evidence for the value of what we do.

Doctors have to live with uncertainty. Much of clinical decision making at the bedside has to proceed without adequate information. Over recent years attempts have been made to put this activity onto a more scientific and rational basis. Some extremely simple but effective methods have been used, such as encouraging individual patients to act as their own controls for particular drug regimes. But much more sophisticated approaches, which include the use of computerized decision-making trees, are being assessed. Complex probability theory, based on Bayesian logic, a form of probability analysis first described in a 1763 essay by the Reverend Thomas Bayes, an English clergyman, is being adapted to help clinicians make more logical decisions. Many doctors are extremely skeptical about all this. However, it is essential that research is pursued along these lines if we are to make better use of our current knowledge.

The same principles apply to a great deal of therapeutics, much of which is simply assumed to be effective. As we saw earlier, although some of the very expensive treatments of modern cardiology have been subjected to adequate clinical trial, many have not. This is true of many routine surgical procedures as well. Every aspect of clinical practice requires close scrutiny, careful audit, and, where necessary, properly designed trial. We must try to persuade the practitioners of alternative medicine to follow the same route. And we must take advantage of modern electronic technology and communications systems to make the results of these investigations available to the medical profession.

We must be equally critical in our approach to preventive medicine. While clinical trials have shown the benefit of stopping cigarette smoking, many of the other changes in lifestyle that are being promoted by Western governments are based on information lacking in solid evidence. It is unpardonable to try to alter the diet of an entire population without sufficient information. If, for example, adequate cholesterol-lowering trials had been carried out years ago, we would now have the information on which to base sound advice. As it is, we have a series of impressions stemming from deficient data, not to mention a completely confused public. Things are no better when we come to assess the different methods for the delivery of health care. Great Britain, with its government-based National Health Service, missed many opportunities for studies of this kind. In the recent reorganization of the National Health Service, it would have been possible to change the pattern of health care delivery in some parts of the country, leave the rest alone, and compare the results. But the new organization was simply assumed to offer the right way forward, and little effort was made to assess the outcome. Measuring the lengths of waiting lists or the numbers of patients processed through the system may tell us something, but little more than whether we are running an efficient supermarket.

But do we have the appropriate analytical tools for assessing the delivery of health care? While the randomized clinical trial and its more recent offspring have formed the basis for much of effective medical practice, they have their limitations. As the American epidemiologist Alvan Feinstein has noted, many of the models, methods, and paradigms of "basic" biomedical research are not suitable for evaluating clinical intervention and the outcome of health care. He believes that completely new approaches to the assessment of clinical practice are required.

Recently efforts have been made to measure the quality of life and the outcomes of treatment from the patient's viewpoint. Much of this work involves the mathematical treatment of information obtained from interviews with patients. A variety of different profiles are being tested, some to assess the quality of life as a whole and others to measure the impact of treatment on particular diseases. This new branch of medical research brings together some strange bedfellows, ranging from psychology and the social sciences to biomathematics. It presents many difficulties, not least the uneasy amalgamation of the relatively "soft" science of interviewing techniques with some fairly sophisticated mathematics. And much of its current jargon is incomprehensible to most doctors. However, its methods are improving and some of its findings are sufficiently robust to suggest that it will yield important benefits for patients in the future. The recent announcement of a $40 million study in the United States, the Patient Outcomes Research Program, which will include an assessment of the best way to treat prostatic hypertrophy, backache, heart attacks, and other common diseases, highlights this interesting new trend in medical research.

Getting to Grips with Our Intractable Diseases: What Can We Expect from the Molecular Sciences?

The problems of trying to predict when the molecular sciences will make a major impact in the clinic are exemplified in a recently published book, *The Code of Codes,* edited by Daniel Kevles and Leroy Hood. While this provocative work is aimed mainly at the ethical and social issues arising from the Human Genome Project, two of its chapters deal at some length with the project's medical applications. The first, on the diagnosis and management of single-gene disorders, concludes that molecular biology and the genome project will transform clinical genetics over the next few years; few would disagree with this conclusion. The other, entitled "Biology and Medicine in the Twenty-first Century," written by Leroy Hood, takes a much broader view of the applications of molecular biology to human disease.

After describing the new technology that we outlined in chapter 9, Hood discusses its application to our common killers. His view of the future is as follows:

The diagnosis of disease-predisposing genes will alter the basic practice of medicine in the twenty-first century. Perhaps in twenty years it will be possible to take DNA from newborns and analyze fifty or more genes for the allelic forms that can predispose the infant to many common diseases— cardiovascular, cancer, autoimmune, or metabolic. *For each defective gene there will be therapeutic regimes that will circumvent the limitations of the defective gene.* Thus medicine will move from a reactive mode (curing patients already sick) to a preventitive mode (keeping people well). Preventative medicine should enable most individuals to live a normal, healthy, and intellectually alert life without disease. (Italics mine)

In summarizing his thoughts on the future of biology and medicine, Hood goes on to say,

I believe that we will learn more about human development and pathology in the next twenty-five years than we have in the past two thousand.

He is not alone in this belief; many molecular biologists are spreading a similar gospel.

While I am sure that Hood and his colleagues are correct in believing that if we are successful, we will learn a great deal about the pathology of our common diseases from their molecular genetics, I am far less confident that we will be successful in controlling them in such a short time. The reasons for my uncertainty will become apparent if I summarize briefly my earlier discussions of these illnesses.

Unlike the infectious diseases that were the important killers of all age groups up until the middle of this century, most of our current diseases do not have a single cause and tend to attack the middle-aged and elderly. This problem is exemplified by the old gentleman with kidney failure whom we met earlier in this chapter. As is often the case in elderly people, the damage to his kidneys may well be due to disease of the walls of the blood vessels that supply them. Vascular disease is rarely localized to one region of the body. Thus he is also likely to have coronary artery disease, a raised blood pressure, impairment of his nervous system as a result of a narrowing of the arteries to the brain, and many other complications of widespread damage to his circulatory system. This may have been caused by his lifestyle and diet, together with the effects of his genetic makeup and the pathology of aging. But many other disease processes, such as high blood pressure or insulin-resistant diabetes, which are common in older populations and which have an equally complex, though different, constellation of causes, may also have been involved in damaging his blood vessels and kidneys.

Many common diseases have complex and multiple pathologies that reflect the effects of both nature and nurture together with the damage that our tissues sustain as we age. We have explored the evolutionary basis for the genetic diversity that may make us more or less susceptible to these diseases. In an attempt to combat malaria over just a few thousand years, natural selection seems to have made use of almost every conceivable genetic variant of our red cells and immune system. Insulin-resistant diabetes shows the diversity of the many genetic polymorphisms that may be entailed in susceptibility or resistance to common disease, reflecting, as they do, selection over a much longer time in many different environments. Because most diseases have multiple and additive pathologies, we are in effect seeing the end result of genetic susceptibility to different environmental agents mediated through many different evolutionary pathways of this kind.

Because of the enormous interactive complexity of the cardiovascular system, and the large number of genes involved, progress in just discovering the major players in vascular diseases, and learning more about their actions and about ways to modify them, is bound to be slow. We already know that genetic susceptibility to coronary artery disease involves many different genes that are part of completely different chemical pathways; a recent count suggests that over two hundred play a role in the control of cholesterol levels in our blood, and this is only one of many factors involved in heart disease. But we are slowly chipping away at the problem, and can already identify some groups of patients who are at particular risk and on whom we should focus our efforts at prevention. There seems little doubt that we will gradually come to understand how these diseases are caused and how better to prevent or treat them. But the multiplicity of routes to a diseased blood vessel makes it unlikely that we will find a single "magic bullet" for preventing or treating vascular disease.

Cancer provides another good example of the problems that will face us. Now that we are much closer to understanding how some cancers arise, we have an inkling of their biological complexity. For instance, the development of colon cancer requires the generation of mutations in at least half a dozen or more oncogenes, which control the way in which cells proliferate, mature, and function. It is likely that environmental carcinogens can cause some of them, although we may be born with one or more, or they may be by-products of the aging process. Of course, some cancers may require many fewer cellular "events," perhaps no more than two. But there is now little doubt that many are the end result of a chain reaction, in which many different sets of genes are involved. We are probably generating mutations of this type all the time, but it is only when a particular pattern occurs in the same tissue at the right time that a cancer develops. In essence, cancer reflects an element of bad luck, damage to our DNA by environmental carcinogens or those that we produce as we age, a particular set of changes in our genes, and, possibly, a failure of our immune mechanisms that may also be related to aging.

What comfort is all this to patients with cancer? The first message is that a great deal more work will be required before we understand how mutations in oncogenes combine to produce a malignant tumor. But there may be practical benefits long before we get that far. Suppose that at least one of the genetic changes is inherited. When we are absolutely sure that this mutation increases the likelihood of developing cancer, we then have a diagnostic handle for identifying persons at particular risk. The most obvious examples are the familial cancers that strike the bowel or the eye. But there may be applications to many other forms of cancer. For example, it appears that *BRCA* I, a gene on chromosome 17, is involved in susceptibility to breast cancer, particularly before the menopause. If it turns out that this is due to a specific variant of this gene, we will have an important tool for identifying women at high risk and hence for focusing our screening programs. Of course, we will then have to find out which other oncogenes and environmental factors contribute to this form of breast cancer; simple family studies tell us that this is not a single-gene disease.

Once we have information about the oncogenes involved, and since methods are now available for analyzing minute amounts of DNA, which means that we require only a few cells from a particular tissue, we will quite likely be able to screen for many common cancers, such as those of the lung, prostate, bowel, and cervix. We should also be able to monitor patients for recurrence of tumors after treatment.

What are these new concepts about the "multiple hit" cause of cancer telling us about treatment in the future? There are two opposing views. The pessimists contend that if cancer of the colon requires mutations in at least six different sets of genes, the whole thing will be far too complicated to reverse. The more optimistic view, and the one to which I subscribe, is that if cancers result from chains of events of this complexity and if each mutation must be in place for a malignant change to occur, we may be able to control the disease by breaking only one link in the chain—that is, by tackling the consequences of one of the mutations. But we should not pin our hopes on the discovery of a blanket "cure" for cancer. It might happen; much more likely though, we will have to tailor our new approaches to diagnosis and treatment to each individual type of cancer.

I suspect that progress toward the prevention and treatment of our other killers and causes of chronic ill health will follow a similar pattern. If we are lucky, we should be able to identify some of the important genes that make us more or less susceptible to these conditions, and as we come to understand their products we will undoubtedly learn more about their underlying pathology. Along the way we will be able to identify some groups of people who are unusually prone to the ill effects of particular environmental agents, and hence we will be able to concentrate our efforts at prevention more effectively. As we come to understand its underlying pathology, the combination of recombinant DNA technology and protein engineering will lead to completely new approaches to drug design and to a more logical way of treating established disease. And, finally, when we have discovered how environmental agents, or those that we generate as we age, mediate their effects at the molecular and cellular level, we will learn how better to prevent or treat the diseases that have puzzled us for so long. Some of them may be curable, but because of their multiple origins and close relation to the aging process, we may have to settle for a combination of partial prevention and improved control.

It would be unwise, therefore, in planning research programs or medical services for the next half century, to rely on a sudden change in the whole face of medical practice; our present patterns of the epidemiological investigation of the causes of disease, and research into better and more humane ways to manage it when it is established, must continue. Work in the molecular sciences will undoubtedly lead to steady improvements in the way we diagnose and treat our patients, but the road to prevention and radical cure for our intractable problems, given their multiple causes and complexity, will be a long one.

THE DEVELOPING WORLD

A recent publication from the World Bank, *World Development Report, 1993,* sets out a blueprint of simple public health measures and essential clinical services that, if applied in the poorer countries, would significantly improve the health of their inhabitants. The public health measures include immunization, school-based health services, family planning, programs to reduce tobacco and alcohol consumption, and AIDS prevention. The essential clinical services consist of a simple package of medical care directed at a reduction in maternal deaths, family planning, tuberculosis control, a lowering in the frequency of sexually transmitted diseases, and the care of serious illnesses of children, particularly diarrhea, respiratory infection, measles, and malaria. The report estimates that a program of this type would cost approximately twelve dollars per capita, or some 3.4 percent of the per capita income of low-income countries. It could reduce the burden of disease in these populations by approximately 30 percent.

While these are admirable objectives, many of which should be attainable, some of the major problems of the developing world will clearly not be solved without the help of the medical sciences. Nowhere is this more true than in the fields of population control and infection.

As we have seen, population growth is one of the major challenges for the future. Despite its popularity, the subject of human sexual behavior has long been proscribed by theologians and neglected by biologists and medical educators. But, as a recent editorial in the medical journal *Lancet* points out, things are changing. Rather optimistically, I suspect, the writer suggests that events of the last few years in animal behavior research resemble those of the Copernican age for astronomers. The new information coming from this rapidly expanding discipline offers the beginnings of a primitive understanding of the evolution and meaning of human sexual activity.

Since humans are closely related to chimpanzees, in evolutionary terms at least, the sexual behavior of chimpanzees is of great interest. It appears that they, like us, engage in many acts of infertile intercourse. Some of them have been observed to indulge in over 3,000 such acts before the first birth, and in up to 2,500 during the rest of their fertile life. This is, in effect, similar to the behavior of women in preliterate societies in which there are relatively few births and long intervals of breast feeding; it has been estimated that women may have 2,000 to 3,000 acts of sexual intercourse in a lifetime, of which perhaps only 200 are potentially fertile.

In light of on these new observations, the *Lancet* editorial goes on to suggest, the key claim of the 1968 papal encyclical, *Humanae vitae*—"every conjugal act must be open to the transmission of life"—must be incorrect. And it points out that much of the evolution of Catholic teaching on intercourse stems from Saint Augustine's misinterpretation of the account of Adam and Eve, as a result of his lack of acquaintance with the Greek language. In fact, the Roman Catholic teaching of natural law, which is based on Aristotelian rather than on biblical thought, does not preclude intercourse in infertile couples or after the menopause. Its central theme is that it is wrong to frustrate procreation by unnatural means, a view the Pope has just reaffirmed.

It is beginning to look as if patterns of sexual activity, in humans and chimpanzees alike, have been shaped by natural selection, but it is not clear, although theories abound, what advantage has been conferred by our unusually active, in terms of comparative biology, sex lives. But at least these findings, together with more enlightened discussions between behavioral scientists and Catholic theologians, have led Cardinal Fiorenzo Angelini of the Vatican to invite theologians and scientists to continue studying the meaning and purpose of human sexual intercourse. If further work leads to a more pragmatic approach to birth control on the part of the Catholic church and other religious bodies, some genuine progress in the control of the world's population may be made.

It is doubtful, however, whether there is time to wait for further advances in the behavioral sciences or a change of attitude on the part of some of the world's religions. It is essential that work in reproductive biology be directed at more effective and simpler methods of contraception. In reviewing this field, a report written by the Commission on Health Research for Development in 1990 lists several areas of research that show great potential for population control. They include the development of a spermicide with antiviral properties, a reliable predictor of ovulation, an improved oral contraceptive for women, reliable and reversible methods for male sterilization, and antifertility vaccines. One of the depressing features of the report, however, is that whereas in 1970 thirteen major pharmaceutical companies were conducting research and development on new contraceptives, by 1987 the number had dwindled to four, with only one in the United States. Presumably this reflects some disillusionment on the part of these companies in the speed of progress in this critically important area of research. But the research must be pursued.

One reason for the pharmaceutical industry's reluctantance to pursue research in contraception is the fear that it may be impossible to produce contraceptive agents that are completely free of harmful side effects. Unlike most medicines, contraceptives are used by healthy people, and therefore any unwanted effects are considered unacceptable. We need to think much more positively about this problem, however. It has been known for a long time that women who take contraceptive pills have slightly reduced risks of ovarian and endometrial cancers, and the much publicized fear of an increased risk of breast cancer has not stood the test of time. Further research could ultimately generate agents that would combine effective contraception with protection against both cancer and menopausal symptoms. As John Aitken, the English reproductive biologist, has pointed out, developments along these lines might lead to a fundamental shift in the way in which contraceptives are viewed by the early years of the next century—no longer as burdensome necessities but as important contributers to a full and healthy lifestyle. The enormous progress made in recent years by the basic biological sciences in understanding the mechanisms of both male and female fertility suggests that this area of research will have a lot to offer the developing world in the future.

Studies of our sexual behavior are equally important if we are to make any headway in controlling the epidemic of AIDS, which presents such a horrifying prospect for the developing world. There have been very few surveys of sexual lifestyles, which are difficult to organize and extremely expensive. A recent study carried out in Great Britain under the auspices of the Wellcome Trust, the medical charity that had to sponsor the work because the government refused, underlines some of these difficulties. It turns out that the lifetime number of sexual partners in the United Kingdom averages 9.9 for men and 3.4 for women, figures consistent with those obtained in a similar survey conducted recently in France. The variance was 6,575 for men, compared with 165 for women, which says that men show a much larger spread in the number of partners. Perhaps the most important finding of this study is that mutually monogamous couples with no risk of sexually transmitted diseases are extremely rare.

Heterosexual transmission of AIDS in the richer countries has been slower than at first feared, but its spread in developing countries has been frightening. Limited information from Africa suggests a swifter partner change there than in Europe. Comparative studies have also shown that the activities of prostitutes are much greater in Thailand than in Europe, whereas condoms are less available and the treatment of sexually transmitted disease is not nearly as advanced in the developing countries. The recent survey in Great Britain brings out the potential problems of containing AIDS in Europe and the United States; from what little we know, these difficulties will be magnified many times in the developing world.

As we saw earlier, although public health measures may do much to reduce death rates from infection in the poorer countries, a great deal of work in the basic sciences will be required if we are to eradicate many of the important infectious diseases. The common parasitic killers serve as a good example. Given that drug-resistant malaria is spreading and that the genetic adaptability of the parasite is always a step ahead of the pharmaceutical industry, we will either have to discover better drugs to treat malaria and other parasitic illnesses or have to develop vaccines against them.

The importance of the development of vaccines does not stop at the parasitic killers. The commonest causes of death in childhood in the tropics are still respiratory and diarrheal illnesses. In addition to making available the vaccines that we have already—for measles, for example—we must continue work on developing new ones for other common infectious illnesses that affect young children, both in the tropics and in richer Western countries. If we are to check the current AIDS epidemic, we will need either a better antiviral agent or an effective vaccine. It may be already too late to help many African countries; if we are to control the disease in Southeast Asia, we will have to move very quickly. Infectious disease is universal, and the need for further basic research and the development of viral chemotherapy, and better treatment for bacterial, viral, and parasitic illnesses, is almost as great now as it was in the period when infectious disease was the leading killer in the West.

Finally, we have seen how, when poorer countries become richer and start to adopt Western lifestyles, they are prone to new patterns of illnesses, similar to those that affect the wealthy industrialized societies. Because of their genetic makeup, some of these populations seem unusually prone. The major epidemic of diabetes in Asians offers a timely example of the problems that may arise. It is essential to try to determine the differences in susceptibility of different racial groups to high-energy Western lifestyles. We need this information not only to help the emerging countries but also to help us understand why these conditions occur at all. Medical research in the developed countries has much to learn from the rest of the world. In the future it must take a global perspective, a lesson that has to be impressed on those who fund it and on the governments of the developed world.

THE CURRENT LIMITATIONS OF MEDICAL RESEARCH

What are the limiting factors in medical research, and why can't we move faster? Questions of this kind are being posed by politicians and the public with increasing vigor. Why, they ask, is all the money that has poured into the basic biological sciences and medical research not eliminating our common killers? By describing the complexity of these illnesses, I hope I have already gone some way to answering this question. But there is a broader issue that we must face as we move into the era of medical research at the molecular and cellular level.

A few years ago a British newspaper asked James Black, the pharmacologist and Nobel laureate, to give his views on the future of his field of science. His reply was characteristically brief and provocative: "The progressive triumph of physiology over molecular biology." This theme has been taken up recently in a book, *The Logic of Life,* edited by the Oxford physiologists Richard Boyd and Dennis Noble. The central message is encapsulated in its subtitle, *The Challenge of Integrative Physiology.* What the authors, who come from many different biological disciplines, are saying, in effect, is that living things are enormously complex, and that once we have taken apart all their molecular building blocks, we will have to evolve a completely new type of physiology, probably based on a new kind of mathematics and computer modeling, in order to put them all together again. It is difficult to fault this thesis; it is the major reason why I have advised caution when trying to anticipate when the fruits of the molecular sciences will be applied in the clinic, particularly with respect to some of our more intractable diseases.

Science has always looked for simple rules that govern the universe. Newton explained how the heavenly bodies behave according to the laws of motion and gravitation. His deterministic view of things received a setback in the early part of this century with the development of quantum mechanics, a theory based on probability and the impossibility of measuring the position and momentum of objects at the same time. The uncertainty principle of quantum theory seems to be inherent in the laws of nature. Nevertheless, quantum mechanics has proved to supply a valuable framework in which to study the properties of matter and the forces that govern the universe. Currently physicists believe that once they are able to arrive at a unifying "theory of everything," it will be possible to explain all natural phenomena. Human beings, it is assumed, will be subject to the same laws of nature that rule galaxies and atoms.

Perhaps. But, for the moment, science cannot even use the laws of nature to predict with certainty what the weather will be like tomorrow morning, let alone to explain the enormous complexities of the workings of even the simplest living organism. Recently, however, research workers in many disciplines have begun to realize that, while there seem to be built-in limits to predicting the future, some mathematical approaches to this problem may have applications in every branch of science. One new field that has raised these hopes is based on "chaos theory."

In introducing a recently published book on this topic for the nonexpert, Nina Hall describes how chaos theory has evolved from a synthesis of imaginative mathematics and readily accessible computer power. It presents a picture of a universe that is deterministic and obeys the fundamental laws of physics, but which has a built-in predisposition for disorder, complexity, and unpredictability. Chaos theory has already been applied to several areas of biology and medicine. It has started to make some limited sense of the complexities of epidemics, the behavior of parasites in populations, hitherto unexplained oscillations in the rates of production of blood cells, and even the disorganized behavior of the heart muscle in states of fibrillation. It is already clear that at least some physiological and pathological systems exhibit patterns that can be interpreted in terms of chaos theory, but it is too early to say how far this exciting new field, and its more recent offspring, will take us in our ability to predict complex events in multicellular organisms. And it will be a long time before this kind of sophisticated mathematics will have a major effect on day-to-day clinical practice.

In *The Logic of Life,* in summarizing the current problems of understanding living organisms, F. Eugene Yates points out that whatever science of complexity finally emerges, it is likely to displace cybernetics, general systems theory, artificial intelligence, information theory, and control theory, all of which, he believes, have failed to account for the stunning diversity of biological structures and functions. He closes by saying that as he was pondering the possible relevance of chaotic dynamics to the problems of self-organization in biology, he came across a short poem by Robert Frost, first published in 1949, which seems to encapsulate our current dilemmas.

Let chaos storm!
Let cloud shapes swarm!
I wait for form.

While I suspect that Yates, like the other authors of this stimulating book, tends to underestimate what we are likely to learn about living things from the study of their genomes, he and his colleagues leave us in little doubt about the difficulties we will encounter as we try to understand the workings of intact organisms, particularly those as complex as human beings. We do not yet have the technology with which to evolve an "integrative physiology," but clearly this is the direction in which we must move if we are to reap the benefits of the molecular sciences.

For the moment, therefore, we are stuck with our reductionist approach of trying to dissect disease at the level of molecules, cells, and organs. Recently, I attended several meetings on the Human Genome Project at which I was surprised to see slides of the beautiful anatomical drawings of Vesalius. They were probably there to liken the remarkable potentials of the genome project to those of the great age of anatomy, but it seemed to me that their message went much further. When we have completed our gene map, or sequence, of the whole of the human genome, we will be in much the same position as the anatomists of the Renaissance. We will understand the anatomy of our genetic makeup but will then face the extraordinarily difficult task of trying to find out how it works. Undoubtedly, we will start to appreciate better the molecular and cellular mechanisms of many of the diseases that affect us, but we will still be a long way from comprehending their protean effects on intact organs, let alone sick human beings.

Once we have worked out the mechanisms of our diseases at the molecular level, we will require a new type of physiology and pathology, based on a branch of mathematics and probability theory that does not yet exist, to help us anticipate and understand their effects on our patients. We may by then have gained some inklings of the physical bases of emotion and social adaptation, but we will still be far from understanding the many different ways in which we are molded by our changing environments and contact with our fellow human beings. Disease has diverse effects not just on our physical well-being but on almost every aspect of our lives and those of our families, friends, and communities. It is the sheer interactive complexity and unpredictability of the behavior of living organisms that sets the limits of the medical sciences, regardless of whether they involve highly sophisticated molecular technology or the simplest observational studies.

PART VI

Do Patients Have
Any Place in Modern
Scientific Medicine?

ELEVEN

Back to Our Patients

I should never have been happy in any profession that did not call
forth the highest intellectual strain, and yet keep me in good warm
contact with my neighbours. There is nothing like the medical profession for that: one can have the exclusive scientific life that
touches the distance, and befriend the old fogies in the parish too.
—*George Eliot (1819–1880)*

Before finishing our journey through the maze of medical science and practice, we must return briefly to a recurring theme—the lot of our patients.
Although it should be self-evident that medicine's long-term goal is to prevent disease and improve the quality of care of sick people, this fact is sometimes neglected in the frenetic excitement of the research laboratory, or in
the quest for efficiency and the financial constraint that dominate the lives of
today's doctors.

In some respects I am skeptical of the views of those who criticize modern
medicine as a dehumanizing, impersonal activity. I know of no evidence that
doctors of the past were so much better at handling people. I still vividly
remember one of my first teaching ward rounds as a medical student, over
thirty years ago. We had just reached the bed of a patient who had been
found to have an inoperable cancer of the lung. The senior physician who
was conducting the round suddenly veered away from the bed and gathered
together the throng of staff and students in the middle of the ward into what
resembled a huddle of American football players planning their next play.
The diagnosis and prognosis were discussed sotto voce, after which we all
returned to the bedside, where a few banalities were exchanged, and we
moved on to the next patient. The young doctors and nurses of today would
never let me get away with this kind of behavior. Nor would they allow me
to tell an impoverished dock laborer, suffering from a duodenal ulcer, to cut
down on the beer, take a little decent claret, and spend his winters in the

Mediterranean—advice offered by another of my teachers, in this case a surgeon.

The last thirty years have brought genuine improvements in communications skills and in the pastoral aspects of medical practice. However, it would be wrong if I left the reader with the idea that all is well with the patient's lot in the current medical scene. In this final chapter we will survey some of their problems and, against this background, return to the question with which we started out: how far has medical practice moved from being an art to being a science, and, perhaps more important, is it inevitable that, as this transition continues, it will have an adverse effect on the caring aspects of medical practice?

SICK MACHINES

Medical research and practice have no doubt grown increasingly reductionist over the years, and there is some basis for the criticism that doctors have tended to become more interested in the workings of diseased organs than in their owners. The reasons for this are complex. A Cartesian view of sick people as sick machines has been encouraged by specialization. As knowledge of human physiology and pathology increased, many doctors found that they could no longer cope with the totality of medical practice and therefore concentrated their activities on a particular organ or organ system. Today many physicians, especially clinician-scientists who wish to remain competent in both the laboratory and the ward, find that they can do so only if they focus their activities in this way. Generalists, that is, physicians or surgeons who will have a go at anything, have become less effective, and it is to the specialist that students and young doctors often turn, both for education for and inspiration.

In many leading American and European teaching centers responsibility for the care of the "whole patient" is delegated to more recently qualified members of the staff, who orchestrate events by calling in one specialist after another to offer an opinion on the piece of the patient that interests him or her. Even though patients, particularly as they get older, are not always obliging enough to package their diseases into only one system, the day of the generalist appears to be over. This is not to say that all specialists are bad doctors; some of them remain proficient in other branches of medicine and are equally effective in the pastoral aspects of care. But there is no doubt that, despite the advances in medical care that have resulted from scientific research and specialization, we have tended to compartmentalize disease in a fashion not always to the patient's benefit.

Of course, the true generalists, primary care clinicians or family doctors, could have made up for these deficiencies. They can assess all of the problems of their patients in the home environment and refer them for advice to specialists in appropriate systems, at the same time making sure that the rest of their needs and fears are not neglected. But in Britain and, I suspect, elsewhere family practice is changing. The role of family doctors as generalists, friends, and family counselors is threatened. They are now expected to function as pioneers in preventive medicine, all-around physicians, financial managers, and committee members. The time that they have to spend with patients is limited, and much of the work of out-of-working-hours home visiting is conducted on a rota, which often involves doctors from outside their practices. Thus, in both research and practice, there is a real danger that patients are looked on as the receptacles of one or more sick organs, and that the effect of disease on other systems, not to mention its impact on their lives, is neglected. The advent of the study of disease at the molecular and cellular level has therefore been seen as yet another phase of high-technology medicine that will lead to even further reductionism. Are we, in effect, simply going to reduce the standing of our patients to the status of a mass of misbehaving molecules?

Many of these concerns have their roots in the broader issue of biological determinism, in the notion that all human accomplishments, talents, and institutions are immutably fixed by our genetic makeup. Some of the main proponents of the Human Genome Project have not helped in this respect. For example, James Watson has described it as "the ultimate tool for understanding ourselves at the molecular level," adding, "We used to think our fate was in our stars. Now we know, in large measure, our fate is in our genes." If this is the way modern biologists are thinking, surely medical science and practice will move in the same direction. If we are depressed and worried about ourselves, it is because our genes have made us that way; there is nothing we can do about it, so why try?

But, as Peter Medawar has reminded us, we should not take such a simplistic view of ourselves. He speaks of two varieties of "heredity": endosomatic, or internal, heredity, for the ordinary or genetic heredity that we have in common with other animals; and exosomatic, or external, heredity, for the nongenetic heredity that is particularly our own—the heredity that is mediated through tradition. By the latter Medawar means the transfer of information through nongenetic routes from one generation to the next. In other words, we are what we are as a result of what we inherit and the environment in which we are raised, with its attitudes and traditions that have been passed down from our forebears by word and deed.

There are several other, more pragmatic reasons for taking a much more optimistic view of the new biology. Paradoxically, the change in emphasis in basic medical research, from whole patients and their diseased organs to their molecules and cells, has the potential for bringing together the disparate sub-specialties of the current medical scene and for breaking down its watertight compartments. As the pattern of research changes in this way, there will be a natural tendency toward the reunification of its activities; cardiologists, chest physicians, psychiatrists, and the rest will all be sharing the same tools to study their particular problems. Disease mechanisms at the molecular and cellular levels may well turn out to be similar for all our organs, and require the same type of approaches for their investigation, whatever system is involved. Furthermore, once we have explored the anatomy of the human genome, we will have to develop a more holistic approach to human biology and medical research if we are to find out how the whole thing works.

Another important principle emerging from the study of disease at the molecular level, particularly the structure of our genes, is that each of us is unique. We must transmit to our students of the future the notion that two people with the same base changes in their DNA, though they have identical pathology, may react to disease in completely different ways. The interactions of their many other variant genes with the complex environments in which they have been raised have combined to make them the individuals that they are. If these simple messages from the "new biology" are interpreted and taught in this manner, there is no reason why it should have a dehumanizing effect on our doctors of the future—just the contrary.

LOSING SIGHT OF OUR OBJECTIVES

There is a growing belief that many of the shortcomings of today's doctors stem from the ossified kind of education that is common to most Western medical schools. Typically, this starts with two or three years of the basic medical sciences—anatomy, physiology, and biochemistry—followed by a course in pathology. Students then spend a further few years acquiring clinical skills, mainly by walking the wards of large teaching hospitals. Is this the best way to train a doctor? Do two years spent in the company of cadavers provide the best introduction to a professional lifetime spent communicating with sick people and their families? Does a long course of pathology, with its emphasis on diseased organs, and exposure to the esoteric diseases that fill the wards of many of our teaching hospitals, prepare students for the very different spectrum of illness they will encounter in the real world? And is the protracted study of the "harder" basic biological sciences, to the detriment of topics like psychology and sociology, the best way to introduce a future doctor to human aspects of clinical practice? In short, have we got it all wrong; is our current preoccupation with diseases rather than with sick people simply a reflection of an educational system that has lost sight of its objectives?

Thinking along these lines has led to a complete reappraisal and reorganization of the medical curricula of many medical schools, both in the United States and in Europe. In Great Britain the General Medical Council, the body that governs the practice of medicine, has set out a new program that reduces the amount of taught material to what it calls a "core" of essential knowledge (it is wise enough not to define what this means), backed up by a much less crowded curriculum with less emphasis on the basic biological sciences and more on the social sciences and communications skills and with an early exposure to patients.

While the motives behind these changes are admirable, it is essential that, while trying to improve the social, pastoral, and communications skills of our future doctors, we do not dilute their scientific education. As we have seen, the basic sciences are of enormous importance to the future development of clinical practice. If this is not impressed on our students, we risk losing that small but vital core of medical scientists who will be able to form the link between the clinical and the basic-science worlds. It also poses the danger that our doctors will not learn how to evaluate new information critically, a facility they require throughout their professional lifetimes. Achieving the balance between a scientifically based medical education and one that introduces these new approaches to training more caring and socially aware doctors—all without overcrowding the curriculum—is the major problem that faces medical education today. It would be less daunting, of course, if we had the faintest idea about how to measure the quality of the end product.

But it is naive to lay all the blame for the shortcomings of our doctors on their education. Remarkable developments in medicine and increasing pressures to contain costs have placed intolerable demands on them; it is not surprising that they have less time to spend talking to patients and their families. After qualifying as a doctor, a little over thirty years ago, I spent some time as an intern in a teaching hospital and then worked with a busy family practice in the slums of Liverpool, a large city in the north of England. Because there was relatively little that we could do for many of our patients, except control their symptoms and make them more comfortable, much of my time was spent talking to them about their illnesses and trying to comfort and support their families. I gained not only extremely valuable clinical experience but also an unforgettable introduction to illness in the real world. The scene that greets those who qualify today is quite different. The frantic pace of life in a modern hospital, and the diverse technical skills required of young doctors in the research laboratory, hospital, and community, allows little time for thought or sleep, let alone talking to patients and their families.

Other facets of the current medical scene are also separating doctors from their patients. One of the most worrying features of medical practice in Great Britain, particularly following the recent reorganization of the National

Health Service, is the large amount of time that doctors are having to spend on committees concerned with business plans, efficiency, and hospital organization; caring for patients, teaching, and research seem to have become secondary issues. In short, our hospitals are beginning to resemble supermarkets. It is not surprising that, in this overcharged atmosphere, the pastoral side of medical care is sometimes forgotten.

And, in searching for the roots of the shortcomings of the current medical scene, we must remember that today's doctors are practicing at a time of extraordinary change and development. Rarely a week goes by without a new advance of potential benefit to their patients. Many of them raise new ethical dilemmas, yet there is increasingly less time to sit back and reflect on how to deal with them. Hence we cannot fully appreciate the reasons for some of the current concerns about the future of the medical sciences without considering a few of the ethical issues that face us.

ETHICAL ISSUES

Those who enjoy browsing in bookshops might be excused for believing that modern scientific medicine has created nothing short of an ethical nightmare. Almost every week brings another new book, often with a lurid cover, proclaiming the dangers and evils of human genetic manipulation. These repetitive works are backed up by a growing body of literature that is critical of the whole ethos of modern science and of the direction in which it is going. While all this activity may reflect a more questioning society, there is a genuine danger that we may regress to a state in which superstition and the activities of quasi-religious movements and other pressure groups combine to retard medical progress; the campaign of the creationists in the United States, with their attack on evolutionary theory and their focus on schoolchildren, is a frightening example of retrograde thinking of this kind.

Not surprisingly, many new ethical issues have arisen from high-technology medicine. Perhaps most important in the long term are the questions, real and perceived, raised by our newfound ability to meddle with our genes and scrutinize our molecules and cells.

High-Technology Medicine and Euthanasia

Many of the ethical problems of modern medical practice revolve around the question of how far we are justified in interfering with natural events—in short, around the accusation that we are "playing God." Every time we swallow an aspirin tablet or buy a pair of spectacles, we are interfering with nature. Modern medical science has simply raised these activities to a level of sophistication at which it is now possible artificially to prolong life, almost indefinitely. Considering medicine's pace of development over recent years, we have managed to cope remarkably well with many of our ethical dilemmas. When it started, organ transplantation seemed to pose insurmountable problems. Yet we have worked out acceptable codes for protecting live donors and eliminating family pressures on related donors and, by sensible legislation, have controlled its commercial exploitation. The same evolutionary process has characterized much of modern high-technology medicine. But we have been left with a few problems with which we still seem ill equipped to deal.

Last year two decisions were made by the courts in Great Britain that may have profound implications for the future of medical practice. Doctors were given permission to stop feeding a brain-damaged youngster who had been maintained on a life-support system for many months. And a doctor who had killed an elderly patient in agonizing pain from rheumatoid arthritis with an intravenous injection of potassium chloride, although found guilty of murder in a court of law, was released and, after a reprimand by the General Medical Council, allowed to continue to practice. Both these events were acclaimed by most of the British press and, except for some objections from assorted so-called right-to-life movements, were accepted by the medical profession and by society as a whole.

Medical science has advanced so much this century that it can take over the activity of almost any damaged organ and maintain brain-damaged patients in a "vegetative state" for as long as it chooses. What at first sight seemed to be a signal triumph for organ-based medical research has left society with a difficult ethical dilemma. What is the status of a patient whose brain is irrevocably damaged but whose heart, lungs, and kidneys are still able to function, albeit with the help of a machine or two? Modern biology tells us that the transition from "life" to "death" reflects gradual changes in the chemistry of our cells that affect individual organs at different rates. But this begs the question of what we mean by "death" in the context of the complex collection of cells and organs that constitute a human being. In the event, we have had to compromise. Committees of doctors and lawyers have produced an arbitrary definition of "brain death" that, though helpful in day-to-day practice, is often confusing and distressing for the relatives of a patient who is being maintained on a life-support system.

The recent decision by the British courts was sensible. It does not reflect a less caring attitude toward the value of life by society. On the contrary, the intense interest and debate that this case generated suggests that we are keenly aware of these questions. It is an excellent example of how, by opening up sensitive issues to fully informed public debate, involving the church, various action groups, and the man in the street, ethical difficulties of this type can be slowly worked through. When this case was resolved, it turned out that there were over one thousand patients in Great Britain in a similar predicament; each one will have to be resolved by close consultation with the family, the courts, and the medical profession, but at least there is now a precedent for acting in a humane and sensible fashion.

The question of euthanasia is even more difficult but, like all ethical issues, is best addressed against a background of knowledge about the extent to which medicine has become a genuine science. In many countries euthanasia is illegal and exposes doctors to the risk of arrest for murder. Right-to-life movements are extremely vigilant, and the days when a kindly doctor could shorten the period of suffering for a patient with a terminal disease, by administering increasingly large doses of sedative, are gone, though there is no doubt that this sort of thing happened in the past. Perhaps it is good that euthanasia has been brought out into the open, but it is important that the public appreciate the difficulties this question poses for doctors.

It has been a recurrent theme in this book that medicine is still very much an art based on limited scientific knowledge. It offers no certainties, and there must be few doctors who have not had the experience of stopping treatment for a patient with a terminal disease and starting a course of strong sedation, only to see him or her improve for several months and spend a period of fulfilling life. Despite the recent outbursts from supporters of euthanasia, there is a world of difference between stopping active treatment for a brain-damaged patient on a life-support system and giving somebody a lethal injection. The state of medical knowledge is simply not sufficiently advanced for doctors to be given these powers; neither they nor their patients have reached the stage at which they can deal with decisions of this kind. I suspect that many of those who press for euthanasia have very little knowledge of the limitations of medical science. Even in what appear to be hopeless situations there may be surprises. Most sensible doctors have long given up announcing a prognosis in terms of weeks or months; they can make some kind of guess, based on experience with many patients with a similar condition, but realize their limitations when pronouncing on the outlook for an individual patient.

There are, however, some reasonable compromises. For example, in Great Britain, it is now possible for an individual to sign documents that make requests about the course to be taken under certain clinical situations, including the vegetative state. These are not legally binding but will provide families, doctors, and courts with some guidance, should these unfortunate circumstances arise. We must move slowly toward a solution of these difficulties, and society must not rush the medical profession into trying to make decisions for which the scientific basis or ethical framework is simply not adequate. Ethical practice, like all good doctoring, must be founded on firm science or, when this is not possible, on a consensus reached only after a completely open debate between doctors, courts, and public.

Greater Choice of Methods of Reproduction

The ethical and social questions that have been raised by scientific advances in the management of infertility are good examples of the problems that can be posed for society when it encounters issues that seem to strike at the very core of what it means to be a human being.

In Great Britain about one in six couples see their physicians with problems of fertility. About 35 percent of the cases result from a failure of ovulation or from damage to the Fallopian tubes. The development of techniques to fertilize eggs outside the body, in vitro fertilization (IVF), offered a valuable new treatment of these forms of infertility. The first "test tube" baby was born in July 1978. Almost overnight, in vitro fertilization, and the kind of research on embryos that made it possible, became a major issue in Great Britain, and the subject for a committee of inquiry chaired by a distinguished philosopher, Dame Mary Warnock.

The Warnock committee took a fairly pragmatic view of this new field, and many of its recommendations seem to have been based on what its members felt might be acceptable to present-day society. Because there is so much public concern about artificial reproduction, this new field was placed under the control of a statutory body. Practices deemed unacceptable, such as the use of surrogate mothers, were proscribed, and it was decided that research on human embryos should be permitted only up to the fourteenth day after fertilization. The latter decision was arbitrary and based on the fact that at about this time an embryo develops what is called a primitive streak, that is, the first beginnings of a nervous system.

In the ensuing years many hundreds of babies were born following in vitro fertilization, and the procedure became an accepted part of medical practice. But at the beginning of 1994 several stories in the press generated another vigorous debate on the whole ethos of artificial reproduction. It was reported that, for reasons of social convenience, a woman aged nearly sixty had given birth to a baby following in vitro fertilization and that a black woman had given birth to a white child. At about the same time, the story broke that it might soon be possible to isolate eggs from women who had died in accidents, or even from aborted fetuses, to be used for in vitro fertilization.

All these new issues called for a cool, extended debate. Unfortunately, before this could happen, a politician put a bill before Parliament making it illegal to use eggs obtained from embryos or people killed in accidents for the treatment of infertile women. Although this sort of precipitous reaction is understandable, it is very important that society takes a long and careful look at new scientific developments of this kind. It could be argued that the use of eggs from donors who have been killed in accidents or from a dead fetus is, in effect, no different from using organs obtained from the same source for transplantation. The concern that children born of such procedures would have no way of identifying their mothers is not new either; it often arises when eggs are donated from others for the treatment of infertility.

As Mary Warnock and her committee realized ten years ago, there are some practices that are repugnant to many people, though their objections are not always logical. In such cases, even though the particular scientific advances may benefit some members of society, they should not be pursued further, or not until they have been subjected to a thorough and open public debate and a consensus is reached that they are acceptable forms of practice.

Genetic Manipulation

What of the ethical issues that arise from our ability to dissect and manipulate human genes? Here there are two sets of problems: those that are already with us, and the more futuristic possibilities posed by genetic engineering—"slippery slope" concerns that are based on "what if" arguments and that, though undoubtedly real, may ultimately have to be sorted out by our great-grandchildren.

Many of the hotly debated problems of human genetic manipulation are not new but have simply been highlighted by the ability of recombinant DNA technology to identify many more single-gene disorders in carriers or in early fetal life. After all, we have had the means to identify mongol babies in mid-pregnancy for many years; modern technology has simply broadened the choice for parents to decide whether or not they will have children with serious genetic disability or congenital malformation. To what extent, if at all, is it appropriate to avoid genetic disease? The extreme view is that under no circumstances should parents be allowed to decide to terminate a pregnancy in which a baby has a serious genetic or congenital disability. Rather, society should look after its handicapped children, many of whom, it is argued, can lead happy and fulfilled lives, if properly cared for.

It would be surprising if these attitudes were not common among doctors and thoughtful members of society, regardless of whether their views are based on religious or humanistic beliefs. After all, doctors are trained to preserve life, and for many the idea of terminating a pregnancy on medical grounds is anathema. And which of them have not come across a happy mongol baby? On the other hand, equally caring physicians and others maintain that if we have the knowledge, parents with the potential for having a genetically abnormal child should be allowed to decide what kind of children they bring into the world. Has society the right to deny them the choice if medical progress has made it possible?

Some of those who believe that termination of pregnancy for serious genetic disability is wrong take their case even further. Surely, they say, physical disability is not always a bad thing. Some of our greatest creative artists suffered in this way. Is it right to terminate a pregnancy and risk losing a Beethoven? I doubt if it is helpful to base our attitudes on the avoidance of genetic disease on such unusual individuals. After all, many outstandingly creative people have remained in rude health throughout their working lives. It is difficult to substantiate the argument that unusual talent, or genius, is seen only in the context of serious disability.

As Bryan Appleyard has recently pointed out, arguments of this kind about the value of life are not always based on entirely rational grounds. He encapsulates this problem in a quotation from G. K. Chesterton, a writer who had an intense fear of the eugenics movement. "Keats died young; but he had more pleasure in a minute than a Eugenicist gets in a month." Appleyard reminds us that a genius with tuberculosis is a phenomenon we like to feel is beyond any rational balancing of probabilities, and that it is natural, therefore, to imagine that an aborted fetus or a pregnancy prevented by genetic counseling might have become a Keats or an Einstein. He is right, of course. We are all irrational, and the world would be a much duller place were we not. But we ought not offer our irrationality as an excuse for not trying to get to grips with some of the ethical issues that face us.

With the exception of certain activist groups and religious movements, most societies seem to be coming around to the view that the question of whether to bring children with serious genetic disabilities into the world should be left to the parents to decide, after careful counseling and discussion with their doctors and families. This is sensible; it is unreasonable to argue that society should prevent them from having normal children. Although I do not see the termination of pregnancy for congenital malformation or serious genetic disease as the ultimate goal for medical genetics, having worked with the families of children with a crippling inherited blood disorder for over thirty years, and witnessed the joy with which they have greeted our newfound ability to enable them to have normal children, I have no doubt that they should be allowed to make these difficult choices.

Our increasing sophistication is extending the number of conditions that can be identified prenatally. While fewer and fewer people contest the rights of parents who wish to terminate pregnancies in which the baby has a serious genetic handicap, the question of milder disability is more controversial. Where do we stop? Until recently most societies agreed that, except on medical grounds, it is inappropriate for parents to choose the sex of their children. But as I write this chapter, the leaders of the British Medical Association have told their members that, with a few trivial provisos, even this is acceptable. Fortunately their advice was disregarded by the rank and file of doctors who belong to the association. But it was a near thing.

And what of mild disability? In parts of China prenatal detection and selective termination of pregnancy are already being carried out for children with relatively mild inherited anemias; what is appropriate for a largely rural society in which one child is the norm may not, of course, be acceptable in conditions where a relatively mild anemia has no effect on well-being and longevity. But these less stringent indications for a termination of pregnancy are not confined to China; similar decisions are being made in the case of inherited blood disease in England. These are worrying trends, particularly since, to my knowledge, they have not been subjected to serious public debate.

And as our ability to analyze our DNA increases, other questions will arise. For instance, what will we do with unexpected information that is of medical or social importance to an individual? Two examples should suffice to bring home some of the difficulties that we may face. For the prenatal detection of genetic disease it is often essential to know that the parents, who claim to be such, are the true biological parents. And yet we find that in about 2 percent of the families referred to us, this is not the case. What do we do with this information? And what of disorders like Huntington's disease, a debilitating disease of the brain that appears only in middle life. Will a healthy teenager want to know that he or she has a gene for a disease that is likely to produce a miserable, lingering death, a long time in the future? Some will welcome the knowledge so that they can evaluate the risks for their own children; others will not wish to live with it.

As we develop screening programs and learn more about the prediction of genetic disease, we will have to set out codes of practice and sensible ways of maintaining confidentiality. The notion of employers' demanding genetic blueprints is causing concern, particularly in the United States. Those who hire train engineers (drivers) might reasonably ask potential employees whether they have inherited a gene for color-blindness. But should they also require to be told that the candidate has a gene that increases his likelihood of having a heart attack at the age of forty-five? We can identify genes for color-blindness with some certainty by a simple test for color vision; with a few exceptions we are still many years away from being able accurately to predict the time of death from coronary artery disease. But since we may get there in the end, it is important that debates on this type of problem begin now. After all, my insurance company demands to know my weight, height, and family history. An accurate prediction of my likelihood of dying young that is based on my genetic makeup may well turn out to be a better guide to my risk than the fact that I am 30 percent overweight.

Currently screening for inherited disorders is confined to populations in which there is a very high frequency of a serious genetic defect—sickle-cell anemia or thalassemia, for example. In the case of other common genetic diseases, such as cystic fibrosis, many societies are still uncertain whether they wish to embark on major screening programs. As often happens, we may be trying to walk before we can run. It is only recently that research has started to inquire into the social and emotional effects of learning that we are carrying a serious genetic defect. If at some future date screening is extended to our likelihood of developing cancer, heart attacks, diabetes, and even psychiatric disease, what will we do with the information? The problem is that we may well go through a long period of having this knowledge and not being able to put it to much practical use.

The German geneticist Benno Müller-Hill, writing in *Nature,* has recently highlighted some of these concerns. Although he sees problems arising from screening for physical disorders, he anticipates the greatest dangers in the field of mental disease. Suppose, he argues, at some time in the near future we find a gene, or genes, which, when they carry particular mutations, make us more likely to develop schizophrenia. If, as seems probable, there is a long gap between obtaining this information and putting it to any practical use, how will we handle it? He outlines several scenarios. First, we may be tempted to ignore the important environmental factors that undoubtedly interact with our genetic makeup to make us prone to serious mental diseases. Hence we will label individuals at high risk for developing schizophrenia without knowing what the risk is. Even worse, we may be tempted to screen all possible ethnic and social groups for these genetic variants. Undoubtedly some ethnic differences will be found, and then, Müller-Hill believes, genetic racial injustice will be a major issue. While I am sure that he underplays the importance of finding these genes for helping us understand the basis of schizophrenia, his concerns about the misuse of information of this kind before we understand its true significance are well founded.

Within a month or two of Müller-Hill's article a paper was published that suggested that there may be a gene on the X chromosome that predisposes toward homosexual behavior. Within days the popular media had picked up this story and proclaimed the possibility that parents could terminate pregnancies carrying potential homosexuals. While the journalists who wrote these stories had clearly never read the original article, which did not say that there might be a single gene for homosexuality, their reaction to what they thought the article was saying serves as a chilly reminder that we must take Müller-Hill's warnings seriously.

The harm that a misguided and premature genetic-screening program can do to society is exemplified by what happened when sickle-cell anemia was rediscovered as a major health problem in the United States in the early

1970s. Blacks were informed that many of them were suffering from this neglected disorder, and a massive screening program was set up, backed by heavy federal funding. Local communities became involved, supported by assorted individuals of diverse backgrounds and limited training, and the entire program became a major political, racial, and social issue. Several states passed laws that made screening mandatory, without any provision for education and counseling. Worst of all, nobody had decided beforehand what should be done with the information after the population had been screened. All that the program achieved was public anxiety, stigmatization, job and health insurance discrimination, and a variety of other undesirable effects.

If we can get into such a mess screening for a simple monogenic disease, the clinical consequences of which we understand reasonably well, we will have to be particularly careful when we start screening for complicated diseases that result from the interaction of several different genes with complex environmental factors. First, we will have to define the precision of our screening tests. We will then have to set up well-designed studies to find out whether the test will enable us to prevent a particular disease or to treat it earlier. And, finally, we must encourage a dialogue between scientists and the public to explain the complexities of tests of this kind. For example, if, as seems likely, we discover in the near future that there is a gene that greatly increases the likelihood of developing breast cancer, before we set up screening programs we must embark on pilot studies to determine the effect of this information on young women. Only a proportion of those who have genes of this kind will ultimately suffer from cancer; in our enthusiasm for preventive medicine we must not forget that we may be exposing many women to a life of unnecessary worry. This may be justified, particularly if the information is transmitted in an open and sensitive way and if it saves lives, but we simply do not know.

Modification of our genetic makeup with the objective of controlling or curing genetic diseases is another subject of public concern. In chapter 9 we discussed the two different approaches to gene therapy. Somatic gene therapy simply involves inserting genes into particular cells to take the place of defective genes; the inserted genes cease to function when the person dies and are not passed on to future generations. This is, in effect, no different from an organ transplant. Where this form of treatment has been subjected to serious ethical scrutiny, it has been agreed that it does not raise any new issues, provided that it is carried out under properly controlled conditions. The other form of gene therapy, which involves placing genes into fertilized eggs so that they are disseminated throughout all the organs of an individual, is quite different. Here, the "foreign" gene would be expressed in germ cells and would be passed on to future generations. Because of the technical uncertainties involved and because we would be altering the genetic makeup of unborn people, who would therefore have had no choice in the matter, this presents a completely different ethical issue. Most societies that have examined this question carefully have decided that germline therapy should not be pursued, certainly not in our current state of knowledge.

There are several practical reasons why germline therapy is quite inappropriate at the present time. First, it is now possible to identify genetic diseases in fertilized human ova. Thus we can obtain several eggs after in vitro fertilization, find out which carry the genetic defect, and replace only those that do not. In this way a family can be assured of having normal children, and there is no need to tamper with the genetic makeup of individuals of the future. Furthermore, we have no idea about the stability over successive generations of genes inserted into fertilized eggs, or about any long-term deleterious effects that they might have. Thus, for the moment, there is no reason to consider this approach in humans. Whether there ever will be must remain for future generations to decide.

These issues and others—including embryo research, genetic modification of animals to produce human therapeutic agents, the possible ill effects of changing the genetic makeup of individuals, ownership of knowledge of human DNA, and the sticky problem of patents—should all be amenable to solution, provided they are subjected to honest debate on the part of scientists and the public. When this approach works well, as in the case of the embryo research debate in Great Britain, it achieves a consensus in society. But it requires complete openness on the part of scientists and politicians, who are the final arbiters in many of these disputes.

"Slippery slope" concerns are more difficult. In the context of human genetic manipulation, many of them are based on the fear of eugenics, a movement started in Great Britain by Francis Galton. Galton invented the word to describe the improvement of the species by selective breeding, something farmers had been doing for centuries. The eugenics movement flourished for a while in Great Britain in the early part of this century and numbered among its followers many of the founder figures of human genetics, including Karl Pearson, Ronald Fisher, Lionel Penrose, and J. B. S. Haldane. It is important to emphasize the quality of these men; they were not cranks but distinguished scientists. However, they became disillusioned with the movement, which tended to die out in Great Britain after the Second World War.

The eugenics movement caught on for a time in many countries. In the United States, under the auspices of John D. Rockefeller, Charles Davenport set up the Eugenics Record Office, in Cold Spring Harbor, to which young men and women came for courses and training in human heredity and field research techniques. Once indoctrinated, they were sent off into the community with a "trait book." The data they collected were returned to the Eugenics

Record Office and duly cataloged. By 1914 at least thirty states had enacted new marriage laws, which restricted marriage among the unfit of various categories. The first state sterilization law was passed as early as 1907, and over the next ten years similar laws were enacted by fifteen or more states. And the movement had wider implications. For example, in 1924 an act based on its ideas was passed and and signed into law by President Calvin Coolidge, who, when vice president, had declared, "America must be kept American. Biological laws show . . . that Nordics deteriorate when mixed with other races."

Although the final story of eugenics in Germany in the 1920s and 1930s, and the influence of British and American eugenicists on Nazi thinking, remains to be told, it is clear that the politicians of Nazi Germany gained considerable encouragement from overseas. However, by the end of the Second World War, and the announcement to the world of the atrocities carried out in Germany in the cause of eugenics, the movement petered out. But it has reared its head again from time to time over the last thirty years. For example, in 1969 Arthur Jensen published an article in the *Harvard Education Review* entitled "How Much Can We Boost IQ and Scholastic Achievement." This paper posed questions about genetic variation in intelligence among different racial groups and suggested that the lack of performance among the American black population might reflect an innate lack of intelligence.

Although these particular brands of nonsense have disappeared, especially after much of the work of the British geneticist Cyril Burt on the inheritance of intelligence was found to have been fraudulent, the emergence of sociobiology and biological determinism has undoubtedly raised fears that a new eugenics movement may yet evolve. Now that we can transfer genes between different people, the idea that this type of technology might be used for the enhancement of human traits, or for even less desirable purposes, is being voiced. These fears are being encouraged by some of the less responsible elements of the media. We are regaled by television programs that show potential parents wandering around gene supermarkets stocking up with the traits that they would like to see in their children. Memories of Aldous Huxley's brave new world, and George Orwell's 1984, a few years later than predicted, are being reawakened.

The article by Benno Müller-Hill that we mentioned earlier concludes by arguing that the isolation of the first gene involved in determining "intelligence" will mark a major turning point in human history. Will, he asks, governments endorse the view of the eugenicists of the 1930s that the carriers of less effective versions of such genes are "bad" and "inferior" and that the eugenicists and their followers, endowed, we presume, with the "high intelligence" variants, are "good" and "superior"? Or will they stress privacy and, wisely, leave selection to market forces? Might they resort to legal measures to speed up the process of the "physical disappearance of the unwanted"? And what of the geneticists? he asks. Perhaps they will simply be relieved to find that their own mental genotypes are "healthy." But will some of them propose ways to eliminate the "bad" genes from others?

Remembering the early successes of the eugenics movement, we cannot ignore these concerns. However, despite the claims of biological determinists, these problems are unlikely to be with us for a very long time to come, if ever. When the Human Genome Project is complete, it will take us many years to learn how the products of one hundred thousand genes are orchestrated. And even when we do, we may still not be able to predict a person's true potential, let alone modify human behavior or achievement by changing a DNA sequence. At least for the foreseeable future, much of the thinking behind these concerns is naive; in particular, it underestimates the importance of our environment in making us what we are. It is important that scientists are completely honest and open about the way in which the biological sciences are moving, and lose no opportunity to debate these issues with the public. The potential benefits of molecular and cell biology for medicine are so great that we must ensure that progress is not retarded by speculations based on ill-founded, "slippery slope" arguments.

Toward a Time of Greater Certainty

Despite all the caveats that I have placed on our expectations about the pace of progress in the molecular sciences, someday, and maybe long before we really understand the functions of all our genes, and how they are orchestrated, we will be able to make reasonably accurate predictions about our likelihood of developing common diseases. Ultimately we may know that we have a very high probability of developing heart disease, cancer, or a stroke, and we may even have a shrewd idea when this is likely to happen. We may be lucky, of course, and by the time this cheerful information is available there may be much better ways of preventing or treating these conditions. But we cannot bank on this.

A particularly important ethical issue will arise from our increasing knowledge of ourselves. Some of us will turn out to be lucky and have rosy genetic profiles. Other will not and may be destined to develop a number of unpleasant diseases by middle life. The availability of information of this type will be difficult enough in a caring society that looks after its sick in a civilized way. But what will happen to those whose genetic profiles suggest a high risk of diseases in middle life, and who live in countries in which health care is based on commercial arrangements between insurance companies and the government and medical profession? The end result could be a gross inequality of the costs of health insurance and care, based simply on the way in which an individual's genetic dice turned up.

It is to be hoped, therefore, that governments take some of these problems into consideration as they think about the future provision of medical services. It would be a great tragedy if, as a by-product of the enormous amount of knowledge that we will gain about susceptibility to disease, we do not at the same time evolve a system of health care that treats everybody equally, regardless of his or her genetic heritage.

Who Sets Priorities?

In the overcharged atmosphere of the ethical debate generated by problems like euthanasia, selective abortion, and genetic engineering, we hear much less about one of the most difficult issues that face society. Recently, several reports in the newspapers and on television about the treatment of patients in the British National Health Service have generated public furor. The events responsible for these outbursts of righteous indignation involved elderly patients who had been refused one or another form of treatment because they were deemed to be too old. I suspect that doctors have been making decisions of this kind for many years and that it is only now, in our new cost-conscious era of health care, that such things are spoken about in public. As medical science and practice progress, and our populations get older, we are rapidly reaching the stage at which it will not be possible to offer every form of advanced medical technology to everybody. Indeed, there are hints that a rationing is already being applied that is based on criteria other than age; there was another recent public outcry in Great Britain when it was discovered that patients were being denied coronary artery surgery because they had refused to stop smoking.

Sooner or later we will have to decide on our priorities. But who will set them? Doctors cannot do this; the whole ethos of medical practice is based on doing everything possible for an individual patient. So who will make the decisions? Will it be left to market forces or government committees? And on what criteria will their judgments be based? It is ludicrous to define particular ages above which certain forms of medical treatment are not available; many eighty-year-old people have the mental agility and, to some extent, the physical capabilities of people half their age, and many thirty-year-olds behave as if they were already brain-dead.

We cannot duck these issues for much longer. Unlike many of the ones we have just discussed, they are with us already. Yet how often do we see them aired in the media? If there is one current ethical issues that requires wide and informed debate, it is the extent to which society wishes its doctors to apply the benefits of high-technology practice to our aging populations. If the message of this book is correct, and the complexities of the important diseases of the elderly are such that medical science is unlikely to prevent or cure them in the near future, the need for this debate is even more urgent.

ART OR SCIENCE?

The British physician Robert Coope, who was unlucky enough to be one of my teachers, compiled a delightful anthology, *The Quiet Art,* published in 1952. The title is taken from Virgil's *Aeneid:* "it was his part to learn the powers of medicines and the practice of healing, and careless of fame, to exercise that quiet art." The book contains a short quotation from a piece in the *Practitioner,* a British journal for general practitioners, written by Sir Arthur Hall in 1941:

> Medicine—however much it develops—must always remain an "applied science" and one differing from all the rest in that the application is to man himself. Were there no sick persons there would be no need for Medicine, either the Science or the Art. So long as there are both, both will be necessary. The application of its Science, to be of value, must be made in such a way that it will produce the maximum of relief to the sick man. This calls for certain qualities in the practising physician which differ entirely from anything required in the practice of the other applied sciences. Herein lies the Art of Medicine. The need for it is as great today as it ever was, or ever will be, so long as human sickness continues.

Sentiments of this kind have been expressed for almost as long as medicine has been practiced. But what is the art of medicine? Is it a reality, or is it a myth created by doctors over the years to add an aura of mystique to their activities and to set themselves onto a higher plane (with a higher salary) than that of ordinary mortals? Although it probably has been used to this end, it has a much less prosaic meaning, which gets to the core of what good doctoring is all about. In essence, it describes the holistic approach to the care of patients. While it includes skills in diagnosis and treatment, it also encompasses the management of every aspect of patients' reactions to their illness and its impact on their lives. It reflects the notion that the outcome of disease is the end result of a host of factors, many of which we still do not understand, but which undoubtedly include our genetic makeup, the environment in which we live, our upbringing, and our individual response to pain and adversity, together with the level of support we receive from our employers, family, and friends.

But if medical practice were based simply on the recognition of the importance of the many facets of disease, together with kindness, thoughtfulness,
and common sense, backed up with a working knowledge of how to communicate and how the body works, it would not have been necessary to make
so much of the art of doctoring. However, the problems of managing sick
people go much further than this. All living organisms are extremely complex, even when they are healthy. Illness adds yet more layers of complexity.
It is this, along with our extremely scanty knowledge of the cause of many of
the diseases that we have to manage, that makes the work of a doctor so
difficult. Even the most straightforward illness, for which we think we know
the cause, course, and treatment, is never exactly the same in any two
patients.

If, as frequently happens in our better teaching hospitals, we try to analyze
all the features of a patient's illness and attempt to interpret them in terms of
our current scientific understanding of disease processes, we always fail. I
suspect that this is because we are only starting to understand the complex
adaptive processes that are brought into action in response to disease, and
we know even less about the mysterious interactions between the mind and
the body that must also be important in modifying its course. Furthermore,
as the age of our population increases, many of our patients have more than
one disease; when one organ fails, a chain reaction often follows. The days
when disease could be ascribed to a single cause, say, a particular infection,
are long past.

An example from my own clinical practice might help the nonmedical
reader appreciate some of these problems. For many years I have looked after
a brother and sister with thalassemia, a disease that causes profound anemia.
Both of them have exactly the same defect in one of the genes that controls
the production of their hemoglobin. Yet the boy is often sick and requires
blood transfusions to correct his anemia, while his sister is a healthy, lively
young lady who suffers no symptoms at all. After many years of struggling
with this conundrum, we now understand a little why the same disease
affects these two so differently; it simply depends on the function of another
gene, which the two do not share and which helps the sister reduce the
severity of her anemia. So much for the reductionist approach. But the
brother's disease is more complicated. Often he is ill and cannot cope with
life; at other times he manages to live with his disease and to function well.
Yet he is always anemic, and his symptoms are not related to changes in the
severity of his anemia. In his case a more holistic approach—simply sitting
and talking to him—uncovers an extraordinary saga: difficulties at work,
financial problems, breakdown of relationships with family members, and a
host of other social problems. As each of these is dealt with, the whole pattern
of his illness changes for a while, and he is able to cope and lead a more

happy life, until the next crisis. And after further delving, and only after talking to him for several years and slowly gaining his confidence, it finally emerges that many of his difficulties are based on long-standing agoraphobia, a fear of open spaces. Thus what at first sight seemed to be a simple genetic disease turns out to be a complex problem, reflecting layer upon layer of social maladjustment set in the background of a completely unexplained and unrelated psychological disorder.

It is, then, the sheer complexity and unpredictability of the manifestations of illness that is responsible for the notion of medicine as an art. Apart from clinical and pastoral skills, good doctoring requires an ability to cut through many of the unexplained manifestations of disease, to appreciate what is important and what can be disregarded, and hence to get to the core of the problem, knowing when scientific explanation has failed and simple kindness and empiricism must take over. This is the real art of clinical practice. It comes naturally to some doctors, but others never quite accomplish the difficult transition from the textbook and lecture hall to practice in the real world.

As we saw in the life and work of Thomas Sydenham, there are deep-seated tensions between the art and science of medical practice. Undoubtedly some doctors exaggerate the role of the art. While this may be no more than a defense mechanism for concealing the empiricism of much of what they do, at its worst it leads them to ignore or even denigrate the scientific basis of medicine. This attitude is epitomized in a 1933 article in the *British Medical Journal* by the pathologist and medical historian Sir Andrew Macphail (1864–1938):

> I am well aware that in these days, when a student must be converted into a physiologist, a physicist, a chemist, a biologist, a pharmacologist, and an electrician, there is no time to make a physician of him. This consumma-tion can only come after he has gone out into the world of sickness and suffering, unless indeed his mind is so bemused by the long process of education in those sciences, that he is forever excluded from the art of medicine. . . . In that case he is destined for the laboratory, the professors chair, or the consultants office. What would have happened to Sydenham had he been put through this machinery is a problem in infinity which no human intelligence is competent to solve.

Was Sir Andrew's tongue planted firmly in his cheek as he wrote this? Though I would like to think so, in truth it reflects a view of medicine that is still very much alive in our teaching hospitals.

Although clinical method and experience are vital for good practice, this kind of exaggeration of the art of medicine can be deleterious. At its most extreme it can lead to ossified practice, based on fashion and anecdote, and an unwillingness to submit each new development to sound scientific inspec-tion. But the fact that is has survived for so long, and through so many different patterns of medical education, suggests that it may reflect a funda-mental incompatibility between the craft of healing and scientific discipline, which generations of doctors have found difficult to bridge.

The principal problem for those who educate our doctors of the future is how, on the one hand, to encourage a lifelong attitude of critical, scientific thinking to the management of illness and, on the other, to recognize that moment when the scientific approach, because of ignorance, has reached its limits and must be replaced by sympathetic empiricism. Because of the dichotomy between the self-confidence required at the bedside and the self-critical uncertainty essential in the research laboratory, it may always be dif-ficult to achieve this balance. Can one person ever combine the two qualities? Possibly not, but this is the goal to which medicine must aspire.

POSTSCRIPT

Current negative attitudes toward science in general and toward the medical sciences in particular must be changed. For, as we have seen, almost all the genuine advances in medical practice over the years have their basis in solid scientific research. Furthermore, the biological and medical sciences are mov-ing into the most exciting phase of their development and have enormous potential for the improvement of practice. And what are the alternatives? On the basis largely, I suspect, of society's disillusionment with modern high-technology medicine and the apparent inability of the basic sciences to yield practical results in the short term, it is sometimes suggested that we already know enough about the causes of many of our major killers. Hence we should apply this knowledge and our limited resources to controlling our environ-ments and lifestyles and forget about fundamental biomedical research. But where would this leave us? In both the developing and the developed world we could undoubtedly clean up our environments and deal with the factors thought to be the root causes of some of our leading killers. But this would not eradicate infectious disease, population growth, and malnutrition. And what kind of lives could those in richer countries look forward to if we leave things as they are?

Granted, stopping smoking, getting a little more exercise, and making some modest modifications in our diet are sensible. But, as things look at the moment, the prospects for life from early middle age on are not too cheerful. We will trek to our doctors at regular intervals to have all our accessible organs poked, prodded, imaged, and scanned; those that are inaccessible will be assessed by batteries of sophisticated tests of our blood, urine, feces, and anything else that is on offer. Depending on what is found, many of us will receive regular medication to control our blood pressure, blood cholesterol level, enlarging prostate, collapsing bones, minor depressions, and a lot more. We may also need drugs to control our failing heart, stop our blood from clotting or make it clot more effectively, control the pain from our rheumatism, and treat all the other ills of increasing age. And this is just for the lucky ones. The rest will be in and out of hospitals, receiving new hearts, blood vessels, joints, kidneys, and all the other wonders of medical technology.

This is not to belittle preventive medicine or some of the technical feats of the last fifty years. But we still know very little about how far it will be possible to prevent our killers and causes of ill health, and the extent to which many of them are the inevitable consequences of aging. A future in which the majority of our elderly population is spending its extra years having health checkups and trying to find enough time in the day to swallow its tablets and adjust its hearing aids sounds far from ideal. The way forward toward genuine health must be through the application of those simple preventive measures that really work, together with research in the basic medical sciences and epidemiology directed to determining the causes, prevention, and cure of the diseases that lead to premature death or to the disabilities of old age. And if we are able to afford to look after ourselves, every aspect of health care must be subjected to scientific scrutiny; massive changes in the way in which we provide medical services to our communities should stem not from the overnight whims of politicians but from careful pilot studies in well-defined populations. In short, in both medical research and the delivery of health care we must take a longer-term view of what we are trying to achieve.

But because of the complexity of the diseases that have been the less fortunate by-products of our increased longevity, we must not expect modern science to produce another "penicillin" to cure cancer or heart disease. Rather, we should look to steady progress in reducing the numbers of premature deaths and improving the quality of the lives of our increasingly aged populations.

The increasingly important role of science in the provision of health care, and the difficult social and ethical issues that will stem from our newfound ability to determine our futures, makes it essential that all of us become more scientifically literate. Our politicians must understand the rudiments of scientific evidence, and society as a whole must be sufficiently well informed to understand how best to achieve a healthy life and to participate in debating the complex issues that will continue to be posed by advances in biological and medical research. This movement toward greater scientific awareness will have to start in schools, with better support for science education and a greater accent on teaching human biology. Modern medicine will not achieve its true potential until all of us have a better appreciation of how we function, and of the wonderful complexity of living things.

I hope that this view of medical science and practice will have persuaded the reader that there has been a modest shift from an art or craft to a more rational and scientifically based discipline. We must pin our hopes for the future on more and better science. But, as I have emphasized, illness is an enormously complicated biological problem that has to be understood at many different levels, ranging from molecules to communities; while there are still patients to treat, medicine is likely to remain very much an art.

Some Original Sources and
Suggestions for Further Reading

CHAPTER 1: INTRODUCTION

Brown, G. Quoted in *Nature* 359 (1992): 175.
"The Future of Medicine: Warning, Doctors Can Damage Your Wealth." *Economist,* October 20, 1990, 25–30.
Rothenberg, L. S. "Molecular Medicine in the USA: Can Three Paradigms Shift Simultaneously?" *Lancet* 341 (1993): 1381–83.
Thomas, L. "The Health Care System." In *The Medusa and the Snail,* 45–50. New York: Viking Press, 1979.

CHAPTER 2: BEGINNINGS

Booth, C. C. "Clinical Research." In *Historical Perspectives on the Role of the MRC,* ed. J. Austoker and L. Bryder, 205–41. Oxford: Oxford Univ. Press, 1989.
———. "History of Science in Medicine." In *Science in Medicine: How Far Has It Advanced?* ed. G. Teeling-Smith and C. Roberts, 11–22. London: Office of Health Economics, 1993.
Bynum, W. "Historical Background: Medicine as a Dogma." In *Science in Medicine: How Far Has It Advanced?* ed. G. Teeling-Smith and C. Roberts, 1–10. London: Office of Health Economics, 1993.
Chadwick, J., and W. N. Mann. *The Medical Works of Hippocrates.* Oxford: Blackwell, 1950.
Cohen, I. B. *Revolution in Science.* Cambridge: Harvard Univ. Press, 1985.
Dampier, W. C. *A History of Science.* 4th ed. Cambridge: Cambridge Univ. Press, 1989.
Debus, A. G. *Man and Nature in the Renaissance.* Cambridge: Cambridge Univ. Press, 1992.

Dewhurst, K. *An Oxford Medical Quartet: Sydenham, Willis, Locke and Lower*, 23–31. Oxford: Oxford Medicine, Sandford Publications, 1970.

Diamond, L. K. "A History of Blood Transfusion." In *Blood, Pure and Eloquent,* ed. M. M. Wintrobe, 659–718. New York: McGraw-Hill, 1980.

Feder, G. "Paradigm Lost: Celebration of Paracelsus on His Quincentenary." *Lancet* 341 (1993): 1396–97.

Hankins, T. L. *Science and the Enlightenment.* Cambridge: Cambridge Univ. Press, 1993.

Jacob, F. *The Logic of Life: A History of Heredity* and *The Possible and the Actual.* London: Penguin Books, 1989.

Keynes, G. *The Life of William Harvey.* Oxford: Oxford Univ. Press, 1981.

Kipple, K. F., ed. *The Cambridge World History of Human Disease.* Cambridge: Cambridge Univ. Press, 1993.

Krebs, H. A. *Reminiscences and Reflections.* Oxford: Oxford Univ. Press, 1981.

Loudon, I. *Death in Childbirth: An International Study of Maternal Care and Maternal Mortality, 1800–1950.* Oxford: Oxford Univ. Press, 1992.

Magner, L. N. *A History of Medicine.* New York: Marcel Dekker, 1992.

Mason, S. F. *A History of the Sciences.* New York: Collier Books, 1962.

Murray, R. "Science and Psychiatry." In *Science in Medicine: How Far Has It Advanced?* ed. G. Teeling-Smith and C. Roberts, 53–62. London: Office of Health Economics, 1993.

Nutton, V. "Healers in the Market Place: Towards a Social History of Greco-Roman Medicine." In *Medicine in Society,* ed. A. Wear, 15–58. Cambridge: Cambridge Univ. Press, 1992.

Pasteur, L. *Etudes sur la bière.* 1877. English trans., London: Macmillan, 1879.

Porter, R. *Mind-Forged Manacles.* London: Penguin Books, 1987.

Sarder, Z. "Conventional Wisdoms." *Nature* 360 (1992): 713–14.

Temkin, O. *The Double Face of Janus and Other Essays in the History of Medicine.* Baltimore: Johns Hopkins Univ. Press, 1977.

Thomas, L. "Medical Lessons from History." In *The Medusa and the Snail,* 158–75. New York: Viking Press, 1979.

———. *The Fragile Species.* New York: Macmillan, 1992.

Wear, A., ed. *Medicine in Society.* Cambridge: Cambridge Univ. Press, 1992.

Weatherall, M. *In Search of a Cure: A History of Pharmaceutical Discovery.* Oxford: Oxford Univ. Press, 1990.

Weindling, P. "From Infection to Chronic Disease: Changing Patterns of Sickness in the Nineteenth and Twentieth Centuries." In *Medicine in Society,* ed. A. Wear, 303–12. Cambridge: Cambridge Univ. Press, 1992.

CHAPTER 3: EARLY SUCCESSES BREED NEW PROBLEMS

Bliss, M. *The Discovery of Insulin.* London: Faber and Faber, 1982.

Castle, W. B. "The Conquest of Pernicious Anemia." In *Blood, Pure and Eloquent,* ed. M. M. Wintrobe, 283–332. New York: McGraw-Hill, 1980.

Dubos, R. *Louis Pasteur.* New York: Da Capo Press, 1960.

Fisher, R. B. *Edward Jenner, 1749–1823.* London: André Deutsch, 1991.

Harvey, A. M. *Science at the Bedside: Research in American Medical Schools, 1905–1945.* Baltimore: Johns Hopkins Univ. Press, 1984.

Hotchkiss, R. D. "From Microbes to Medicine: Gramicidin, René Dubos, and the

Rockefeller." In *Launching the Antibiotic Era,* ed. C. L. Moberg and Z. A. Cohn, 1–18. New York: Rockefeller Univ. Press, 1990.

Lewis, T. "Clinical Science." *British Medical Journal* 2 (1933): 717–22.

Loudon, I. "Puerperal Fever, the Streptococcus, and the Sulphonamides, 1911–1945." *British Medical Journal* 293 (1987): 485–90.

MacFarlane, G. *Howard Florey: The Making of a Great Scientist.* Oxford: Oxford Univ. Press, 1975.

———. *Alexander Fleming: The Man and the Myth.* London: Chattow and Windus, 1984.

Magner, L. N. *A History of Medicine.* New York: Marcel Dekker, 1992.

Maurois, A. *The Life of Alexander Fleming.* Harmondsworth: Penguin Books, 1959.

Medawar, P. B. *The Threat and the Glory: Reflections on Science and Scientists.* Ed. D. Pyke. Oxford: Oxford Univ. Press, 1990.

Nutton, V. "The Reception of Fracastoro's Theory of Contagion, the Seeds That Fell among Thorns?" *Osiris,* 2nd ser., 6 (1990): 196–234.

Rothstein, W. G. *American Medical Schools and the Practice of Medicine.* Oxford: Oxford Univ. Press, 1987.

Smith, A. D. M. "Megaloblastic Madness." *British Medical Journal* ii (1960): 1840–1845.

Thomas, L. *The Youngest Science.* Oxford: Oxford Univ. Press, 1985.

Williams, T. I. *Howard Florey: Penicillin and After.* Oxford: Oxford Univ. Press, 1984.

CHAPTER 4: A GLIMPSE OF HIGH-TECHNOLOGY MEDICAL PRACTICE

Anderson, H. V., and J. T. Willerson. "Current Concepts: Thrombolysis in Acute Myocardial Infarction." *New England Journal of Medicine* 329 (1993): 710–14.

Antman, M. E., et al. "A Comparison of Results of Meta-analysis: A Randomized Control Trial and Recommendations of Clinical Experts' Treatment for Myocardial Infarction." *Journal of the American Medical Association* 268 (1992): 240–48.

Beeson, P. B. "The Ways of Academic Clinical Medicine in America since World War II." *Man and Medicine* 11 (1975): 65–79.

———. "Changes in Medical Therapy during the Past Half Century." *Medicine* 59, no. 2 (1980): 79–85.

Booth, C. C. "Clinical Research." In *Historical Perspectives on the Role of the MRC,* ed. J. Austoker and L. Bryder, 205–41. Oxford: Oxford Univ. Press, 1989.

Cochrane, A. L. *Effectiveness and Efficiency: Random Reflections on Health Services.* Cambridge: Cambridge Univ. Press, 1989.

Comroe, J. H., Jr., and R. D. Dripps. "Scientific Basis for the Support of Biomedical Science." *Science* 192 (1976): 105–11.

Feinstein, A. R. "Models, Methods, and Goals." *Journal of Clinical Epidemiology* 42 (1989): 301–8.

Foster, V. "Coronary Thrombolysis: A Perspective for the Practicing Physician." *New England Journal of Medicine* 329 (1993): 723–25.

"ISIS 3 (Third International Study of Infarct Survivors)." *Lancet* 339 (1993): 753–70.

Jenicek, M. "Meta-analysis in Medicine: Where We Are and Where We Want to Go." *Journal of Clinical Epidemiology* 42 (1989): 35–44.

Julian, D. "Where Is Heart Disease Going? Away?" In *Future Trends in Medicine,* ed. A. Ashton, 8–12. London: Royal Society of Medicine, 1993.

Kanbrocki, E. L., et al. "A Quest for the Relief of Atherosclerosis: Potential Role of Intrapulmonary Heparin—A Hypothesis." *Quarterly Journal of Medicine* 83 (1992): 259–82.

Lawrence, C. "Moderns and Ancients: The 'New' Cardiology in Britain, 1880–1930." *Medical History,* suppl. 5 (1985): 1–33.

Lewis, T. *The Mechanism of the Heart Beat.* London: Shaw and Sons, 1911.

———. "Clinical Science." *British Medical Journal* 2 (1933): 717–22.

Mackenzie, J. *Diseases of the Heart.* London: Oxford Univ. Press, 1908.

MacMahon, S., et al. "Blood Pressure, Stroke, and Coronary Artery Disease. Part 1: Prolonged Differences in Blood Pressure; Perspective Observational Studies Corrected for the Regression Dilution Bias." *Lancet* 335 (1990): 765–73.

Silverman, W. A., and I. Chalmers. "Sir Austin Bradford Hill: An Appreciation." *Controlled Clinical Trials* 13 (1992): 100–105.

Topol, E. J. "Which Thrombolytic Agents Should One Choose?" *Progress in Cardiovascular Disease* 34 (1991): 165–78.

White, K. L. "Health Care Research: New Wine in Old Bottles." *The Pharos* 56 (1993): 12–17.

CHAPTER 5: HOW MUCH HAS BEEN ACHIEVED?

Birth to Old Age. Book 5 of *Health and Disease.* Milton Keynes: Open Univ. Press, 1985.

Braudel, F. *The Structures of Everyday Life: The Limits of the Possible.* London: Collins, 1981.

Cook, A. M. "Fringe Medicine, Cults and Quackery." In *Oxford Companion to Medicine,* ed. J. Walton, P. B. Beeson, and R. Bodley Scott, 416–21. Oxford: Oxford Univ. Press, 1986.

Commission on Health Research for Development. *Health Research: Essential Link to Equity in Development.* Oxford: Oxford Univ. Press, 1990.

Dubos, R. *The Mirage of Health.* New York: Harper Colophon, 1979.

Feachem, R. G., and D. T. Jamison, eds. *Disease and Mortality in Sub-Saharan Africa.* Oxford: Oxford Univ. Press, 1991.

Feachem, R. G., and et al. *The Health of Adults in the Developing World.* Oxford: Oxford Univ. Press, 1992.

Fulder, S. "Alternate Therapists in Britain." In *Alternative Medicine in Britain,* ed. M. Saks, 166–98. Oxford: Oxford Univ. Press, 1992.

Gray, A., ed. *World Health and Disease.* Buckingham, U.K.: Open Univ. Press, 1993.

Haines, A. "Global Warming and Health." *British Medical Journal* 302 (1991): 669–70.

Hill, A. B. "The Clinical Trial." *British Medical Bulletin* 7 (1951): 278–82.

Inequalities of Health: The Black Report. Ed. P. Townsend and N. Davidson. And *The Health Divide.* Ed. M. Whitehead. London: Penguin Books, 1990.

King, M. 'Health Is a Sustainable State." *Lancet* 336 (1990): 664–67.

Malthus, T. *An Essay on the Principle of Population.* London: Penguin Classics, 1985.

Our Planet, Our Health: Report of the WHO Commission on Health and Environment. Geneva: WHO, 1992.

Papas, G., et al. "The Increasing Disparity in Mortality between Socioeconomic Groups in the United States." *New England Journal of Medicine* 329 (1993): 103–9.

Saks, M., ed. *Alternative Medicine in Britain.* Oxford: Oxford Univ. Press, 1992.

Sen, A. K. *Poverty and Famines: An Essay on Entitlement and Deprivation.* Oxford: Oxford Univ. Press, 1981.

United Nations Development Programme. *Human Development Report, 1992.* Oxford: Oxford Univ. Press, 1992.

Whitehead, M., and G. Dahlgran. "What Can Be Done about Inequalities in Health." *Lancet* 338 (1991): 1059–91.

Wilson, J., ed. *Disability Prevention: The Global Challenge.* Oxford: Oxford Univ. Press, 1983.

World Bank. *The World Development Report, 1984.* Oxford: Oxford Univ. Press, 1984.

———. *World Development Report, 1993: Investing in Health.* Oxford Univ. Press, 1993.

CHAPTER 6: NEW WAYS OF THINKING ABOUT DISEASE

Ames, B. N. "Dietary Carcinogens and Anticarcinogens." *Science* 221 (1983): 1256–63.

Barker, D. J. P., et al. "Fetal Nutrition and Cardiovascular Disease in Adult Life." *Lancet* 341 (1993): 938–41.

Burnet, M. *Genes, Dreams and Realities.* Harmondsworth: Penguin Books, 1973.

Cochrane, A. L. "1931–1971: A Critical Review with Particular Reference to the Medical Profession." In *Medicines for the Year 2000,* ed. G. Teeling-Smith and N. Wells, 2–12. London: Office of Health Economics, 1979.

Doll, R. "Smoking: The Past 40 Years and the Next 40: Twice the Killer We Thought." *Health Summary* 10 (1993): 5–9.

Doll. R., and R. Peto. *The Causes of Cancer.* Oxford: Oxford Univ. Press, 1981.

Dunniga, M. G. "The Problem with Cholesterol: No Light at the End of This Tunnel?" *British Medical Journal* 306 (1993): 1355–56.

Feather, I. H., J. A. Warner, and S. T. Holgate. "Cohabiting with Domestic Mites." *Thorax* 48 (1993): 5–9.

Henderson, P. E., R. K. Ross, and M. C. Pyke. "Towards the Primary Prevention of Cancer." *Science* 254 (1991): 1131–37.

Illich, I. *Limits to Medicine: Medical Nemesis: The Expropriation of Health.* Harmondsworth: Penguin Books, 1977.

Julian, D. "Where Is Heart Disease Going? Away?" In *Future Trends in Medicine,* ed. A. Ashton, 8–12. London: Royal Society of Medicine, 1993.

McGovern, D., G. Harrison, and G. R. Glover. "Ethnic Factors in Psychosis." In *Ethnic Factors in Health and Disease,* ed. J. K. Cruickshank and D. G. Beevers, 190–203. London: Wright, 1989.

McKeown, T. *The Role of Medicine.* Oxford: Blackwell, 1979.

———. *The Origins of Human Disease.* Oxford: Blackwell, 1988.

McPherson, K. "Health Gain and the Health of the Nation." In *Future Trends in Medicine,* ed. D. Ashton, 31–35. London: Royal Society of Medicine, 1993.

Morris, J. N., et al. "Exercise in Leisure Time: Coronary Attack and Death Rates." *British Heart Journal* 63 (1990): 325–34.

Peto, R. "Statistics of Chronic Disease Control." *Nature* 356 (1992): 557–58.

Peto, R., et al. "Mortality from Tobacco in Developed Countries: Indirect Estimate from National Vital Statistics." *Lancet* 339 (1992): 1268–78.

Smith, D. G., F. Song, and T. A. Sheldon. "Cholesterol Lowering and Mortality: The Importance of Considering Initial Level of Risk." *British Medical Journal* 306 (1993): 1367–72.

Steinberg, D. "Antioxidant Vitamins and Coronary Heart Disease." *New England Journal of Medicine* 328 (1993): 1486–87.

Ulbricht, T. V. L., and D. A. T. Southgate. "Coronary Artery Disease: Seven Dietary Factors." *Lancet* 338 (1991): 985–92.

White, K. L. "Health Care Research: New Wine in Old Bottles." *The Pharos* 56 (1993): 12–17.

Wood, D. "Coronary Disease in Populations." In *Future Trends in Medicine,* ed. D. Ashton, 18–20. London: Royal Society of Medicine, 1993.

CHAPTER 7: NATURE, NURTURE, AND AGING

Ames, B. N. "Dietary Carcinogens and Anticarcinogens." *Science* 221 (1983): 1256–63.

Ames, B. N., M. K. Shigenaya, and T. M. Hagen. "Oxidants, Antioxidants and the Degenerative Disease of Aging." *Proceedings of the National Academy of Sciences of the United States of America* 90 (1993): 7915–22.

Beeson, P. B. "Changes in Medical Therapy during the Past Half Century." *Medicine* 59, no. 2 (1980): 79–85.

Black, F. L. "Infectious Disease in Primitive Societies." *Science* 187 (1975): 515–18.

Blackwell, C. C., et al. "Non-secretion of ABO Antigens Predisposing to Infection by *N. meningitidis* and *S. pneumoniae.*" *Lancet* ii (1986): 284–85.

Blaser, M. J., and J. Parsonnet. "Parasitism by the 'Slow' Bacterium *Helicobacter pylori* Leads to Altered Gastric Homeostasis and Neoplasia." *Journal of Clinical Investigation* 94 (1994): 4–8.

Bodmer, W. F., ed. *Inheritance of Susceptibility to Cancer.* Oxford: Oxford Univ. Press, 1982.

Bodmer, W. F., and L. L. Cavalli-Sforza. *Genetics, Evolution, and Man.* San Francisco: W. H. Freeman, 1976.

Brown, M. S., and J. L. Goldstein. "Receptor-mediated Pathway for Cholesterol Homeostasis." *Science* 232 (1986): 34–47.

Burnet, M. "Biomedical Research: Changes and Opportunities." *Perspectives in Biology and Medicine* 24 (1981): 511–24.

Butterfield, J. "The Contribution of Modern Medicine." In *Medicines for the Year 2000,* ed. G. Teeling-Smith and N. Wells, 2–12. London: Office of Health Economics, 1979.

Cavalli-Sforza, L. L., P. Menozzi, and A. Piazza. "Demic Expansions and Human Evolution." *Science* 259 (1993): 639–46.

Crawford, M. H. "When Two Worlds Collide." *Human Biology* 64 (1992): 271–79.

Cruickshank, J. K., and D. G. Beevers, eds. *Ethnic Factors in Disease.* London: Wright, 1989.

Diamond, J. M. "Diabetes Running Wild." *Nature* 357 (1992): 362–63.

Flint, J., et al. "The Population Genetics of the Haemoglobinopathies." *Clinics in Haematology* 6, no. 1 (1993): 215–62.

Garrod, A. E. *Inborn Errors of Metabolism.* London: Frowd, Hodder and Stoughton, 1909.

———. *The Inborn Factors in Disease: An Essay.* 1931. Reprint, ed. C. R. Scriver and B. Childs, Oxford: Oxford Univ. Press, 1989.

Gibbon, A. "Gerontology Research Comes of Age." *Science* 250 (1990): 622–25.

Greenberg, J. H. *Language in the Americas.* Stanford: Stanford Univ. Press, 1987.

Haldane, J. B. S. "The Rate of Mutation of Human Genes." *Proceedings of the Eighth International Congress of Human Genetics (Hereditas* suppl. 35) (1948): 267–73.

Harman, D. "The Aging Process: Major Risk Factor for Disease and Death." *Proceedings of the National Academy of Sciences of the United States of America* 88 (1991): 5360–63.

Iseman, M. D. "The Evolution of Drug-Resistant Tuberculosis: A Tale of Two Species." *Proceedings of the National Academy of Sciences of the United States of America* 91 (1994): 2428–29.

Lewin, R. *Human Evolution.* Oxford: Blackwell Scientific Publications, 1989.

McKiegue, P. M., G. J. Miller, and M. G. Marmot. "Coronary Heart Disease in South Asians Overseas: A Review." *Journal of Clinical Epidemiology* 42 (1989): 597–609.

Mascie-Taylor, C. G. N., and G. W. Lasker. *Biological Aspects of Human Migration.* Cambridge: Cambridge Univ. Press, 1988.

Medawar, P. "In Defence of Doctors." In *The Threat and the Glory: Reflections on Science and Scientists,* ed. D. Pyke, 259–70. Oxford: Oxford Univ. Press, 1990. (This book review was first published in the *New York Review of Books,* May 15, 1980.)

Miller, L. H. "Impact of Malaria on Genetic Polymorphism and Genetic Diseases in Africans and African Americans." *Proceedings of the National Academy of Sciences of the United States of America* 91 (1994): 2415–19.

Mourant, A. E. *Blood Relations: Blood Groups and Anthropology.* Oxford: Oxford Univ. Press, 1983.

Mourant, A. E., A. C. Kopec, and K. Domaniewska-Sobczak. *Blood Groups and Disease.* Oxford: Oxford Univ. Press, 1978.

Neel, J. V. "Diabetes Mellitus: A 'Thrifty' Genotype Rendered Detrimental by Progress." *American Journal of Human Genetics* 14 (1962): 353–62.

———. "Lessons from a 'Primitive' People." *Science* 170 (1970): 815–22.

———. "A Thrifty Genotype Revisited." In *The Genetics of Diabetes Mellitus,* ed. J. Kobberling and R. Tattersall, 283–94. London: Academic Press, 1982.

O'Brien, S. J. "Ghetto Legacy." *Molecular Evolution* 1 (1991): 209–11.

O'Dea, K. "Diabetes in Australian Aborigines: Impact of the Western Diet and Lifestyle." *Journal of Internal Medicine* 232 (1992): 103–17.

Olshansky, S. J., B. A. Carnes, and C. Kassel. "In Search of Methuselah: Estimating the Upper Limits to Human Longevity." *Science* 250 (1990): 634–40.

Rogers, R. A., L. A. Rogers, and L. D. Martin. "How the Door Opened: The Peopling of the New World." *Human Biology* 64 (1992): 281–302.

Rusting, R. L. "Why Do We Age?" *Scientific American,* Dec. 1992, 86–95.

Templeton, A. R. "The 'Eve' Hypothesis: A Genetic Critique and Reanalysis." *American Anthropologist* 95 (1993): 51–72.

Thomas, L. "Medical Lessons from History." In *The Medusa and the Snail*, 158–75. New York: Viking Press, 1979.

Vogel, F., and A. G. Motulsky. *Human Genetics*. New York: Springer-Verlag, 1982.

Weatherall, D. J. *The New Genetics and Clinical Practice*. 3rd ed. Oxford: Oxford Univ. Press, 1991.

———. "The Role of Nature and Nurture in Common Disease: Garrod's Legacy." The Harveian Oration, Royal College of Physicians, London, 1992.

Weatherall, D. J., et al. "Genetic Factors as Determinants of Infectious Disease Transmission in Human Communities." *Philosophical Transactions of the Royal Society of London*, ser. B, 321 (1988): 327–48.

Weiss, K. M. *Genetic Variation and Human Disease: Principles and Evolutionary Approaches*. Cambridge: Cambridge Univ. Press, 1993.

Weiss, K. M., R. E. Ferrell, and C. L. Harris. "A New World Syndrome of Metabolic Disease with a Genetic and Evolutionary Basis." *Yearbook of Physical Anthropology* 27 (1984): 153–78.

Williams, G. C., and R. M. Nesse. "The Dawn of Darwinian Medicine." *Quarterly Review of Biology* 66 (1991): 1–22.

Wills, C. *Genetic Variability*. Oxford: Oxford Univ. Press, 1981.

———. *The Wisdom of the Genes*. Oxford: Oxford Univ. Press, 1991.

Zimmet, P., et al. "The Epidemiology and Natural History of NIDDM: Lessons from the South Pacific." *Diabetes/Metabolism Reviews* 6 (1990): 91–124.

CHAPTER 8: THE NEW REVOLUTION IN THE MEDICAL SCIENCES

Beadle, G. W. "Genes and Chemical Reactions in Neurospora." In *Nobel Lectures including Presentation Speeches and Laureates' Biographies: Physiology or Medicine, 1942–1962*, 587–99. Amsterdam: Elsevier, 1964.

Bearn, A. G. *Archibald Garrod and the Individuality of Man*. Oxford: Oxford Univ. Press, 1993.

Berg, P., and M. Singer. *Dealing with Genes: The Language of Heredity*. Mill Valley, Calif.: Univ. Science Books, 1992.

Bodmer, W. F., ed. *Inheritance of Susceptibility to Cancer*. Oxford: Oxford Univ. Press. 1982.

Bowler, P. J. *The Mendelian Revolution*. Baltimore: Johns Hopkins Univ. Press, 1989.

Conley, C. L. "Sickle Cell Anemia: The First Molecular Disease." In *Blood, Pure and Eloquent*, ed. M. M. Wintrobe, 320–59. New York: McGraw-Hill, 1980.

Crick, F. H. C. "Recent Research in Molecular Biology: Introduction." *British Medical Bulletin* 21 (1965): 183–86.

Galton, F. *Hereditary Genius*. London: Julian Friedman, 1979. (Text based on the 1869 ed. with a preface from the 1892 ed.)

Garrod, A. E. *Inborn Errors of Metabolism*. London: Frowd, Hodder and Stoughton, 1909.

Glover, D. M., and B. D. Hames. *Oncogenes*. Oxford: Oxford Univ. Press, 1989.

Gribbin, J. *In Search of the Double Helix*. London: Corgi Books, 1985.

Haldane, J. B. S. *Possible Worlds and Other Essays*. London: Chatto and Windus, 1927.

Hobbs, H. H., M. S. Brown, and J. L. Goldstein. "Molecular Genetics of the LDL

Receptor Gene in Familial Hypercholesterolemia." *Human Mutation* 1 (1992): 445–66.

Jacob, F. *The Logic of Life: A History of Heredity* and *The Possible and the Actual.* London: Penguin Books, 1989.

Judson, H. F. *The Eighth Day of Creation: Makers of the Revolution in Biology.* New York: Simon and Schuster, 1979.

Kevles, D. J. *In the Name of Eugenics.* Berkeley: Univ. of California Press, 1985.

Kevles, D. J., and L. Hood. *The Code of Codes.* Cambridge: Harvard Univ. Press, 1992.

Lewin, B. *Genes IV.* Oxford: Oxford Univ. Press, 1990.

Olson, M. V. "The Human Genome Project." *Proceedings of the National Academy of Sciences of the United States of America* 90 (1993): 4338–44.

Orel. V. *Mendel.* Oxford: Oxford Univ. Press, 1984.

Schrödinger, E. *What Is Life?* and *Mind and Matter.* 11th reprint. Cambridge: Cambridge Univ. Press, 1989.

Tsui, L.-C. "Probing the Basic Defect in Cystic Fibrosis." *Current Biology* 1 (1991): 4–10.

Watson, J. D. "The Human Genome Project: Past, Present, and Future." *Science* 248 (1990): 44–49.

Watson, J. D., and F. H. C. Crick. "Molecular Structure of Nucleic Acids: A Structure for Deoxyribose Nucleic Acid." *Nature* 171 (1953): 737–38.

Weatherall, D. J. *The New Genetics and Clinical Practice.* 3rd ed. Oxford: Oxford Univ. Press, 1991.

———. "Towards an Understanding of the Molecular Biology of Some Common Inherited Anemias: The Story of Thalassemia." In *Blood, Pure and Eloquent,* ed. M. M. Wintrobe, 373–407. New York: McGraw-Hill, 1980.

CHAPTER 9: WHAT MIGHT WE EXPECT FROM BASIC MEDICAL RESEARCH IN THE FUTURE?

Ames, B. N. "Endogenous Oxidative DNA Damage, Ageing and Cancer." *Free Radical Research Communications* 7 (1989): 121–28.

Balkwill, F. "Messenger Molecules." *Science and Public Affairs.* Summer ed., 10–16. London: Royal Society, 1992.

Basic Molecular and Cell Biology. 2nd ed. London: B. M. J. Publishing Group, 1993.

Brown, M. S., and J. L. Goldstein. "Receptor-Mediated Pathway for Cholesterol Homeostasis." *Science* 232 (1986): 34–47.

Coghlan, A. "Engineering the Therapies of Tomorrow." *New Scientist* 138 (1993): 26–32.

Djerassi, C. "New Contraceptives: Utopian or Victorian." *Science and Public Affairs* 6 (1991): 5–17.

Halliday, R. G., et al. "A Decade of Global Pharmaceutical R and D Expenditure (1981–1990)." *Pharmaceutical Medicine* 6 (1992): 281–96.

Hamer, D. H., et al. "A Linkage between DNA Markers on the X Chromosome and Male Sexual Orientation." *Science* 261 (1993): 321–27.

Jacob, H. J., et al. "Genetic Mapping of a Gene Causing Hypertension in the Stroke-Prone Spontaneously Hypertensive Rat." *Cell* 67 (1991): 213–24.

Johnson, R. T. "Prion Disease." *New England Journal of Medicine* 326 (1992): 486–87.

Kosik, K. S. "Alzheimer's Disease: A Cell Perspective." *Science* 256 (1992): 780–83.

Leckie, B. J. "High Blood Pressure: Hunting the Genes." *BioEssays* 14 (1992): 37–41.

MacCluer, J. W., and C. M. Kemmerer. "Dissecting the Genetic Contribution to Coronary Artery Disease." *American Journal of Human Genetics* 42 (1991): 1139–44.

McKusick, V. A. *Mendelian Inheritance in Man.* 9th ed. Baltimore: Johns Hopkins Univ. Press, 1992.

Mundy, G. R. "Boning Up on Genes." *Nature* 367 (1994): 216–17.

Murray, R. M., et al. "Genes, Viruses, and Neurodevelopmental Schizophrenia." *Journal of Psychiatric Research* 26 (1992): 225–35.

Newsom-Davis, J., and D. J. Weatherall, eds. *Health Policy and Technological Innovation.* London: Chapman and Hall, 1993.

Perutz, M. F. *Protein Structure: New Approaches to Disease and Therapy.* London: W. H. Freeman, 1992.

Prockop, D. J. "Mutations in Collagen Genes as a Cause of Connective Tissue Disease." *New England Journal of Medicine* 326 (1992): 540–46.

Roses, A., M. Alberts, and W. Strittmatter. "Alzheimer's Disease: Reassessing the Data." *Current Biology* 2 (1992): 7–9.

Ross, R. "The Pathogenesis of Atherosclerosis: An Update." *New England Journal of Medicine* 314 (1986): 488–99.

Scott, J. "Nature, Nature and Hypercholesterolaemia." *Lancet* 341 (1993): 1312–13.

———. "Apolipoprotein E and Alzheimer's Disease." *Lancet* 342 (1993): 697.

Sham, P. C., et al. "Schizophrenia following Prenatal Exposure to Influenza Epidemics between 1939 and 1960." *British Journal of Psychiatry* 160 (1992): 461–66.

Steinberg, D., et al. "Beyond Cholesterol: Modifications of Low-Density Lipoprotein That Increase Its Atherogenicity." *New England Journal of Medicine* 320 (1989): 915–24.

Swales, J. D. "The ACE Gene: A Cardiovascular Risk Factor." *Journal of the Royal College of Physicians of London* 27 (1993): 106–8.

Weatherall, D. J. *The New Genetics and Clinical Practice.* 3rd ed. Oxford: Oxford Univ. Press, 1991.

CHAPTER 10: HOPES AND REALITIES

Ashton, D., ed. *Future Trends in Medicine.* London: Royal Society of Medicine, 1993.

"ASSF Investigation of AIDS and Sexual Behaviour in France." *Nature* 360 (1992): 407–9.

Boyd, C. A. R., and D. Noble, eds. *The Logic of Life: The Challenge of Integrative Physiology.* Oxford: Oxford Univ. Press, 1993.

Caper, P. "Solving the Medical Care Problem." *New England Journal of Medicine* 318 (1988): 1535–37.

Feinstein, A. R. "Models, Methods, and Goals." *Journal of Clinical Epidemiology* 42 (1989): 301–8.

Frost, R. *The Complete Poems of Robert Frost,* 407. New York: Holt, Rinehart and Winston, 1962.

Hall, N. *The New Scientist Guide to Chaos.* London: Penguin Books, 1991.

Harman, D. "The Aging Process: Major Risk Factor for Disease and Death." *Proceedings of the National Academy of Sciences of the United States of America* 88 (1991): 5360–63.

Johnson, A. M., et al. "Sexual Lifestyles and HIV Risk." *Nature* 360 (1992): 416–20.

Kevles, D. J., and L. Hood. *The Code of Codes.* Cambridge: Harvard Univ. Press, 1992.

Lewin, R. *Complexity: Life at the Edge of Chaos.* London: J. M. Dent, 1993.

"Mapping Sexual Lifestyles." Editorial. *Lancet* 340 (1992): 1441–42.

Newsom-Davis, J., and D. J. Weatherall, eds. *Health Policy and Technological Innovation.* London: Chapman and Hall, 1993.

Olshansky, S. J., B. J. Carnes, and C. K. Cassel. "The Aging of the Human Species." *Scientific American,* April 1993, 18–25.

Sackett, D. L. "Inference and Decision at the Bedside." *Journal of Clinical Epidemiology* 42 (1989): 309–16.

Verkuyl, D. A. "Two World Religions and Family Planning." *Lancet* 342 (1993): 473–75.

Weiss, J. N., et al. "Chaos and Chaos Control in Biology." *Journal of Clinical Investigation* 93 (1994): 1355–60.

Wenzel, R. P., ed. *Assessing Quality Health Care: Perspectives for Clinicians.* Baltimore: Williams and Wilkins, 1992.

Yates, F. E. "Self-Organising Systems." In *The Logic of Life: The Challenge of Integrative Physiology,* ed. C. A. R. Boyd and D. Noble, 189–218. Oxford: Oxford Univ. Press, 1993.

CHAPTER 11: BACK TO OUR PATIENTS

Adams, M. B., ed. *The Wellborn Science: Eugenics in Germany, France, Brazil and Russia.* Oxford: Oxford Univ. Press, 1990.

Appleyard, B. "Our Plunge in the Gene Pool." *The Independent,* May 12, 1993.

Coope, R. *The Quiet Art.* Edinburgh: E. & S. Livingstone, 1952.

Davis, B. D., ed. *The Genetic Revolution.* Baltimore: Johns Hopkins Univ. Press, 1991.

Glover, J. *Causing Death and Saving Lives.* London: Penguin Books, 1990.

Human Genetic Information: Science, Law and Ethics. Ciba Foundation Symposium 149. New York: John Wiley, 1990.

Jones, S. *The Language of the Genes.* London: Harper Collins, 1993.

Jonsen, D. R. *The New Medicine and the Old Ethics.* Cambridge: Harvard Univ. Press, 1990.

Kennedy, I. *Treat Me Right: Essays in Medical Law and Ethics.* Oxford: Clarendon Press, 1988.

Kevles, D. J. *In the Name of Eugenics.* Berkeley: Univ. of California Press, 1985.

Kevles, D. J., and L. Hood. *The Code of Codes.* Cambridge: Harvard Univ. Press, 1992.

Macphail, A. "The Source of Modern Medicine." *British Medical Journal* i (1933): 443–47.

Medawar, P. "The Future of Man." In *The Threat and the Glory: Reflections on Science and Scientists,* ed. D. Pyke, 110–96. Oxford: Oxford Univ. Press, 1990.

Müller-Hill, B. *Murderous Science.* Trans. G. R. Fraser. Oxford: Oxford Univ. press, 1988.

———. "The Shadow of Genetic Injustice." *Nature* 362 (1993): 491–92.

Rose, R., L. J. Camin, and R. C. Lewontin. *Not in Our Genes.* Harmondsworth: Penguin Books, 1984.

Watson, J. D. Quoted by J. P. Swazey. "Those Who Forget Their History: Lessons from the Recent Past for the Human Genome Quest." In *Gene Mapping: Using Law and Ethics as Guides,* ed. G. J. Annes and S. Ellis, 43–56. Oxford: Oxford Univ. Press, 1992.

Weatherall, D. J. *The New Genetics and Clinical Practice.* 3rd ed. Oxford: Oxford Univ. Press, 1991.

———. "Science and the Health of the Nations." *Science and Public Affairs,* Autumn 1993, 3–6.

Weindling, P. "The Survival of Eugenics in 20th Century Germany." *American Journal of Human Genetics* 52 (1993): 643–49.

Wilkie, T. *Perilous Knowledge.* London: Faber and Faber, 1993.

Index

Index 371

intractable diseases, 275–77
  cancer, 291–94
  degenerative diseases of brain, 290–91
  dementia, 287–89
  diabetes, 277–80
  of heart and blood vessels, 280–85
  joint and bone diseases, 294–95
  mental handicaps, 289–90
  molecular sciences applied to, 312–16
  psychiatric, 285–87
introns, 253
in vitro fertilization (IVF), 333, 339
Ireland, 142–43
Islamic science, 31
Itano, Harvey, 250

Jacob, François, 45
Jansen, Zacharias, 37
Japan, 249, 292
Jeffreys, Alec, 297
Jenner, Edward, 43, 70–71
Jensen, Arthur, 340
Johannsen, Wilhelm, 233
joint diseases, 176, 294–95

Kevles, Daniel, 312
kidney disease, 63
King, Maurice, 141
Kirkwood, Thomas, 219
Koch, Robert, 44, 71, 72, 87
Köhler, Georges, 301
koro, 178
Kraepelin, Emil, 47
Krebs, Hans, 36
Kunkel, Louis, 257
kuru, 291
kwashiorkor, 144

Laennec, René Théophile Hyacinthe, 46, 96
Langerhans, Paul, 59
Lap-Chee Tsui, 257
Lavoisier, Antoine, 42
Leeuwenhoek, Anton van, 37–38
  microscope of, *39*
Legionnaires disease, 133
leukemia, 120–21, 223
Lewis, Sir Thomas, 51, 98, *100*
life expectancy, 126, *127,* 217
  coronary artery bypass surgery and, 113

  in developing countries, 145
  GNP correlated with, 306–7
lifestyles, *see* environment
Lind, James, 65
Linnaeus, Carolus, 42
lipoprotein (a), 281, 282
Lister, Joseph, 46, 53
Locke, John, 38
Logic of Life, The (Boyd and Noble, eds.), 320, 321
Logic of Life (Jacob), 45
London Epidemiological Society, 160
Long, Andrew, 126
Loudon, Irvine, 43, 76
Louis, Pierre Charles Alexandre, *49,* 56, 110
  on bloodletting, 50
  influence in America of, 160
  statistical analysis developed by, 47–48
Lower, Richard, 38–40, 52, 263
lung cancer, 162
lungs
  diseases of, 117
  heart diseases related to, 95
  transplantations of, 106
Luther, Martin, 32
Lyme disease, 133
lymphocytes, 278, 301
lysozyme, 77–80

McCarty, Maclyn, 241
McCollum, E. V., 65
Mackenzie, Sir James, 97, 98, 99
McKeown, Thomas, 179
  on change in emphasis in medicine, 181
  on genetically-caused diseases, 188
  on human evolution, 191–93
  on nature and nurture in disease, 187
  on origins of disease, 157–59
  on psychiatric disorders, 177
McKusick, Victor, 249, 269
MacLeod, Colin, 241
Macleod, John, 61, 63
Macphail, Sir Andrew, 345
McPherson, Klim, 174
Madness of George III, The (Bennett), 26
Magner, Louis, 87
malaria, 70, 146–49, *147,* 222
  distribution of, *198*
  drug-resistant strains of, 185, 319